Quantifiers, Propositions and Identity

Many systems of quantified modal logic cannot be characterised by Kripke's well-known possible worlds semantic analysis. This book shows how they can be characterised by a more general 'admissible semantics', using models in which there is a restriction on which sets of worlds count as propositions. This requires a new interpretation of quantifiers that takes into account the admissibility of propositions.

The author sheds new light on the celebrated Barcan Formula, whose role becomes that of legitimising the Kripkean interpretation of quantification. The theory is worked out for systems with quantifiers ranging over actual objects, and over all possibilia, and for logics with existence and identity predicates and definite descriptions. The final chapter develops a new admissible 'cover semantics' for propositional and quantified relevant logic, adapting ideas from the Kripke-Joyal semantics for intuitionistic logic in topos theory.

This book is for mathematical or philosophical logicians, computer scientists and linguists.

ROBERT GOLDBLATT is Professor of Pure Mathematics at the Victoria University of Wellington, New Zealand, and a Fellow of the Royal Society of New Zealand. He has served as the Co-ordinating Editor of The Journal of Symbol Logic, and has been a Managing Editor of Studia Logica for the past two decades.

LECTURE NOTES IN LOGIC

A Publication for The Association for Symbolic Logic

This series serves researchers, teachers, and students in the field of symbolic logic, broadly interpreted. The aim of the series is to bring publications to the logic community with the least possible delay and to provide rapid dissemination of the latest research. Scientific quality is the overriding criterion by which submissions are evaluated.

More information, including a list of the books in the series, can be found at
http://www.aslonline.org/books-lnl.html

Quantifiers, Propositions and Identity

Admissible Semantics for Quantified Modal and Substructural Logics

ROBERT GOLDBLATT
Victoria University of Wellington, New Zealand

ASSOCIATION FOR SYMBOLIC LOGIC

CAMBRIDGE
UNIVERSITY PRESS

CAMBRIDGE UNIVERSITY PRESS
Cambridge, New York, Melbourne, Madrid, Cape Town,
Singapore, São Paulo, Delhi, Tokyo, Mexico City

Cambridge University Press
The Edinburgh Building, Cambridge CB2 8RU, UK

Published in the United States of America by Cambridge University Press, New York

www.cambridge.org
Information on this title: www.cambridge.org/9781107010529

Association for Symbolic Logic
Richard A. Shore, Publisher
Department of Mathematics, Cornell University, Ithaca, NY 14853
http://www.aslonline.org

First published 2011

Printed in the United Kingdom at the University Press, Cambridge

A catalogue record for this publication is available from the British Library

ISBN 978-1-107-01052-9 Hardback

CONTENTS

INTRODUCTION AND OVERVIEW ix

CHAPTER 1. LOGICS WITH ACTUALIST QUANTIFIERS 1
1.1. Syntax 1
1.2. Logics 4
1.3. Incompleteness and Admissibility 11
1.4. Some History of the Quantifiers 19
1.5. Model Structures 22
1.6. Premodels and Models 25
1.7. Soundness 33
1.8. Infinitely Many Constants 43
1.9. Canonical Models and Completeness 45
1.10. Completeness and Canonicity for QS 53
1.11. Kinds of Incompleteness 59
1.11.1. Incompleteness for Kripkean Models 59
1.11.2. Kripkean S-frame Incompleteness 61
1.11.3. Non-Canonical S-frame Incompleteness 63

CHAPTER 2. THE BARCAN FORMULAS 67
2.1. Logics with CBF 68
2.2. Contracting Domains for All 70
2.3. Constant Domains for CBF 74
2.4. One Universal Domain 79
2.5. The Deductive Role of Commuting Quantifiers 84
2.6. Completeness with CBF and BF 89
2.7. Completeness with UI and BF 95
2.8. S-frame Incompleteness Revisited 98

CHAPTER 3. THE EXISTENCE PREDICATE 103
3.1. Axiomatising Existence 104
3.2. Completeness for Kripkean $\mathcal{E}$-Models 108
3.3. Necessity of (Non)Existence 116
3.4. Independence of BF from NNE 119

3.5. What is the Role of the Barcan Formula? 122

CHAPTER 4. PROPOSITIONAL FUNCTIONS AND PREDICATE SUBSTITUTION . 127
4.1. Functional Model Structures 128
4.2. Predicate Substitution: Notation and Terminology 136
4.3. The Anatomy of Predicate Substitution 141
4.3.1. Changing Bound Variables 141
4.3.2. Sufficient Freeness 141
4.3.3. Defining $\varphi(\psi_P/Px_P : P \in \Pi)$ in General 142
4.3.4. Strongly Free Substitution 143
4.3.5. Parameterless Substitution 145
4.4. Soundness of Predicate Substitution 146
4.5. Functional Canonicity 152

CHAPTER 5. IDENTITY 159
5.1. Intension versus Extension 159
5.2. Syntax 161
5.3. Identity and Rigid Functions 163
5.4. Model Structures and Models 165
5.5. Validity of Substitutivity of Identicals 169
5.6. Axiomatisation and Completeness 176
5.7. Kripkean Models 187
5.8. Definite Descriptions 193

CHAPTER 6. COVER SEMANTICS FOR RELEVANT LOGIC 203
6.1. Routley-Meyer Models for R 203
6.2. Admissible Semantics for RQ 209
6.3. Local Truth and Covers 214
6.4. Relevant Cover Systems 218
6.5. Cover System Completeness for R 225
6.6. Modelling Fusion 230
6.7. Cover System Characterisation of RQ 231
6.8. Heyting Implication for Full Systems 239
6.9. Localic Cover Systems for HR and HRQ 245

REFERENCES 251

INDEX 261

INTRODUCTION AND OVERVIEW

This book is about the possible-worlds semantic analysis of systems of logic that have quantifiers binding individual variables. Our approach is based on a notion of "admissible" model that places a restriction on which sets of worlds can serve as propositions. We show that admissible models provide semantic characterisations of a wide range of logical systems, including many for which the well-known model theory of Kripke [1963b] is incomplete. The key to this is an interpretation of quantification that takes into account the admissibility of propositions.

This is a subject that bristles with choices and challenges. Should terms be treated as rigid designators, or should their denotations vary from world to world? Should individual constants and variables be treated the same in this respect, or differently? Should each world have its own domain of existing individuals over which the quantifiable variables range, or should there be just a single domain of individuals? If there are varying domains, how should they relate to each other? Can any function from worlds to individuals be regarded as the "meaning" of some individual concept? Should an arbitrary mapping from individuals to propositions be admissible as a propositional function? Can we deductively axiomatise the class of valid formulas determined by each answer to these questions?

One striking phenomenon arises in quantified modal logic: there are axiomatically defined logics that cannot be characterised by the kind of possible-worlds semantics introduced by Kripke, even though the propositional fragments of those logics *are* characterised by their models as in [Kripke 1963a]. The same can happen with relational semantics for non-modal logics, including quantified relevant logic [Fine 1989] and some extensions of intuitionistic first-order logic [Ono 1973]. This failure of the completeness under Kripke semantics to lift from the propositional to the quantificational level raises a natural question. Is there some similar but more general kind of relational model theory that can characterise these "incomplete" logics?

We give a positive answer to that question, using two key ideas. The first is to take seriously the view that, whereas any proposition can be identified with a set of worlds—namely the set of all worlds in which it is true—the converse can fail. An arbitrarily chosen set of worlds need not correspond to a proposition. There may be no common property of the members of the set that would allow us to say that there is a single proposition that is true in exactly those members. Only for certain *admissible* sets of worlds need there be such a property. Our models will have a designated collection *Prop* of sets of worlds, called the *admissible propositions* of the model, from which the interpretations of formulas are to be selected. *Prop* will be closed under set-theoretic operations corresponding to the logical connectives. It forms a lattice under the partial ordering $\subseteq$ of set inclusion, which serves as the *entailment* relation between propositions.

We use the term "admissible semantics" generally to refer to semantical theories that have this sort of restriction on the type of entity that can be used to interpret formulas, and other expressions, such as terms denoting individuals. There will be some general type of entity that could be used for interpretations, and then it will be required that these interpretations be limited to some nominated set of entities of this type that may not include all of them. The nominated entities are said to be "admissible".

Admissibility has been used effectively to characterise *propositional* logics that are incomplete for their Kripke semantics. But now we have the new challenge of how to interpret the universal and existential quantifiers, $\forall$ and $\exists$, relative to a specified collection *Prop* of admissible propositions. For $\forall$ we can appeal to the intuition that a sentence $\forall x\varphi$ is semantically equivalent to the *conjunction* of the sentences $\varphi(a/x)$ for all $a \in U$, where U is the universe of all possible individuals over which the variable x ranges. This is the *possibilist* interpretation of the quantifier. It suggests that the proposition $|\forall x\varphi|$ expressed by $\forall x\varphi$ should be the conjunction of the collection $\{|\varphi(a/x)| : a \in U\}$ of admissible propositions. Here we invoke the second key idea: we interpret this conjunction to mean that $|\forall x\varphi|$ is to be an admissible proposition that entails all of the $|\varphi(a/x)|$'s, and is the weakest such proposition to do so, i.e. it is entailed by any other admissible proposition that entails all of the $|\varphi(a/x)|$'s. In other words, $|\forall x\varphi|$ is the *meet*, or greatest lower bound, of the $|\varphi(a/x)|$'s in the lattice of admissible propositions.

Our notion of possibilist model will require that *Prop* contains sufficiently many meets to ensure that this *admissible conjunction* of $|\varphi(a/x)|$'s always exists. Sometimes it is just the set-theoretic intersection $\bigcap_{a\in U} |\varphi(a/x)|$, but not always. It may be a proper subset of this intersection.

For the possibilist interpretation of the existential quantifier, $\exists x\varphi$ is identified with the disjunction of the $\varphi(a/x)$'s, and an admissible disjunction in *Prop* is a *join*, or least upper bound, of admissible propositions. Sometimes this is their set-theoretic union, but not in general. It can be larger than the union.

The interpretation of the quantifiers by meets and joins has a long and important history in algebraic logic. An account of this tradition is given in

Section 1.4. What novelty there is here results from combining that interpretation with the use of admissible propositions-as-sets-of-worlds. Moreover, for quantified modal logics we are in fact concerned, not just with the "single universe" possibilist interpretation discussed above, but more generally with models having *varying domains* containing individuals that may exist at some worlds and not others. The universe U of possibilia is taken to include all these domains of "actual" individuals. The set Ea of worlds whose domain contains the individual a forms a proposition expressing "a exists". We use these *existence propositions*, together with admissible conjunctions and Boolean propositional implication $\Rightarrow$, to interpret $\forall x\varphi$ as the conjunction of all the assertions "if a exists then $\varphi(a/x)$" for a ranging over U. This is the *actualist* interpretation of the quantifiers.

On this approach, the general criterion for $\forall x\varphi$ to be true at a world w is that there should be some admissible proposition X that is true at w and entails all of the propositions $Ea \Rightarrow |\varphi(a/x)|$ (see page 27 for details).

The theme running through the book is the use of admissible model theory to investigate metalogical properties that clarify the nature of quantification and its relation to other logical concepts. Here now is a brief abstract of each of the chapters.

Chapter 1 explains how our interpretation works for systems having the axioms for actualist quantification of [Kripke1963b]. A general completeness theorem is obtained that gives a semantic characterisation of the quantified logic QS generated by any propositional modal logic S. This involves constructing a canonical model structure $\mathcal{S}_{QS}$ and showing that QS is characterised by validity in all models on this structure. A special analysis is given in the important case that S is *canonical*, meaning that it is validated by its own canonical Kripke frame $\mathcal{F}_S$. In that case we show that the Kripke frame underlying $\mathcal{S}_{QS}$ is isomorphically embedded into $\mathcal{F}_S$ in a way that ensures that it inherits all modally expressible properties from $\mathcal{F}_S$. The chapter includes an extensive discussion of cases of incompleteness with respect to various conditions, designed to show that our use of admissibility is unavoidable.

Chapter 2 studies the famous Barcan Formula $\forall x\Box\varphi \to \Box\forall x\varphi$, often associated with the semantic condition of *contracting domains*, meaning that if world u is accessible from world w, then the domain of actual individuals of u is included in that of w. The converse of this Formula is associated with the reverse condition of *expanding* domains. However, we show that every logic of the form QS is characterised by admissible models that have contracting domains, and those logics that include the Converse Barcan Formula are characterised by models with *constant* domains, i.e. accessible worlds have the same domain. This may seem surprising, and even contrary to received wisdom. But we provide a new perspective on the Barcan Formula, demonstrating that its real role in admissible models is to ensure that the quantifier $\forall$ gets its standard Kripkean interpretation, in which the lattice meet defining the proposition $|\forall x\varphi|$ is given by the set-theoretic intersection operation $\bigcap_{a\in U}$. It is also noteworthy that the scheme $\forall x\forall y\varphi \to \forall y\forall x\varphi$, expressing commutativity of quantification, is not derivable in some of the logics of Chapter 1, and indeed there are admissible models that falsify it. But it is valid under the Kripkean set-theoretic interpretation of $\forall$, and is needed to axiomatise the logics characterised by this interpretation on constant-domain structures. The precise deductive role played by the commutativity scheme is explained in Section 2.5.

Chapter 3 adds a monadic existence predicate $\mathcal{E}$ to our formal language, interpreting it by the existence propositions Ea. This leads to simplified axiomatisations and model constructions, allows a complete axiomatisation of the logics determined by the Kripkean interpretation of the quantifiers, and provides further clarification of the role of the Barcan Formulas, both deductively and semantically.

Chapter 4 enriches our model structures with a designated set *PropFun* of admissible *propositional functions*. These are functions assigning an admissible proposition to each valuation of the individual variables. In any model on such a *functional model structure*, each formula is interpreted as an admissible propositional function. We show that such models validate an inference rule that licenses substitution of formulas for predicate letters, and provide a detailed syntactic analysis of this rule. We then construct for each quantified modal logic L a canonical functional model structure $\mathfrak{S}_{\mathrm{L}}$, and show that L is characterised by validity in all models on $\mathfrak{S}_{\mathrm{L}}$ if, and only if, it is closed under this rule of substitution for predicate letters.

Chapter 5 adds an identity predicate $\approx$ to the language and considers the vexed subject of existence and identity of intensional and extensional entities, as exemplified by such philosophical chestnuts as "the Morning Star is the Evening Star", "Scott is the author of Waverley","the number of planets is 9", etc. Models now have a designated set of admissible *individual concepts*, which are taken to be partial functions from worlds to individuals, and a designated subset of admissible *objects*, which are taken to be concepts that are *rigid*, i.e. have the same value in accessible worlds. We use two sorts of quantifiable variable, ranging over these two sorts of entity. An identity statement $\tau \approx \tau'$ may involve terms of different sort, and is interpreted at a world in a strict sense as asserting that both terms are defined and denote the same individual. The existence predicate becomes definable, by taking $\mathcal{E}\tau$ to be the self-identity formula $\tau \approx \tau$ whose corresponding proposition is the domain of the partial function interpreting τ. We axiomatise the logics determined by these models, using a new inference rule that allows deduction of assertions of non-existence ($\neg\mathcal{E}\tau$), and treat separately the case of models giving the Kripkean interpretation to the quantifiers. Then we give a semantic analysis of Russelian definite description terms $\imath x.\varphi$ in this context, and show how canonical models and axiomatisations can be obtained for logics in languages that have these description terms as well as the identity predicate.

Chapter 6 applies admissible semantics to an area of non-modal non-Boolean logic, studying the propositional relevant logic R and its notoriously intractable quantified extension RQ. A new kind of model structure is introduced, called a *cover system*, motivated by topological ideas about "local truth" from the Kripke-Joyal semantics for intuitionistic logic in topos theory. These provide the universal quantifier with the standard set-theoretic semantics over a single universe, but give a non-classical interpretation of disjunction and existential quantification, as in Beth's intuitionistic semantics, that is also reminscent of neighbourhood semantics in modal logic. We combine these notions with a modelling of negation by a binary world-relation of orthogonality, or incompatibility, and an operation of "fusion" of worlds to interpret relevant implication. Characteristic model systems for R have an algebra *Prop* of admissible propositions, while those for RQ have a set *PropFun* of admissible propositional functions as well. By adding an intuitionistic implication connective we obtain logics for which the models can be confined to ones in which all possible propositions are admissible. In the case of R the addition is conservative, so R itself is characterised by such models.

The range of logics that we characterise by admissible semantics is indicated by the long list of them that appears under the entry "characterisation of" on the second page of the Index.

Acknowledgements. I am indebted to Ed Mares, whose initial idea it was to apply admissibility to quantifiers in relevant logic. Together we worked this up into a complete possibilist semantics for RQ[1] (different to the cover semantics given here) and then adapted it to certain modal logics with possibilist quantification[2]. My development of the present material has been much helped by discussion with Ed, over countless lunches. I am grateful to Ian Hodkinson for his input to the project, particularly in relation to the analysis of commutativity of quantifiers[3], and to Greg Restall for encouragement and helpful discussions. Thanks to Max Cresswell for many stimulating conversations over the years about quantified modal logic, and for the benefit of his contributions to the subject. Thanks also to Hajnal Andréka, Ian Hodkinson, András Máté and István Németi for hospitality enjoyed in London and Budapest during parts of the writing.

The research undertaken for this book was supported by a grant from the Marsden Fund of the Royal Society of New Zealand.

[1] Mares and Goldblatt [2006]

[2] Goldblatt and Mares [2006]

[3] Goldblatt and Hodkinson [2009]

Chapter 1

LOGICS WITH ACTUALIST QUANTIFIERS

In the possible-worlds models for quantified modal logic introduced by Kripke [1963b], each world w is assigned a set Dw, thought of as the domain of individuals that exist, or are *actual*, in w. Such models are said to have *varying domains*. A universal quantifier $\forall x$ is interpreted at w by taking the variable x to range over the domain Dw. This is called the *actualist* interpretation of quantification. It does not validate the *Universal Instantiation* scheme[4]

$$\forall x\varphi \to \varphi(y/x), \quad \text{where } y \text{ is free for } x \text{ in } \varphi,$$

because the value of variable y may not exist in a particular world. Kripke proposed instead to use the scheme

$$\forall y(\forall x\varphi \to \varphi(y/x)),$$

which is valid under the actualist interpretation, and which we will call *Actual Instantiation*. This chapter explores logics that have this axiom, and develops a new kind of "admissible" semantics for them.

The first two sections present the syntax of languages and logics. Section 3 is motivational, reviewing the way that admissible semantics has been developed to overcome incompleteness for propositional modal logics, and explaining the ideas behind our adaptation of this approach to logics with quantifiers. Section 4 reviews some of the historical background to quantifier notation and its algebraic interpretation. The next few sections set out the formal semantics and prove soundness and completeness theorems. The last section gives some examples of incompleteness, designed to show that the use of admissibility is unavoidable.

1.1. Syntax

To begin with, some notation will be established for the syntax of modal predicate logic with quantification of individual variables. We take as fixed

[4]We use "scheme" generically to mean a set of formulas that comprises all instances of a particular syntactic form.

a denumerable set $\mathsf{InVar} = \{v_0, \dots, v_n, \dots\}$ of such variables. Symbols like $x, y, z, x_1, x_1', \dots$ will often be used for arbitrary members of InVar.

A *signature* is a set $\mathcal{L}$ of individual constants c, predicate symbols P, and function symbols F. Each predicate or function symbol comes with an assigned positive integer, its *arity*, so is *n-ary* for some n. We write $\mathsf{Con}_{\mathcal{L}}$, or just Con, for the set of individual constants in $\mathcal{L}$.

Now we define the *terms* of a signature, for which we typically use the symbol τ, possibly with subscripts or superscripts. An *$\mathcal{L}$-term* is any variable from InVar, any constant c from $\mathsf{Con}_{\mathcal{L}}$, or inductively any expression $F\tau_1 \cdots \tau_n$ where F is an n-ary function symbol from $\mathcal{L}$, and $\tau_1, \dots, \tau_n$ are $\mathcal{L}$-terms. A *closed* term is one that has no variables, and so contains only constants and (possibly) function letters.

An *atomic $\mathcal{L}$-formula* is any expression $P\tau_1 \cdots \tau_n$ where P is an n-ary predicate symbol from $\mathcal{L}$, and $\tau_1, \dots, \tau_n$ are $\mathcal{L}$-terms. The set of *$\mathcal{L}$-formulas* is generated from the atomic ones and a constant formula F (Falsum) in the usual way, using the connectives $\wedge$ (conjunction), $\neg$ (negation), the modality $\Box$ and universal quantifiers $\forall x$ for each $x \in \mathsf{InVar}$. Other connectives $\vee$ (disjunction), $\to$ (implication), $\leftrightarrow$ (biconditional), $\Diamond$ (dual to $\Box$), and the existential quantifiers $\exists x$ are introduced by standard definitions:[5]

$$\varphi \vee \psi = \neg(\neg\varphi \wedge \neg\psi).$$
$$\varphi \to \psi = \neg(\varphi \wedge \neg\psi).$$
$$\varphi \leftrightarrow \psi = (\varphi \to \psi) \wedge (\psi \to \varphi).$$
$$\Diamond\varphi = \neg\Box\neg\varphi.$$
$$\exists x\varphi = \neg\forall x\neg\varphi.$$

The constant T (Verum) is defined to be $\neg\mathsf{F}$. Formulas are denoted by the symbols $\varphi, \psi, \chi, \dots$.

The size of the set of $\mathcal{L}$-formulas depends on the size of the signature $\mathcal{L}$. Even when $\mathcal{L} = \emptyset$ there are countably infinitely many formulas that can be generated from F. On the other hand, if $\mathcal{L}$ contains uncountably many constants, and at least one predicate symbol, it will have uncountably many atomic formulas. In general, for a signature with at least one predicate symbol, the number of formulas is equal to the maximum of $\aleph_0$ and the size of $\mathcal{L}$.

We also need to consider *propositional* modal logics, whose language we take to be generated by an infinite set $\mathsf{PropVar}$ of *propositional* variables, for which we use symbols like $p, q, p_1, q_1', \dots$. Formulas are constructed from these and F by the connectives $\wedge$, $\neg$ and $\Box$. We may use the letters $A, B, \dots$ for such *propositional modal formulas*. An $\mathcal{L}$-formula φ of quantified modal logic will be called a *substitution-instance* of some propositional formula A if φ

[5] Technically, we could define F as some formula of the form $\varphi \wedge \neg\varphi$, but it is convenient to have it as a primitive.

can be obtained by uniform substitution of $\mathcal{L}$-formulas for the propositional variables of A. In particular, an $\mathcal{L}$-formula φ will be called a *Boolean tautology* if it is a substitution-instance of some propositional formula that is valid in the two-valued semantics of Boolean propositional logic.

It is common to assume that the set $\mathsf{PropVar}$ of propositional variables is *countably* infinite, but the theory also allows it to be of any uncountable size. We make use of that option in Section 1.10, where we assume that there are at least as many propositional variables as there are $\mathcal{L}$-formulas for some possibly uncountable signature $\mathcal{L}$.

Extensive use will be made in this book of the operation of substituting terms for free occurrences of variables. This needs to be handled carefully, and we take some trouble over its notation now. *Free* and *bound* occurrences of a variable in a formula are defined in the usual way, and a *sentence* is a formula with no free variables. The symbol $\varphi(\tau/x)$ will be used for the formula obtained by replacing every free occurrence of the variable x in φ by the term τ. This standard notation conveys the idea of τ *overwriting* free x in φ.

We also need to consider the operation of making *simultaneous* substitutions for a number of free variables at once. The notation

$$(\tau_0/v_0, \dots, \tau_n/v_n, \dots)$$

will be used for the substitution operator that uniformly substitutes the term τ_n for all free occurrences of the variable v_n, and does this simultaneously for all $n \geq 0$. This operator can be applied to any formula φ to give a formula

$$\varphi(\tau_0/v_0, \dots, \tau_n/v_n, \dots).$$

It can also be applied to a term τ, to form $\tau(\tau_0/v_0, \dots, \tau_n/v_n, \dots)$. The notation may be abbreviated to

$$(\tau_{n_0}/v_{n_0}, \dots, \tau_{n_p}/v_{n_p})$$

to indicate that the substitution alters only $v_{n_0}, \dots, v_{n_p}$, i.e. $\tau_n = v_n$ for all $n \notin \{n_0, \dots, n_p\}$. We will make particular use of single-substitution operators of the type (c/x) that replace all free occurrences of the variable x by the constant c, leaving all other variables unchanged.

Note that simultaneous substitution may produce a different result to sequential compositions of single substitutions. Thus if φ is the atomic formula Pxy, then $\varphi(y/x, z/y)$ is Pyz, whereas applying (y/x) to φ, and then (z/y) to the result, gives Pyy and then Pzz. So $\varphi(y/x, z/y) \neq \varphi(y/x)(z/y)$. Also, this shows that composition of substitutions need not commute, as $\varphi(z/y)(y/x) \neq \varphi(y/x, z/y)$ in this example. However, if the substituting terms are closed, then the order of substitution is immaterial. Thus if τ, τ' are closed, then

$$(\tau/x, \tau'/x') = (\tau/x)(\tau'/x') = (\tau'/x', \tau/x).$$

This also holds more generally if x' is not in τ and x is not in τ'.

As is customary, τ is said to be *freely substitutable for x in φ*, or more briefly *free for x in φ*, if no free occurrence of x in φ is within the scope of a quantifier $\forall y$ where y is any variable occurring in τ. If this condition holds, then no variable in τ becomes bound in $\varphi(\tau/x)$ within the occurrences of τ that replace free x in φ.

Another kind of operator we will use is substitution for constants, writing $\varphi(\tau/\mathsf{c})$ for the result of replacing *every* occurrence of the constant c in φ by the term τ. τ is *free for* c *in* φ if no occurrence of c in φ is within the scope of a quantifier $\forall y$ where y is any variable occurring in τ. We will also use simultaneous substitutions $\varphi(\tau_1/\mathsf{c}_1, \dots, \tau_n/\mathsf{c}_n)$ of this kind.

Further substitution technology will be introduced later in Section 4.2.

1.2. Logics

The axiom schemes we need are as follows:

K: $\Box(\varphi \to \psi) \to (\Box\varphi \to \Box\psi)$

AI: $\forall y(\forall x\varphi \to \varphi(y/x))$, where y is free for x in φ.

UD: $\forall x(\varphi \to \psi) \to (\forall x\varphi \to \forall x\psi)$

VQ: $\varphi \to \forall x\varphi$, where x is not free in φ.

K, which is named for Kripke, is the basic axiom for propositional modal logics. AI is the scheme of *Actual Instantiation*, named for the *actualist* interpretation of the quantifiers, in a given world, as ranging over the individuals that actually exist in that world. UD is *Universal Distribution*, and VQ is *Vacuous Quantification*.

Inference rules will be displayed in the form

$$\frac{\varphi_1, \dots, \varphi_n}{\varphi}$$

A set of formulas is *closed under* such a rule if it contains the *conclusion* φ whenever it contains all of the corresponding *premisses* $\varphi_1, \dots, \varphi_n$.

The inference rules we use are

MP: $\dfrac{\varphi,\ \varphi \to \psi}{\psi}$ *Modus Ponens*

N: $\dfrac{\varphi}{\Box\varphi}$ *Necessitation*

UG: $\dfrac{\varphi}{\forall x\varphi}$ *Universal Generalisation*

TI: $\dfrac{\varphi}{\varphi(\tau/x)}$, if τ is free for x in φ. *Term Instantiation*

GC: $\dfrac{\varphi(\mathsf{c}/x)}{\varphi}$, if c is not in φ. *Generalisation on Constants*

For a given signature $\mathcal{L}$, a *quantified modal logic*, or more briefly a *logic*, is defined to be any set L of $\mathcal{L}$-formulas that includes all Boolean tautologies and instances of the schemes K, AI, UD and VQ, and is closed under the rules MP, N, UG, TI and GC. A member φ of L is called an L-*theorem*, which we indicate by writing $\vdash_{\mathrm{L}} \varphi$. Sometimes we write this as $\mathrm{L} \vdash \varphi$. We may also just write $\vdash \varphi$ if the logic in question is understood. If we need to be specific about the background signature involved, we may say that L is a logic *over* $\mathcal{L}$.

In working with a logic we may write "by PC", for "by Propositional Calculus", meaning that a result follows by reasoning available in Boolean propositional logic.

It is worth being aware from the outset that our definition of logics does not include the scheme

CQ: $\forall x \forall y \varphi \to \forall y \forall x \varphi$

of *Commuting Quantifiers*. We will have a good deal to say later about the role of this scheme as a further axiom.

The Term Instantiation rule would be derivable, using UG, in any logic that included the Universal Instantiation scheme

UI: $\forall x \varphi \to \varphi(\tau/x)$, where τ is free for x in φ.

But we are not assuming that axiom, which relates to the possibilist interpretation of the quantifiers, and will not analyse its role for some time (see Section 2.4). Neither of our rules TI and GC have been commonly taken as primitive in quantified modal logics, but we will see that both are sound for the semantics of varying-domain models, and both are needed for the construction of a characteristic model for an arbitrary logic. They are in fact derivable in many standard logics that are axiomatised by specified modal axioms, as we show at the end of this section. A number of important basic properties of logics do not depend on these two rules, as we show now.

LEMMA 1.2.1. *In any quantified modal logic* L, *the following can be shown without using* TI *or* GC.

(1) L *is closed under the following* $\forall$-*rules and* $\exists$-*rules*:

$\forall$-*Monotonicity*: $\dfrac{\varphi \to \psi}{\forall x \varphi \to \forall x \psi}$

$\forall$-*Equivalence*: $\dfrac{\varphi \leftrightarrow \psi}{\forall x \varphi \leftrightarrow \forall x \psi}$

$\forall$-*Introduction*: $\dfrac{\varphi \to \psi}{\varphi \to \forall x \psi}$, *if* x *is not free in* φ.

$\exists$-*Monotonicity*: $\dfrac{\varphi \to \psi}{\exists x \varphi \to \exists x \psi}$

$\exists$-*Equivalence*: $$\frac{\varphi \leftrightarrow \psi}{\exists x\varphi \leftrightarrow \exists x\psi}$$

$\exists$-*Elimination*: $$\frac{\varphi \to \psi}{\exists x\varphi \to \psi}, \quad \textit{if } x \textit{ is not free in } \psi.$$

(2) $\vdash_L \forall x(\varphi \wedge \psi) \leftrightarrow \forall x\varphi \wedge \forall x\psi$.

(3) Replacement of Provable Equivalents:
If $\vdash_L \varphi \leftrightarrow \psi$, *then* $\vdash_L \chi[\varphi] \leftrightarrow \chi[\psi]$, *where* $\chi[\varphi]$ *and* $\chi[\psi]$ *differ only in that* $\chi[\varphi]$ *has* φ *in some places that* $\chi[\psi]$ *has* ψ.

(4) *If* x, y *are distinct variables with* y *not free in* φ *but freely substitutable for* x *in* φ, *then* $\vdash_L \forall x\varphi \leftrightarrow \forall y\varphi(y/x)$.

(5) Relettering of Bound Variables:
If φ *and* ψ *differ only in that* φ *has free* x *exactly where* ψ *has free* y, *then* $\vdash_L \forall x\varphi \leftrightarrow \forall y\psi$.

PROOF. (1)–(3) are standard, with the $\exists$-rules of (1) following by PC from the $\forall$-rules. For (4), from Actual Instantiation by Universal Distribution and PC we get

$$\vdash \forall y\forall x\varphi \to \forall y\varphi(y/x).$$

But $\vdash \forall x\varphi \to \forall y\forall x\varphi$ by Vacuous Quantification as y is not free in $\forall x\varphi$, so by PC this implies $\vdash \forall x\varphi \to \forall y\varphi(y/x)$. For the converse implication, the hypotheses on x and y ensure that x is free for y in $\varphi(y/x)$, and that $\varphi(y/x)(x/y) = \varphi$, so as another instance of AI we have

$$\vdash \forall x(\forall y\varphi(y/x) \to \varphi).$$

Using UD and VQ again, this leads by PC to $\vdash \forall y\varphi(y/x) \to \forall x\varphi$, and hence to (4).

For (5), the assumptions mean that ψ is $\varphi(y/x)$, and (5) follows directly from (4). ⊣

The last results of this Lemma allow us to take an occurrence of a formula $\forall x\varphi$ as a subformula of some formula ψ and replace it by $\forall y\varphi(y/x)$ with y not occurring in $\forall x\varphi$. The result is a new formula, provably equivalent to ψ, in which these bound occurrences of x have been replaced by y. Any formula ψ' obtained from ψ by a sequence of such reletterings of bound variables is a *bound alphabetic variant* of ψ. Another term for this is "congruence". More precisely, formulas ψ and ψ' will be called *congruent* if there is a bijection $x_i \mapsto x_i'$ between their sets of bound variables such that ψ and ψ' are identical except that ψ has bound ocurrences of x_i exactly where ψ' has bound ocurrences of x_i' and vice versa. Using the above scheme, and the principle of replacement of provable equivalents (Lemma 1.2.1(3)), it can be shown, as in [Kleene 1952, Lemma 15b], that congruent formulas are provably equivalent, i.e:

LEMMA 1.2.2. *If* ψ *and* ψ' *are congruent, then the formula* $\psi \leftrightarrow \psi'$ *is derivable in any quantified modal logic.* ⊣

The rules TI and GC are not needed to show that this is so. We can use the result to follow the common procedure, when confronted by a term τ that is not free for some variable z in ψ, of replacing ψ by a bound alphabetic variant ψ' with no variable of τ being bound in ψ'. Then τ is free for z in ψ'. We discuss this further in Section 4.3.

Next we consider some consequences of the rules TI and GC.

LEMMA 1.2.3. *Any logic* L *is closed under the following rules.*

TI*: $\dfrac{\varphi}{\varphi(\tau_1/x_1,\dots,\tau_n/x_n)}$, *if each* τ_i *is free for* x_i *in* φ.

GC*: $\dfrac{\varphi(\mathsf{c}_1/x_1,\dots,\mathsf{c}_n/x_n)}{\varphi}$, *if the* c_i *are distinct and not in* φ.

Sub: $\dfrac{\varphi}{\varphi(\tau/\mathsf{c})}$, *if* τ *is free for* c *in* φ.

Sub*: $\dfrac{\varphi}{\varphi(\tau_1/\mathsf{c}_1,\dots,\tau_n/\mathsf{c}_n)}$, *if each* τ_i *is free for* c_i *in* φ.

∀GC: $\dfrac{\varphi\to\psi(\mathsf{c}/x)}{\varphi\to\forall x\psi}$, *if* c *is not in* φ *or* ψ.

PROOF. For TI*, note that if none of the variables x_i occur in any of the terms τ_j (for instance if the τ_j's are closed), then the conclusion of the rule can be obtained from φ by performing the single substitutions $(\tau_1/x_1),\dots,(\tau_n/x_n)$ sequentially. In this case the conclusion of TI* is obtained from φ by n applications of TI. For the general case, we apply this observation by first choosing fresh variables $y_1,\dots,y_n$ that do not occur in any of $\tau_1,\dots,\tau_n$, and then applying the $2n$ single substitutions $(y_1/x_1),\dots,(y_n/x_n),(\tau_1/y_1),\dots,(\tau_n/y_n)$ sequentially to reach the same conclusion (this argument can be found in Church 1956, p. 84).

For GC*, note first that since the c_i are closed, the substitutions involved in forming the premiss can be performed simultaneously, or sequentially in any order. We prove the result by induction on n. The case $n=1$ is just GC itself. Assuming inductively the result for n, then if $\mathsf{c}_1,\dots,\mathsf{c}_{n+1}$ are distinct and not in φ, it follows that c_{n+1} is not in $\varphi(\mathsf{c}_1/x_1,\dots,\mathsf{c}_n/x_n)$, so if $\vdash\varphi(\mathsf{c}_1/x_1,\dots,\mathsf{c}_{n+1}/x_{n+1})$ then $\vdash\varphi(\mathsf{c}_1/x_1,\dots,\mathsf{c}_n/x_n)$ by GC, hence $\vdash\varphi$ by induction hypothesis. So the result holds for $n+1$. Thus it holds for all n by induction.

Sub is just the case $n=1$ of Sub*, and we derive the latter directly. Suppose that $\vdash\varphi$ and τ_i is free for c_i in φ for all $i\leq n$. Take distinct new variables $x_1,\dots,x_n$ that do not occur in φ. Then $\varphi(x_1/\mathsf{c}_1)(\mathsf{c}_1/x_1)=\varphi$, as x_1 is not in φ, so $\vdash\varphi(x_1/\mathsf{c}_1)(\mathsf{c}_1/x_1)$. From this we get $\vdash\varphi(x_1/\mathsf{c}_1)$ by GC, since c_1 is not in $\varphi(x_1/\mathsf{c}_1)$. Repeating this, from $\vdash\varphi(x_1/\mathsf{c}_1)$ we get $\vdash\varphi(x_1/\mathsf{c}_1)(x_2/\mathsf{c}_2)$, since x_2 is not in $\varphi(x_1/\mathsf{c}_1)$ and c_2 is not in $\varphi(x_1/\mathsf{c}_1)(x_2/\mathsf{c}_2)$. Sequentially applying the substitutions (x_i/c_i) in this way we arrive at $\vdash\psi$, where $\psi = \varphi(x_1/\mathsf{c}_1)(x_2/\mathsf{c}_2)\cdots(x_n/\mathsf{c}_n)$. In fact $\psi = \varphi(x_1/\mathsf{c}_1,\dots,x_n/\mathsf{c}_n)$, by the distinctness of the x_i's and the c_j's.

Now for each $i \leq n$, the term τ_i is free for x_i in ψ, since it is free for c_i in φ and ψ has x_i exactly in the places where φ has c_i. So by rule TI* we get $\vdash \psi(\tau_1/x_1,\dots,\tau_n/x_n)$. But $\psi(\tau_1/x_1,\dots,\tau_n/x_n)$ is

$$\varphi(x_1/\mathsf{c}_1,\dots,x_n/\mathsf{c}_n)(\tau_1/x_1,\dots,\tau_n/x_n),$$

which is $\varphi(\tau_1/\mathsf{c}_1,\dots,\tau_n/\mathsf{c}_n)$, again as the x_i's do not occur in φ. This completes the proof that L is closed under Sub*.

For $\forall$GC, suppose $\vdash \varphi \to \psi(\mathsf{c}/x)$ with c not in φ or ψ. Take a fresh variable y that does not occur in φ or ψ. By Sub, $\vdash (\varphi \to \psi(\mathsf{c}/x))(y/\mathsf{c})$. But this last formula is equal to $\varphi \to (\psi(y/x))$, so from it we can derive $\varphi \to \forall y\psi(y/x)$ by $\forall$-Introduction, as y is not free in φ (Lemma 1.2.1(1)). Since by Lemma 1.2.1(4) we have $\vdash \forall y\psi(y/x) \to \forall x\psi$, by PC this yields $\vdash \varphi \to \forall x\psi$ as required. ⊣

A *propositional modal logic* is a set S of propositional modal formulas that includes all such formulas that are Boolean tautologies or instances of the scheme K, and is closed under Modus Ponens and Necessitation.[6] Often such an S is presented as the *smallest*, or *least*, propositional modal logic that includes some specific set S_{ax} of formulas - typically the set of all instances of some axiom scheme(s), like $\Box A \to A$. In other words, S is the intersection of all propositional modal logics that include S_{ax}. When $\mathrm{S}_{ax} = \emptyset$, then S is the smallest of all propositional modal logics, and is commonly known as K. In general, theoremhood in S can be characterised by the existence of a finite proof-sequence in the usual way:

THEOREM 1.2.4. *Let* S *be the smallest propositional modal logic that includes some specified set* S_{ax} *of propositional modal formulas. Then* $\vdash_{\mathrm{S}} A$ *iff there exists a finite sequence* $A_1,\dots,A_n = A$ *of propositional modal formulas such that each* A_i *is either a Boolean tautology; or an instance of the scheme* K; *or a member of* S_{ax}; *or follows from earlier members of the sequence by one of the rules* MP *and* N.

PROOF. This proceeds in a standard fashion by defining S′ to be the set of all propositional modal formulas A for which there exists a sequence $A_1,\dots,A_n = A$ as described in the statement of the Theorem, and then observing that

(i) S′ is a propositional modal logic;
(ii) S′ includes S_{ax}; and
(iii) S′ is included in any propositional modal logic that satisfies (ii).

Then (i) and (ii) imply that $\mathrm{S} \subseteq \mathrm{S}'$, while (iii) ensures that $\mathrm{S}' \subseteq \mathrm{S}$. ⊣

[6] Actually this defines what is commonly called a *normal* propositional modal logic. But we will not be discussing non-normal systems in this book.

We will be particularly interested in logics that are quantified extensions of propositional ones. If S is any set of propositional modal formulas, we use the name QS for the *smallest* quantified modal logic that contains every $\mathcal{L}$-formula that is a substitution-instance of a member of S. In other words, QS is the intersection of all such quantified logics. Theoremhood in QS can also be characterised by the existence of a finite proof-sequence of a suitable type.

THEOREM 1.2.5. *Let* S *be any set of propositional modal formulas.*

(1) $\vdash_{\mathrm{QS}} \varphi$ *iff there exists a finite sequence* $\varphi_1, \ldots, \varphi_n = \varphi$ *of* $\mathcal{L}$*-formulas such that each* φ_i *is either an instance of a tautology or one of the schemes* K, AI, UD *and* VQ; *or a substitution-instance of a member of* S; *or follows from earlier members of the sequence by one of the rules* MP, N, UG, TI *and* GC.

(2) *If* S *is the smallest propositional modal logic that includes some set* S_{ax} *of propositional modal formulas, then* $\mathrm{QS} = \mathrm{QS}_{ax}$.

PROOF. (1) is proved in an analogous manner to Theorem 1.2.4.

For (2), since $\mathrm{S}_{ax} \subseteq \mathrm{S}$, it is immediate that $\mathrm{QS}_{ax} \subseteq \mathrm{QS}$. For the converse inclusion it is enough to show that QS_{ax} contains all $\mathcal{L}$-formulas that are substitution-instances of members of S, since QS is the least such logic to do so.

So let A be a member of S with propositional variables $p_1, \ldots, p_n$, and φ an $\mathcal{L}$-formula obtained by uniform substitution of some $\mathcal{L}$-formula φ_i for p_i in A, for all $i \leq n$. Let $(\varphi_1/p_1, \ldots, \varphi_n/p_n)$ denote the substitution operator that substitutes each φ_i for p_i, and replaces any other propositional variables by F. Thus $\varphi = A(\varphi_1/p_1, \ldots, \varphi_n/p_n)$.

Now let $A_1, \ldots, A_n = A$ be a finite proof-sequence establishing $\vdash_{\mathrm{S}} A$ as given by Theorem 1.2.4. Then

$$A_1(\varphi_1/p_1, \ldots, \varphi_n/p_n), \ldots, A_n(\varphi_1/p_1, \ldots, \varphi_n/p_n) = \varphi$$

is a finite sequence of $\mathcal{L}$-formulas each of which is (i) a substitution-instance of either a tautology or an instances of the scheme K or a member of S_{ax}; or (ii) follows from earlier members of the sequence by one of the rules MP and N, as substitution operators maps instances of MP and N to instances of the same rules. By part (1) of this Theorem, it follows that $\mathrm{QS}_{ax} \vdash \varphi$ as required. ⊣

The smallest quantified modal logic is QK, where K is the smallest propositional modal logic, and theoremhood in QK is characterised by proof-sequences as per Theorem 1.2.5(1) in which S is the set of instances of the axiom scheme K.

The adjective "propositional" will usually be used when referring to propositional modal logics, while the term "logic" used by itself can be taken to mean a quantified modal logic, unless the context clearly indicates otherwise.

Use of the letters "S" and "L" should help avoid confusion about which kind of logical system we are discussing.

The notation $QS + \Sigma$ will often be used for the smallest quantified modal logic extending QS that includes a scheme Σ, as in QS + CQ, QS + UI etc.

The deductive machinery introduced by Kripke [1963b] for the logic of varying domain model-structures consists essentially of the axioms and rules introduced here, with the exception of Term Instantiation and Generalisation on Constants. But the systems considered in that paper are built on propositional logics S in a similar manner to QS, and are of a kind in which these two extra rules are derivable.

To put it another way, we could have defined QS from S without using the rules TI and GC, and then shown that these rules are derivable.[7] To see how this works, write $Q'S \vdash \varphi$ to mean that there exists a finite sequence $\varphi_1, \dots, \varphi_n = \varphi$ of $\mathcal{L}$-formulas such that each φ_i is either an instance of a tautology or one of the schemes K, AI, UD and VQ; or a substitution-instance of a member of S; or follows from earlier members of the sequence by one of the rules MP, N, and UG only. Call such a sequence a Q′S-*proof sequence*. It is immediate that $Q'S \vdash \varphi$ implies $QS \vdash \varphi$. To show the converse is true, it suffices to prove that $\{\varphi : Q'S \vdash \varphi\}$ is closed under TI and GC. We outline a proof of this, leaving its fine details to the interested reader.

For TI, suppose $\varphi_1, \dots, \varphi_n = \varphi$ is a Q′S-proof sequence showing that $Q'S \vdash \varphi$, and that τ is free for x in φ. By systematically relettering bound variables, we can turn this into a sequence $\varphi'_1, \dots, \varphi'_n$ that is still a Q′S-proof sequence such that each φ'_i is congruent to φ_i and τ is free for x in φ'_i. Then $\varphi'_1(\tau/x), \dots, \varphi'_n(\tau/x)$ will also be a Q′S-proof sequence, showing that $Q'S \vdash \varphi'_n(\tau/x)$. Now we observed earlier that congruent formulas can be proved equivalent without using TI or GC (see Lemma 1.2.2). Since φ'_n and φ are congruent and τ is free for x in both, it follows that $\varphi'_n(\tau/x)$ and $\varphi(\tau/x)$ are congruent, hence $Q'S \vdash \varphi'_n(\tau/x) \leftrightarrow \varphi(\tau/x)$. Therefore $Q'S \vdash \varphi(\tau/x)$, as required by the rule TI.

For GC, let $\varphi_1, \dots, \varphi_n = \varphi(\mathsf{c}/x)$ be a Q′S-proof sequence showing that $Q'S \vdash \varphi(\mathsf{c}/x)$, where c is not in φ. Take a new variable y that does not occur in this sequence or in φ. Then $\varphi_1(y/\mathsf{c}), \dots, \varphi_n(y/\mathsf{c})$ will also be a Q′S-proof sequence, showing that $Q'S \vdash \varphi_n(y/\mathsf{c})$. Since neither c nor y occur in φ, the formula $\varphi_n(y/\mathsf{c})$, i.e. $\varphi(\mathsf{c}/x)(y/\mathsf{c})$, is just $\varphi(y/x)$, so $Q'S \vdash \varphi(y/x)$. But x is freely substitutable for y in $\varphi(y/x)$, so $Q'S \vdash \varphi(y/x)(x/y)$ by the rule TI just derived. Since $\varphi(y/x)(x/y) = \varphi$, this gives $Q'S \vdash \varphi$ as required by the rule GC.

[7]Technically, the machinery of [Kripke 1963b] required the axioms to be sentences (no free variables), and the set of axioms to be closed under prefixing of quantifiers $\forall x$ as well as $\Box$. Only Modus Ponens needed to be taken as an inference rule, since the rules N and UG are then derivable using the axioms K and UD respectively.

1.3. Incompleteness and Admissibility

The possible-worlds semantics for propositional modal logic is based on structures of the form $\mathcal{F} = (W, R)$, where W is set and R is a binary relation on W. Members of W are thought of as possible worlds, and R as a relation of accessibility. If wRu, then world u is accessible from w, or is a conceivable alternative to w. We will follow the common practice of calling $\mathcal{F}$ a *Kripke frame* or just a *frame*.

A *model* $\mathcal{M}$ on a frame is given by a function $|-|^{\mathcal{M}}$ that assigns to each propositional variable p a subset $|p|^{\mathcal{M}}$ of W, called the *truth set of* p, and thought of as the set of worlds at which p is true. This assignment is then extended to define truth sets $|A|^{\mathcal{M}}$ for all propositional formulas by induction on formula formation, as follows:

$$\begin{aligned} |\mathsf{F}|^{\mathcal{M}} &= \emptyset. \\ |A \wedge B|^{\mathcal{M}} &= |A|^{\mathcal{M}} \cap |B|^{\mathcal{M}}. \\ |\neg A|^{\mathcal{M}} &= W - |A|^{\mathcal{M}}. \\ |\Box\varphi|^{\mathcal{M}} &= [R]|\varphi|^{\mathcal{M}}. \end{aligned} \tag{1.3.1}$$

Here $[R]$ is the operation on the powerset $\wp W$ of W defined by putting, for each subset X of W,

$$[R]X = \{w \in W : \forall u \in W(wRu \text{ implies } u \in X)\}. \tag{1.3.2}$$

The truth sets of $\mathcal{M}$ determine a relation $\mathcal{M}, w \models A$ between worlds and formulas, which is read "A is true (or satisfied) at w in $\mathcal{M}$". This is just defined to mean that $w \in |A|^{\mathcal{M}}$. From the definition of $|A|^{\mathcal{M}}$ we get the following properties of this truth relation:

- $\mathcal{M}, w \models p$ iff $w \in |p|^{\mathcal{M}}$.
- $\mathcal{M}, w \not\models \mathsf{F}$, i.e. not $\mathcal{M}, w \models \mathsf{F}$.
- $\mathcal{M}, w \models A \wedge B$ iff $\mathcal{M}, w \models A$ and $\mathcal{M}, w \models B$.
- $\mathcal{M}, w \models \neg A$ iff $\mathcal{M}, w \not\models A$.
- $\mathcal{M}, w \models \Box A$ iff for all $u \in W(wRu$ implies $\mathcal{M}, u \models A)$.

Alternatively, and equivalently, we could have first inductively defined the truth relation $\mathcal{M}, w \models A$ by these clauses, and then defined truth sets in general by putting

$$|A|^{\mathcal{M}} = \{w \in W : \mathcal{M}, w \models A\}.$$

A formula A is *true in the model* $\mathcal{M}$, symbolised $\mathcal{M} \models A$, if $|A|^{\mathcal{M}} = W$, i.e. if A is true at every member of W. A is *valid in the frame* $\mathcal{F}$, symbolised $\mathcal{F} \models A$, if A is true in every model based on $\mathcal{F}$. A is *valid in a class* $\mathscr{C}$ of frames if it is valid in each member of $\mathscr{C}$.

For any model $\mathcal{M}$, the set $\mathrm{S}_{\mathcal{M}} = \{A : \mathcal{M} \models A\}$ of propositional formulas true in $\mathcal{M}$ is a propositional modal logic, as defined in the previous section.

Similarly, for each frame $\mathcal{F}$ the set $S_{\mathcal{F}} = \{A : \mathcal{F} \models A\}$ of propositional formulas valid in $\mathcal{F}$ is a propositional modal logic, while each class $\mathscr{C}$ of frames determines the logic $S_{\mathscr{C}} = \{A : \mathscr{C} \models A\}$. When $\mathscr{C}$ is the class of all frames, $S_{\mathscr{C}}$ is the smallest propositional modal logic K. When $\mathscr{C}$ is the empty class, $S_{\mathscr{C}}$ is the *inconsistent* logic for which all formulas are theorems.

Every propositional logic of the form $S_{\mathcal{F}}$ or $S_{\mathscr{C}}$ is closed under the rule of uniform substitution of formulas for propositional variables. But the logic $S_{\mathcal{M}}$ defined by a model need not be closed under this rule, since it may be that $|p|^{\mathcal{M}} = W$, and so $\mathcal{M} \models p$, for some variable p, but $\mathcal{M} \not\models A$ for other formulas A. So substituting A for p does not preserve truth in $\mathcal{M}$. This helps to explain why the notion of validity in a frame is important in propositional modal logic.

One of the reasons for the great success of Kripke semantics is that a number of well-studied axioms were shown to correspond to natural properties of the binary relation of these simple structures. For instance, the scheme $\Box A \to A$ is valid in a frame $\mathcal{F}$ iff its relation R is reflexive. The semantics provided some of the more important modal logics with characterisations in terms of validity in frames that were easy to deal with. Two famous examples:

- $\vdash_{S4} A$ iff A is valid in all frames for which R is reflexive and transitive. Thus S4 is precisely the logic $S_{\mathscr{C}}$ where $\mathscr{C}$ is the class of all reflexive and transitive frames.
- S5 is the logic $S_{\mathscr{C}}$ where $\mathscr{C}$ is the class of all frames whose R is an equivalence relation.

Here S4 is the smallest logic including the axiom schemes $\Box A \to A$ (corresponding to reflexivity) and $\Box A \to \Box\Box A$ (transitivity), while S5 extends S4 by the additional axiom $A \to \Box\Diamond A$ (symmetry).

Many other logics were shown similarly to be characterised by validity in a class of frames, and it was even conjectured at an early stage by Lemmon [1977, p. 74] that any propositional modal logic that is closed under uniform substitution for propositional variables must be equal to the logic $S_{\mathscr{C}}$ for some class $\mathscr{C}$. This turned out to be far from true. There are cases where S is *incomplete* for its frame semantics, because the class $\mathscr{C}_S = \{\mathcal{F} : \mathcal{F} \models S\}$ of all S-*frames*, i.e. frames validating S, defines a logic $S_{\mathscr{C}_S} = \{A : \mathscr{C}_S \models A\}$ which is different to S. It is true by definition that $S \subseteq S_{\mathscr{C}_S}$, but it is possible that there is a formula that is not a theorem of S, and yet is valid in all the S-frames, so belongs to $S_{\mathscr{C}_S}$. When this happens, S cannot be equal to $S_{\mathscr{C}}$ for *any* class $\mathscr{C}$.

The first example found to exhibit this *incompleteness* was not in fact a modal logic, but a finitely axiomatised propositional tense logic devised by S. K. Thomason [1972]. It is a deductively consistent logic S for which $\mathscr{C}_S$ is empty—there are no frames at all validating S. That cannot happen in the modal case: Makinson [1971] showed that every consistent propositional modal logic is valid in some one-element frame. Nonetheless the incompleteness phenomenon is pervasive for modal logics. It was shown by Blok [1980] that if a propositional modal S includes the scheme $\Box A \to A$, then there are *uncountably many* other logics S′ that are valid in exactly the same frames as S, i.e. have $\mathscr{C}_{S'} = \mathscr{C}_S$. Only one of these logics S′ can be characterised by a class of Kripke frames. A survey of results on incompleteness can be found in [Goldblatt 2006b, Section 6.1].

The source of incompleteness in frame semantics can be traced to the way it treats propositional variables. In specifying a model $\mathcal{M}$ we can arbitrarily choose the truth set $|p|^{\mathcal{M}}$ of a variable to be any subset of W. The assertion that a formula is valid in frame $\mathcal{F}$, i.e. true in all models on $\mathcal{F}$, amounts to asserting that the formula is true no matter what subsets of W we assign to its variables. In effect this treats propositional variables as set variables ranging over the full powerset of W. Indeed the validity relation "$\mathcal{F} \models A$" can be expressed in a *second-order* language with quantifiable set variables ranging over all of $\wp W$. Now the language of second-order logic is very powerful when given its *standard* interpretation, in which set variables are taken as ranging over the full powerset of its models. It can express the Peano postulates categorically and this leads, via Gödel's Incompleteness Theorem for arithmetic, to an incompleteness theorem for the deductive system of second-order logic: it has formulas that are valid under the standard interpretation but not derivable as theorems of the deductive system. Many systems of propositional modal logic exhibit similar behaviour under the frame semantics.

In order to obtain a complete semantics for second-order logic, Henkin [1950] introduced the wider class of *general* models, in which the range of set variables is restricted to some collection of subsets that is designated as part of the definition of the model, and which may not be the full powerset. A similar approach was taken in modal logic in response to the discovery of the incompleteness of frame semantics. The notion of frame was adapted to allow restriction on the possible choices of the truth sets $|p|^{\mathcal{M}}$ of propositional variables. There are also philosophical motivations for such a restriction. The semantics identifies a proposition with a set of worlds (the set of worlds in which it is true), but it is questionable whether an arbitrarily chosen collection of worlds need correspond to a proposition. There may be an "incoherent" set of worlds whose members lack sufficient commonality for us to be able to say that there is a single proposition that is true in exactly those worlds. So only some sets of worlds are *admissible* as propositions.

Formally, a *general frame*[8] is a structure $\mathcal{G} = (W, R, Prop)$, where (W, R) is a Kripke frame and *Prop* is a non-empty collection of subsets of W that

[8] This notion was introduced by S. K. Thomason [1972], who called it a "first-order structure". The name "general frame" is due to van Benthem [1977], [1979], adapting the terminology of Henkin [1950].

is closed under the Boolean set operations and the operation $[R]$ defined by (1.3.2). Thus $Prop \subseteq \wp W$ and

- $X, Y \in Prop$ implies $X \cap Y \in Prop$;
- $X \in Prop$ implies $W - X \in Prop$;
- $X \in Prop$ implies $[R]X \in Prop$

(the set complement $W - X$ will usually be written just as $-X$). The members of *Prop* are the *admissible propositions* of $\mathcal{G}$. We may also say that a subset of W is *admissible*[9] to indicate that it belongs to *Prop*.

A general frame is *full* if $Prop = \wp W$. The Kripke frame (W, R) can be viewed as the full general frame $(W, R, \wp W)$, in which every set of worlds is admissible.

A model on a general frame is given by a variable assignment $|-|^{\mathcal{M}}$ such that $|p|^{\mathcal{M}} \in Prop$ for every propositional variable p. The closure conditions on *Prop*, together with the equations (1.3.1), then ensure that every formula is interpreted in the model as an admissible proposition: $|A|^{\mathcal{M}} \in Prop$ for all propositional modal A. A formula is *valid* in a general frame if it is true in all such models on the frame that provide admissible interpretations.

It can be shown that validity in general frames provides a sound and complete semantics for any propositional modal logic S. We have S $\vdash A$ iff A is valid in every general frame that validates S. A proof of this may be found in [Blackburn, de Rijke and Venema 2001, Section 5.5].

When it comes to quantified versions of modal logics there are other sources of incompleteness. Even if a propositional logic S is characterised by its class $\mathscr{C}_S = \{\mathcal{F} : \mathcal{F} \models S\}$ of validating Kripke frames, it may be that QS is incomplete for validity in its standard models based on those frames. There may be a formula, involving quantifiers as well as modalities, that is not a theorem of QS, but cannot be falsified in any model based on an S-frame.

The first example of this kind was announced by Kripke [1967].[10] It takes S to be S4M, the extension of S4 by the McKinsey axiom $\Box\Diamond A \to \Diamond\Box A$. A proof of the incompleteness of QS4M can be found in [Hughes and Cresswell 1996, p. 283]. Strictly speaking this proof is for an extension of this logic, but the argument also gives the result for QS4M itself. The extension of QS4M has the Universal Instantiation scheme $\forall x\varphi \to \varphi(\tau/x)$, from which the *Converse Barcan Formula*

$$\text{CBF:} \quad \Box\forall x\varphi \to \forall x\Box\varphi$$

[9] Hughes and Cresswell [1984] call this an "allowable" set. The term "admissible" is employed by Blackburn, de Rijke, and Venema [2001], and has been used in this way by the logicians of Amsterdam at least since the early 1990's.

[10] The first published proof of the phenomenon appears to be that of Ono [1983], not for modal logic, but giving examples of intermediate propositional logics that are complete for their intuitionistic Kripke semantics, while their quantificational extensions are not.

can be deduced. For systems with these schemes, a standard model over a frame (W, R) has a universe U of *possible individuals* and a specification for each world w of a set $Dw \subseteq U$, thought of as the domain of individuals that are actual, or exist, in w. Validity of CBF requires the *expanding domain* condition that wRu implies $Dw \subseteq Du$. The truth condition for a sentence of the form $\forall x\varphi$ is

$$\mathcal{M}, w \models \forall x\varphi \text{ iff for all } a \in Dw, \mathcal{M}, w \models \varphi(a/x) \tag{1.3.3}$$

(treating members of Dw as individual constants).

It was shown by Cresswell [2001] that the quantified logic of expanding-domain models based on S4M-frames can be axiomatised by adding one more scheme, namely

$$\Diamond\forall x_1 \cdots \forall x_n(\varphi \to \Box\varphi), \tag{1.3.4}$$

where $x_1, \ldots, x_n$ are all the free variables of φ. In this way, the incompleteness of quantified S4M can be "repaired". Validity of the extra axiom (1.3.4) depends on a special feature of S4M frames that is also needed for validity of the McKinsey axiom, namely that each world is R-related to some world that in turn is R-related only to itself (see Subsection 1.11.2).

An interesting case of incompleteness that cannot be repaired concerns GL, the propositional modal logic extending K by the axiom

$$\Box(\Box A \to A) \to \Box A.$$

This logic, named for "Gödel-Löb", is also known variously as G, KW, and K4W. It axiomatises the interpretation of $\Box$ as "it is provable in Peano arithmetic that" (see Boolos 1993). The GL-frames are precisely those (W, R) in which R is transitive and has no "infinite R-chains" $w_0Rw_1R\cdots$. Montagna [1984] showed that the system QGL + Universal Instantiation is incomplete for validity in expanding-domain models on GL-frames. Then Cresswell [1997] proved that, unlike the S4M case, this incompleteness cannot be repaired by adding further explicit axiom schemes. Provided that the signature $\mathcal{L}$ includes predicate symbols to express properties of addition, multiplication and identity, then the set of formulas valid in all expanding domain models on GL-frames can encode first-order arithmetic, and is not recursively axiomatisable. The proof does not depend on the expanding-domain condition, and also shows that the set of formulas valid in all varying-domain models on GL-frames is not recursively axiomatisable, so cannot be equal to QGL or any of its recursively axiomatisable extensions.

In Section 1.11 we will use the logics QS4M and QGL to illustrate various notions of incompleteness for the semantics we develop.

There are by now quite a number of incompleteness results in quantified modal logic (see also Ghilardi [1989], [1991], Shehtman and Skvortsov [1990],

Cresswell [1995], as well as Hughes and Cresswell [1996]). Some of these involve systems that have the *Barcan Formula*[11]

$$\text{BF:}\quad \forall x\Box\varphi \to \Box\forall x\varphi.$$

Standard models for systems with BF have just a single universe U of individuals, with the semantics of a sentence $\forall x\varphi$ being

$$\mathcal{M}, w \models \forall x\varphi \ \text{ iff } \ \text{for all } a \in U,\ \mathcal{M}, w \models \varphi(a/x). \tag{1.3.5}$$

In effect this takes the domain Dw of each world to be U itself. One example of incompleteness with BF concerns the propositional logic S4.2, the extension of S4 by the axiom $\Diamond\Box A \to \Box\Diamond A$. The S4.2-frames are those whose relation R is reflexive, transitive, and *convergent* in the sense that

$$\text{if } wRu \text{ and } wRu', \text{ then for some } t,\ uRt \text{ and } u'Rt.$$

The system QS4.2 + UI + BF is incomplete for validity in single-universal-domain models on S4.2-frames [Cresswell 1995]. It is not known if this is repairable. On the other hand, there are logics including BF which are known to be irreparably incomplete. For instance, the logic of quantificational models on GL-frames with a single domain equal to the universe is not recursively axiomatisable [Cresswell 1997].

In fact it was the more expressive tense logic, rather than modal logic, that provided the first examples of irreparable incompleteness. In 1967 Dana Scott (unpublished) showed this for the quantified temporal logic of continuous time, the logic characterised by models on the temporal frame $(\mathbb{R}, <)$ of the real numbers. It is not recursively axiomatisable. Likewise for discrete time, in which $(\mathbb{R}, <)$ is replaced by the frame $(\mathbb{Z}, <)$ of integers (a demonstration of these results can be found in Gabbay, Hodkinson, and Reynolds [1994, Section 4.6]). By contrast, the propositional tense logic of each of these frames is *finitely* axiomatisable [Bull 1968] and decidable [Gabbay, Hodkinson, and Reynolds 1994, p. 576].

On the one hand there are cases, like S=S4M, where the semantically defined quantified logic over S-frames can be axiomatised, but only by using special axioms like (1.3.4) that relate to particular features of S-frames. On the other hand there are cases, like S=GL, where the semantically defined quantified logic over S-frames cannot be recursively axiomatised at all. In either case we are left with the question of how to interpret a deductively defined logic, of

[11] BF and CBF are named after Ruth C. Barcan (later Ruth Barcan Marcus), whose article [Barcan 1946a] was the first axiomatic study of a logical system combining modality with quantification. This was an extension of C. I. Lewis's system S2, with $\Diamond$ as a primitive symbol and strict implication $\prec$ defined by $\varphi \prec \psi := \neg\Diamond(\varphi \wedge \neg\psi)$. It included the axiom $\Diamond\exists x\varphi \prec \exists x\Diamond\varphi$, deductively equivalent to $\forall x\Box\varphi \prec \Box\forall x\varphi$. That is derivable from BF by the Necessitation rule, and conversely has BF as a deductive consequence in systems in which $\Box\varphi \to \varphi$ is derivable, such as the quantified extension of S4 of Barcan [1946b]. The description 'the Barcan formula' is due to Prior [1956], in a paper showing that BF is derivable in a quantified version of S5.

the form QS with the possible addition of general principles like the Barcan Formula and its converse, when this logic is incomplete for validity in the standard models over S-frames.

Our answer is to use the *admissibility* approach that solved the problem for propositional logics. We develop models for quantified modal logic that include a designated set *Prop* of admissible propositions that are to be the interpretations of formulas. The issue that this raises is how the quantifiers are to be interpreted. In the case of single-domain models, the semantic clause (1.3.5) can be expressed, using set-theoretic intersection, as

$$|\forall x\varphi|^{\mathcal{M}} = \bigcap_{a\in U} |\varphi(a/x)|^{\mathcal{M}}.$$

But even if all the truth sets $|\varphi(a/x)|^{\mathcal{M}}$ are admissible, it does not follow that their intersection is admissible, since *Prop* is, or has been so far, only required to be closed under binary intersections, in order to interpret conjunctions $\varphi \wedge \psi$ as admissible propositions. Requiring *Prop* to always be closed under arbitrary intersections would seem to be a severe restriction on the structure of sets of propositions, and not one that can be readily enforced by familiar model constructions.

Instead we take a more abstract approach, motivated by the intuition that the sentence $\forall x\varphi$ expresses the conjunction of all the sentences $\varphi(a/x)$. So we need an operation that forms the conjunction of suitable collections of propositions. We use the notion that *entailment*, or logical implication, between propositions-as-sets-of-worlds is expressed by their set inclusion. If $X \subseteq Y$, then proposition X entails proposition Y, in the sense that Y is true in every world in which X is true. We also view Y as being *weaker* than X, and X as being *stronger* than Y. Now the conjunction of a collection $\{X_i : i \in I\}$ of admissible propositions is to be an *admissible* X that (i) entails all of the the X_i's, and (ii) does no more than that. Here (ii) means that X is weaker than, i.e. is entailed by, any other admissible proposition that entails all of the X_i's. In other words,

(i) $X \subseteq X_i$ for all $i \in I$, and
(ii) if $Z \in$ *Prop* and $Z \subseteq X_i$ for all $i \in I$, then $Z \subseteq X$.

So the conjunction is to be the *weakest admissible* proposition that entails all of the X_i's. It will be a subset of $\bigcap_{i\in I} X_i$ by (i), but it need not be equal to $\bigcap_{i\in I} X_i$ because the latter may not be admissible. In general this conjunction will be the weakest (=largest) admissible subset of $\bigcap_{i\in I} X_i$, and will be denoted by $\sqcap_{i\in I} X_i$ when it exists.

Conditions (i) and (ii) together state that X is the *greatest lower bound*, or *meet*, of the X_i's in the partially ordered set $(\mathit{Prop}, \subseteq)$, the set of admissible propositions under the partial ordering of entailment/set inclusion. Our single-domain models will define $|\forall x\varphi|^{\mathcal{M}}$ to be the admissible conjunction of all the propositions $|\varphi(a/x)|^{\mathcal{M}}$, i.e. to be their meet. In general *Prop* need not be closed under arbitrary meets, but our models will require it to contain sufficiently many meets for this interpretation of $\forall$ to be possible.

For varying-domain models, which are our main interest, the semantic clause (1.3.3) for truth of $\forall x\varphi$ at a world w relativises the range of the quantifier to the domain Dw of individuals existing in w. This clause can be rewritten as

$$\mathcal{M}, w \models \forall x\varphi \text{ iff for all } a \in U, a \in Dw \text{ implies } \mathcal{M}, w \models \varphi(a/x), \quad (1.3.6)$$

so $\forall x\varphi$ expresses the conjunction of the assertions "if a exists then $\varphi(a/x)$" for all $a \in U$. To formalise this, let

$$Ea = \{w \in W : a \in Dw\},$$

so that Ea represents the proposition "a exists". We also need the standard Boolean implication operation $X \Rightarrow Y = (W - X) \cup Y$ on subsets of W.[12] Then (1.3.6) states that

$$|\forall x\varphi|^{\mathcal{M}} = \bigcap_{a \in U} \left(Ea \Rightarrow |\varphi(a/x)|^{\mathcal{M}}\right).$$

Again, even if all of the sets $Ea \Rightarrow |\varphi(a/x)|^{\mathcal{M}}$ are admissible, it need not follow that their intersection is admissible. Our varying-domain models will define $|\forall x\varphi|^{\mathcal{M}}$ to be the admissible conjunction of the sets $Ea \Rightarrow |\varphi(a/x)|^{\mathcal{M}}$, to be denoted

$$\bigsqcap_{a \in U} \left(Ea \Rightarrow |\varphi(a/x)|^{\mathcal{M}}\right).$$

Should the "existence propositions" Ea be admissible? It turns out that this is not required for validity of the basic axioms defining quantified modal logics. But admissibility of the Ea's does have consequences. For instance it ensures validity of the commutativity axiom $\forall x\forall y\varphi \to \forall y\forall x\varphi$ (see Theorem 1.7.16).

Our treatment of the existential quantifier $\exists$ will be dual to that of $\forall$. In a single-domain model, the standard semantics of a sentence $\exists x\varphi$ is

$$\mathcal{M}, w \models \exists x\varphi \text{ iff for some } a \in U, \mathcal{M}, w \models \varphi(a/x).$$

Set-theoretically this states that

$$|\exists x\varphi|^{\mathcal{M}} = \bigcup_{a \in U} |\varphi(a/x)|^{\mathcal{M}}.$$

Intuitively, $\exists x\varphi$ expresses the disjunction of all the sentences $\varphi(a/x)$, and single-domain models will define $|\exists x\varphi|^{\mathcal{M}}$ to be the "admissible disjunction" of all the propositions $|\varphi(a/x)|^{\mathcal{M}}$. Formally, this is the *least upper bound*, or *join*, of the $|\varphi(a/x)|^{\mathcal{M}}$'s in the partially ordered set $(Prop, \subseteq)$, and may be thought of as the *strongest admissible proposition* that is entailed by all the

[12] This connects to the notion of logical implication as set inclusion by the fact that $X \subseteq Y$ iff $X \Rightarrow Y = W$.

$|\varphi(a/x)|^{\mathcal{M}}$'s. We use the symbol $\bigsqcup$ for this join operation. In general, $\bigsqcup_{i\in I} X_i$ will include the set-union $\bigcup_{i\in I} X_i$, but may be larger.

In varying-domain models, $\exists x\varphi$ expresses the disjunction of the assertions "a exists and $\varphi(a/x)$" for all $a \in U$, and so $|\exists x\varphi|^{\mathcal{M}}$ will be the admissible disjunction of the sets $Ea \cap |\varphi(a/x)|^{\mathcal{M}}$, which we write as the join

$$\bigsqcup_{a\in U} \left(Ea \cap |\varphi(a/x)|^{\mathcal{M}}\right).$$

Before developing the details of this approach, we will review some more background.

1.4. Some History of the Quantifiers

Frege [1879] created the first rigorous formal logic with variable-binding quantification, using a two-dimensional graphical notation to present axioms and inference rules. Independently, Peirce [1885] gave a treatment of quantifiers as operators on expressions of Boolean algebra, an idea that he credited to O. H. Mitchell, his "gifted student". Mitchell had devised the notation F_1 for "All U is F", and F_u for "Some U is F", where U is the "universe of class terms". He represented the Aristotelean forms in this symbolism, with $(\bar{a}+\bar{b})_1$ expressing *No a is b*, $(a\bar{b})_u$ expressing *Some a is not b*, and so on. He observed that $\overline{(F_1)} = \overline{F}_u$, which we recognise as the equivalence of $\neg\forall x$ and $\exists x\neg$, and stated many other such laws that these operators obey [Mitchell 1883, pp. 74ff.].

Peirce, in referring to "the distinction of *some* and *all*", wrote that

> All attempts to introduce this distinction into the Boolian algebra were more or less complete failures until Mr. Mitchell showed how it was to be effected. His method really consists in making the whole expression of the proposition consist of two parts, a pure Boolian expression referring to an individual and a Quantifying part saying what individual this is [1985, p. 194].

Peirce illustrated with the expressions *Any*(x) and *Some*(y), noting that from these we may infer *Some*(xy), and then wrote:

> Here, in order to render the notation as iconical as possible we may use Σ for *some*, suggesting a sum, and Π for *all*, suggesting a product. Thus $\Sigma_i x_i$ means that x is true of some one of the individuals denoted by i or
>
> $$\Sigma_i x_i = x_i + x_j + x + x_k + \text{etc.}$$
>
> In the same way, $\Pi_i x_i$ means that x is true of all these individuals, or
>
> $$\Pi_i x_i = x_i x_j x_k, \text{etc.}$$

Peirce had used Σ as a symbol for a logical sum (i.e. disjunction) as early as in a paper of 1867, and had subsequently used it and Π (for product/conjunction) to express forms of quantification. Tellingly, in this 1885 paper he says [p. 195]

> It is to be remarked that $\Sigma_i x_i$ and $\Pi_i x_i$ are only similar to a sum and a product; they are not strictly of that nature, because the individuals of the universe may be innumerable.

Thus Peirce viewed logical summation and multiplication as acting on finitely many terms. Mitchell took the step of simply attaching symbols marking quantification to a Boolean expression, in his case as subscripts. This allowed Peirce to see that his own Σ_i and Π_i prefixes could perform this role, liberating them from their original service as symbols for sums and products.[13]

More than sixty years later, Mostowski [1948] introduced the interpretation of quantifiers as joins and meets. In the interim, lattice theory had become well established, and from a logical perspective joins and meets in a lattice are abstractions of disjunctions and conjunctions. Mostowski worked with partial orderings that are *complete*, i.e. every set of elements has a join and a meet, so that his interpretation of the quantifiers is always available. In fact he used complete Brouwerian lattices, which are algebraic models of intuitionistic logic, because his purpose was to give a method of obtaining counter models to show that certain formulas are not deducible in Heyting's intuitionistic first-order logic. On such a lattice, each formula was interpreted as a "functional", a function of a certain kind taking values in the lattice. We may view the lattice itself as an algebra of propositions partially ordered by logical implication, and the functionals as "propositional functions".

The key to this counter model technique is soundness: all theorems of intuitionistic first-order logic are valid in these models. Mostowski ending his paper by asking if the converse was true: is Heyting's logic semantically complete for validity in order-complete Brouwerian models? He also asked if there is a single such model that characterises the logic. Both questions were answered affirmatively by Rasiowa [1951], in her doctoral thesis supervised by Mostowski. She also answered those questions for a first-order version of the modal logic S4. For each logic, its *Lindenbaum algebra* was used. The elements of this algebra are the equivalence classes $\|\varphi\|$ of formulas φ, under the equivalence relation of deducible biconditional equivalence, i.e. $\|\varphi\| = \|\psi\|$ iff the formula $\varphi \leftrightarrow \psi$ is deducible in the logic. These elements are partially ordered by putting $\|\varphi\| \leq \|\psi\|$ iff the implication $\varphi \to \psi$ is

[13]This point appears to have been undervalued by Dipert [1994, p. 529], who claims that Mitchell's contribution was "much less significant than Peirce's remarks might indicate". Brady [2000, pp. 93–94] describes Peirce's account of Mitchell's method as "a generous reading", and "extravagant praise" that "suggests that Peirce may have understood Mitchell as having made a greater advance that he actually did". But she also acknowledges that Peirce "may still have thought of Σ and Π as sums and products, and not yet had the logical ideas, independent of algebraic ideas, that he was attributing to Mitchell in 1885".

deducible. The Lindenbaum algebra is not order-complete, but does have sufficiently many joins and meets to characterise the quantifiers. For each formula φ and variable x, the set

$$\{\|\varphi(y/x)\| : y \text{ is any variable}\} \tag{1.4.1}$$

has a join, namely $\|\exists x\varphi\|$, and a meet, namely $\|\forall x\varphi\|$. Frege's postulates for the quantifiers are just what is needed to prove this.

Rasiowa used the order-completion construction of MacNeille [1937] and results of McKinsey and Tarski [1944] to embed the Lindenbaum algebra of quantified S4 into a complete S4-algebra, thereby showing that the logic is characterised by algebraic models involving functionals taking values in order-complete algebras. Further results of McKinsey and Tarski [1946] on the relationship between Brouwerian lattices and S4-algebras were used to provide a parallel construction answering Mostowski's questions about intuitionistic logic.

Rasiowa and Sikorski [1950] used this methodology in their algebraic proof of the completeness theorem for classical (non-modal) first-order logic. Instead of taking the MacNeille completion of the Lindenbaum algebra, they applied to it their now famous Lemma about the existence of prime ideals preserving countably many prescribed joins.[14] This allowed them to characterise the logic by functional models whose value algebra is the (order-complete) two-element Boolean algebra. Such models are identifiable with the Tarskian relational structures that are the models for the usual semantics of classical first-order logic.[15]

In the late 1940's, Tarski initiated a study of algebraic laws of quantification that evolved into an extensive theory of *cylindric* algebras [Henkin, Monk and Tarski, 1971, 1985; Henkin, Monk, Tarski, Andréka, and Németi, 1981]. These are Boolean algebras that have operators c_n abstracting the quantifying operators $\|\varphi\| \mapsto \|\exists v_n\varphi\|$ in a Lindenbaum algebra. One of the principal motivating examples of a cylindric algebra is the Boolean algebra of subsets of a set of the form U^ω. A member of U^ω may be thought of as a sequence $a_0, a_1, \dots$ of elements of U which serves as a variable-assignment giving the value a_n to the n-th variable v_n. U^ω is viewed as a Cartesian product space, and c_n is the geometric operation of cylindrification, taking each $X \subseteq U^\omega$ to the cylinder generated by translating X parallel to the n-th coordinate axis. If X is the set of assignments that satisfy some formula φ in a relational structure based on U, then this cylinder $\mathsf{c}_n X$ is the set of assignments that satisfy $\exists v_n\varphi$.

---

[14]This Lemma has been adapted to a variety of forms, often stated in terms of the existence of ultrafilters preserving countably many meets; or in terms of $\{0, 1\}$-valued homomorphisms preserving joins and meets; or, in the context of forcing in set theory, in terms of the existence of generic filters on partially ordered sets of forcing conditions.

[15]A comprehensive exposition of this whole theory about quantified classical, intuitionistic and S4-modal logics can be found in the book of Rasiowa and Sikorski [1963].

A cylindric algebra is *dimension-complemented* if for each of its members b there is some n with $\mathsf{c}_n b = b$. Any Lindenbaum algebra is dimension-complemented, because for each formula φ, if v_n is not free in φ then $\|\exists v_n \varphi\| = \|\varphi\|$. Dimension-complemented algebras satisfy the equation

$$\mathsf{c}_n b = \Sigma_m \mathsf{s}^n_m b, \tag{1.4.2}$$

expressing $\mathsf{c}_n b$ as a join [Henkin, Monk and Tarski 1971, 1.11.6], where s^n_m is a definable operation abstracting the operator $\|\varphi\| \mapsto \|\varphi(v_m/v_n)\|$ induced by the single-variable substitution (v_m/v_n). Equation (1.4.2) generalises the fact that

$$\|\exists v_n \varphi\| = \Sigma_m \|\varphi(v_m/v_n)\|$$

in a Lindenbaum algebra (see (1.4.1)).

What we have been discussing are cylindric algebras of *dimension* ω. The general theory replaces ω here by an arbitrary ordinal α. An α-dimensional cylindric algebra has cylindric operations c_n for all $n < \alpha$. Even more general are the *polyadic algebras* of Halmos [1954] (see Halmos [1962] for the collection of his papers on algebraic logic). A polyadic algebra has operators for quantification of whole sets of variables simultaneously, and other operators for arbitrary substitutions, not just single-variable ones. The main motivating example is the notion of a *functional polyadic algebra* as a collection of propositional functions of the form $U^\omega \to \boldsymbol{B}$, taking values in a Boolean algebra $\boldsymbol{B}$. Again this value algebra need not be order-complete, but must have enough joins and meets for the functional algebra to be closed under operations representing the quantifiers [Halmos 1962, p. 104].

We will be introducing propositional functions of the form $U^\omega \to \mathit{Prop}$ shortly (Section 1.6), and making them into a functional algebra in a later chapter (Section 4.1). One view of our overall approach is that it marries this tradition of interpreting quantifiers as joins and meets with the use of value algebras consisting of admissible propositions-as-sets-of-worlds. It is by giving up the requirement that the joins and meets be set unions and intersections, as well as the requirement that every set be admissible, that we obtain a more general notion of model, one that is able to semantically characterise logics that fail to be characterised by the set-theoretic interpretation of the quantifiers.

With this background in mind, and with the motivation given at the end of the last section, we turn to the formal development of our admissible model theory.

1.5. Model Structures

A *model structure* is a system $\mathcal{S} = (W, R, \mathit{Prop}, U, D)$ such that:

- W is a set, and R is a binary relation on W.

- *Prop* is a non-empty subset of the powerset $\wp W$ of W that is closed under binary intersections $X \cap Y$ and complements $-X$, hence under binary unions $X \cup Y$ and Boolean implications $X \Rightarrow Y = (-X) \cup Y$.
- *Prop* is closed under the operation $[R]$ defined by

$$[R]X = \{w \in W : \forall v \in W(wRv \text{ implies } v \in X)\}.$$

- U is a set, called the *universe* of $\mathcal{S}$; and
- D is a function assigning to each $w \in W$ a subset Dw of U, called the *domain of* w.

The Kripke frame (W, R) is called the *underlying frame*, or just *frame*, of the structure $\mathcal{S}$. (W, R, Prop) is the underlying *general frame* of $\mathcal{S}$. A subset of W is *admissible* if it belongs to *Prop*. Members of *Prop* are also called the *admissible propositions* of $\mathcal{S}$. *Prop* itself is a Boolean subalgebra of the powerset algebra $\wp W$.

Prop is also closed under the operation $\langle R\rangle$, dual to $[R]$, that has $\langle R\rangle X = -[R]-X$, or equivalently,

$$\langle R\rangle X = \{w \in W : \exists v \in W(wRv \text{ and } v \in X)\}.$$

For each $a \in U$ we define $Ea = \{w \in W : a \in Dw\}$. Sets of the form Ea may be referred to as "existence sets" or "existence propositions". They are not required to be admissible.

Using *Prop* we define, for each $X \subseteq W$,

$$X\!\downarrow = \bigcup\{Y \in \mathit{Prop} : Y \subseteq X\},$$
$$X\!\uparrow = \bigcap\{Y \in \mathit{Prop} : X \subseteq Y\}.$$

Then $X\!\downarrow \subseteq X \subseteq X\!\uparrow$. The sets $X\!\downarrow$ and $X\!\uparrow$ need not belong to *Prop*, but if they do, then $X\!\downarrow$ is the largest admissible subset of X, and $X\!\uparrow$ the smallest admissible superset.[16] So if $X \in \mathit{Prop}$, then $X\!\downarrow = X\!\uparrow = X$.

Operations $\sqcap$ and $\sqcup$ on collections of subsets of W are defined by putting, for each $\mathcal{Z} \subseteq \wp W$,

$$\sqcap\mathcal{Z} = (\bigcap\mathcal{Z})\!\downarrow = \bigcup\{Y \in \mathit{Prop} : Y \subseteq \bigcap\mathcal{Z}\},$$
$$\sqcup\mathcal{Z} = (\bigcup\mathcal{Z})\!\uparrow = \bigcap\{Y \in \mathit{Prop} : \bigcup\mathcal{Z} \subseteq Y\}.$$

Note that we do not require $\mathcal{Z} \subseteq \mathit{Prop}$ in these definitions. $\sqcap\mathcal{Z}$ and $\sqcup\mathcal{Z}$ are defined for arbitrary $\mathcal{Z} \subseteq \wp W$, and neither need be admissible in general, even when we do have $\mathcal{Z} \subseteq \mathit{Prop}$ (see Example 1.5.2 below). The following facts clarify the nature of these operations. The proofs are left to the reader.

LEMMA 1.5.1. (1) *If* $X \in \mathit{Prop}$, *then* $X \subseteq \sqcap\mathcal{Z}$ *iff* $X \subseteq \bigcap\mathcal{Z}$, *and* $\sqcup\mathcal{Z} \subseteq X$ *iff* $\bigcup\mathcal{Z} \subseteq X$.

[16] For the topologically minded: *Prop* is a base of clopen sets for topology, under which $X\!\downarrow$ and $X\!\uparrow$ are the interior and closure of X, respectively.

(2) *If $\mathcal{Z} \subseteq Prop$ and $\bigsqcap \mathcal{Z} \in Prop$, then $\bigsqcap \mathcal{Z}$ is the* greatest lower bound, *or* meet, *of $\mathcal{Z}$ in the partially-ordered set $(Prop, \subseteq)$, i.e. the largest admissible set included in every member of $\mathcal{Z}$.*
(3) *If $\mathcal{Z} \subseteq Prop$ and $\bigsqcup \mathcal{Z} \in Prop$, then $\bigsqcup \mathcal{Z}$ is the* least upper bound, *or* join, *of $\mathcal{Z}$ in the partially-ordered set $(Prop, \subseteq)$, i.e. the smallest admissible set that includes every member of $\mathcal{Z}$.*
(4) *If $\bigcap \mathcal{Z}$ is admissible, then $\bigsqcap \mathcal{Z} = \bigcap \mathcal{Z}$.*
(5) *If $\bigcup \mathcal{Z}$ is admissible, then $\bigsqcup \mathcal{Z} = \bigcup \mathcal{Z}$.* ⊣

EXAMPLE 1.5.2. It is quite possible that $\bigsqcap \mathcal{Z}$ is admissible while $\bigcap \mathcal{Z}$ is not, even when $\mathcal{Z} \subseteq Prop$. For example, let $W = \omega =$ the set $\{0, 1, \dots\}$ of natural numbers, and let *Prop* be the Boolean algebra consisting of all the finite subsets of $\{n : n > 0\} = \{1, 2, \dots\}$ and their complements in W. Then $\{0\}$ is not itself admissible, and its only admissible subset is $\emptyset$, i.e. $\{0\}{\downarrow} = \emptyset$. Let

$$\mathcal{Z} = \{W - \{n\} : n > 0\} \subseteq Prop.$$

Then $\bigcap \mathcal{Z} = \{0\} \notin Prop$, while $\bigsqcap \mathcal{Z} = \emptyset \in Prop$.

Illustrating further, if $\mathcal{Z}' = \{W - \{n\} : n \text{ is odd}\}$, then again $\mathcal{Z}' \subseteq Prop$, while $\bigcap \mathcal{Z}'$ is the set of even numbers, and $\bigsqcap \mathcal{Z}'$ is the set of *positive* even numbers, neither of which is admissible. This confirms that *Prop* need not be closed under the operation $\bigcap$.

The example does not depend on R, U or D, but to make it complete we can let U and D be arbitrary, and R any relation such that *Prop* is closed under $[R]$. For instance, if $R = \emptyset$, then $[R]X = W \in Prop$ for all X; or if R is the identity relation, then $[R]X = X$ in general.

We develop this further in Example 1.6.6. ⊣

We now record some facts about the operation $\bigsqcap$ that are useful for proving soundness results. These involve the Boolean set implication operation $\Rightarrow$, whose main property we use is that $Z \subseteq X \Rightarrow Y$ iff $Z \cap X \subseteq Y$.

In the following Lemma, X_i, Y_i, X_{ij} are subsets of W; $\mathcal{Z}$ is a subset of $\wp W$; and $\bigsqcap_{i \in I} X_i$ is $\bigsqcap\{X_i : i \in I\}$.

LEMMA 1.5.3. (1) *If $X_i \subseteq Y_i$ for all $i \in I$, then $\bigsqcap_{i \in I} X_i \subseteq \bigsqcap_{i \in I} Y_i$.*
(2) $\bigsqcap_{i \in I} \bigsqcap_{j \in J} X_{ij} = \bigsqcap_{j \in J} \bigsqcap_{i \in I} X_{ij}$.
(3) *If X is admissible, then $X \Rightarrow \bigsqcap \mathcal{Z} = \bigsqcap_{Y \in \mathcal{Z}}(X \Rightarrow Y)$.*
(4) *If $\{Y_i : i \in I\} \subseteq Prop$, then $\bigsqcap_{i \in I}(X_i \Rightarrow Y_i) = \bigsqcap_{i \in I}(X_i{\uparrow} \Rightarrow Y_i)$.*
(5) *(De Morgan)* $\bigsqcap_{i \in I} -X_i = -\bigsqcup_{i \in I} X_i$ *and* $\bigsqcup_{i \in I} -X_i = -\bigsqcap_{i \in I} X_i$.

PROOF. (1) $\bigcap_{i \in I} X_i \subseteq \bigcap_{i \in I} Y_i$, and the operation $\downarrow$ is $\subseteq$-monotonic.
(2) (N.B: the X_{ij}'s need not be admissible here.)

Let X be an admissible subset of $\bigsqcap_{i \in I} \bigsqcap_{j \in J} X_{ij}$. Then $X \subseteq X_{ij}$ for all $(i, j) \in I \times J$. So, for a given $j_0 \in J$ we have $X \subseteq X_{ij_0}$ for all $i \in I$, hence $X \subseteq \bigsqcap_{i \in I} X_{ij_0}$ because $X \in Prop$. Since this holds for every $j_0 \in J$,

$X \subseteq \bigsqcap_{j\in J}\bigsqcap_{i\in I} X_{ij}$, again as X is admissible. Symmetrically, each admissible subset of $\bigsqcap_{j\in J}\bigsqcap_{i\in I} X_{ij}$ is a subset of $\bigsqcap_{i\in I}\bigsqcap_{j\in J} X_{ij}$. Hence $\bigsqcap_{i\in I}\bigsqcap_{j\in J} X_{ij} = \bigsqcap_{j\in J}\bigsqcap_{i\in I} X_{ij}$, since both are unions of admissible subsets.

(3) (N.B: the members of $\mathcal{Z}$ need not be admissible.)

Since $Y \subseteq (X \Rightarrow Y)$, $\bigsqcap\mathcal{Z} \subseteq \bigsqcap_{Y\in\mathcal{Z}}(X \Rightarrow Y)$ by (1). Also, because $W - X \subseteq (X \Rightarrow Y)$, and also $W - X \in \mathit{Prop}$ since $X \in \mathit{Prop}$, we have $W - X \subseteq \bigsqcap_{Y\in\mathcal{Z}}(X \Rightarrow Y)$. Altogether then,

$$X \Rightarrow \bigsqcap\mathcal{Z} = (W - X) \cup \bigsqcap\mathcal{Z} \subseteq \bigsqcap_{Y\in S}(X \Rightarrow Y).$$

For the converse inclusion it is enough to show that any admissible subset of $\bigcap_{Y\in\mathcal{Z}}(X \Rightarrow Y)$ is a subset of $X \Rightarrow \bigsqcap\mathcal{Z}$. But if $Z \in \mathit{Prop}$ has $Z \subseteq \bigcap_{Y\in\mathcal{Z}}(X \Rightarrow Y)$, then for all $Y \in \mathcal{Z}$, $Z \subseteq (X \Rightarrow Y)$, so $Z \cap X \subseteq Y$. Hence $Z \cap X \subseteq \bigsqcap\mathcal{Z}$ as $Z \cap X \in \mathit{Prop}$. Therefore $Z \subseteq X \Rightarrow \bigsqcap\mathcal{Z}$.

(4) (N.B: the X_i need not be admissible.)

First, since $X_i \subseteq X_i{\uparrow}$, we have $(X_i{\uparrow} \Rightarrow Y_i) \subseteq (X_i \Rightarrow Y_i)$, for all $i \in I$. Hence $\bigsqcap_{i\in I}(X_i{\uparrow} \Rightarrow Y_i) \subseteq \bigsqcap_{i\in I}(X_i \Rightarrow Y_i)$ by (1).

For the converse inclusion, let Z be any admissible subset of the meet $\bigsqcap_{i\in I}(X_i \Rightarrow Y_i)$. Then for all $i \in I$, $Z \subseteq X_i \Rightarrow Y_i$, hence $X_i \subseteq Z \Rightarrow Y_i$. But $Z \Rightarrow Y_i$ is admissible (by admissibility of Z and Y_i), and so $X_i{\uparrow} \subseteq Z \Rightarrow Y_i$, implying that $Z \subseteq X_i{\uparrow} \Rightarrow Y_i$. Hence $Z \subseteq \bigsqcap_{i\in I}(X_i{\uparrow} \Rightarrow Y_i)$.

(5) (Again, the X_i need not be admissible.)

If $w \in (\bigcap_{i\in I} -X_i){\downarrow}$, then $w \in Y \subseteq \bigcap_{i\in I} -X_i$ for some admissible Y. Then $\bigcup_{i\in I} X_i \subseteq -Y$ and $w \notin -Y$. Since $-Y$ is also admissible, it follows that $w \notin (\bigcup_{i\in I} X_i){\uparrow}$. This shows $(\bigcap_{i\in I} -X_i){\downarrow} \subseteq -(\bigcup_{i\in I} X_i){\uparrow}$. The converse inclusion is shown similarly, giving the first De Morgan law. The other follows from the first by

$$-\bigsqcap_{i\in I} X_i = -\bigsqcap_{i\in I} -- X_i = --\bigsqcup_{i\in I} -X_i = \bigsqcup_{i\in I} -X_i.$$

⊣

We will adopt the convention that $\Rightarrow$ binds more strongly that $\bigsqcap$. Thus $\bigsqcap_{i\in I} X_i \Rightarrow Y_i$, will mean $\bigsqcap_{i\in I}(X_i \Rightarrow Y_i)$.

1.6. Premodels and Models

Let $\mathcal{L}$ be a signature. A *premodel* $\mathcal{M} = (\mathcal{S}, |-|^{\mathcal{M}})$ for $\mathcal{L}$, based on a model structure $\mathcal{S} = (W, R, \mathit{Prop}, U, D)$, is given by an interpretation function $|-|^{\mathcal{M}}$ on $\mathcal{L}$ that assigns:

- to each individual constant $\mathsf{c} \in \mathcal{L}$ an element $|\mathsf{c}|^{\mathcal{M}}$ of the universe U.
- to each n-ary function symbol $F \in \mathcal{L}$ an n-ary function $|F|^{\mathcal{M}}$ on the universe U, i.e. $|F|^{\mathcal{M}} : U^n \to U$.
- to each n-ary predicate symbol $P \in \mathcal{L}$ a function $|P|^{\mathcal{M}} : U^n \to \wp W$.

Intuitively, $|P|^{\mathcal{M}}(a_1, \dots, a_n)$ represents the proposition that the predicate P holds of the n-tuple $(a_1, \dots, a_n)$. The original semantics for quantified modal logic of [Kripke 1963b] interprets an n-ary relation symbol P as a function

$$\Phi(P, -) : W \to \wp(U^n)$$

assigning to each world w an n-ary relation $\Phi(P, w) \subseteq U^n$. From such a Φ we can define $|P| : U^n \to \wp W$ by

$$w \in |P|(a_1, \dots, a_n) \quad \text{iff} \quad (a_1, \dots, a_n) \in \Phi(P, w).$$

Alternatively, this can be viewed as a definition of Φ, given $|P|$, so the two methods are equivalent. For our approach, the use of the "proposition-valued" functions $|P|^{\mathcal{M}}$ will provide an effective way of handling the restriction to admissible propositions, and leads to a natural interpretation of all formulas as propositional functions, as we explain shortly.

It is worth emphasising that this kind of model theory allows relations and properties to hold of non-existent objects (e.g. Pegasus has wings). Thus it is *not* required that $\Phi(P, w) \subseteq (Dw)^n$; equivalently, it is not required that

$$|P|^{\mathcal{M}}(a_1, \dots, a_n) \subseteq Ea_1 \cap \dots \cap Ea_n.$$

In order for a premodel $\mathcal{M}$ to assign a truth set to a formula, the free individual variables of the formula must be given values in the universe U. Since the variables are presented as a sequence $v_0, v_1, \dots$, it is natural to take a variable-assignment to be a sequence $a_0, a_1, \dots$ of elements of U, or equivalently, a function from the set ω of natural numbers into U. Thus the set of variable-assignments is the set U^ω of all functions $f : \omega \to U$. Such an f assigns to each $\mathcal{L}$-term τ a value $|\tau|^{\mathcal{M}} f \in U$, so overall $\mathcal{M}$ interprets τ as a function $|\tau|^{\mathcal{M}} : U^\omega \to U$. The inductive definition of $|\tau|^{\mathcal{M}} f$ is:

- $|x|^{\mathcal{M}} f = fn$, if x is the variable $v_n \in \mathsf{InVar}$.
- $|\mathsf{c}|^{\mathcal{M}} f = |\mathsf{c}|^{\mathcal{M}}$.
- $|F\tau_1 \cdots \tau_n|^{\mathcal{M}} f = |F|^{\mathcal{M}}(|\tau_1|^{\mathcal{M}} f, \dots, |\tau_n|^{\mathcal{M}} f)$.

If term τ is closed, and so it built entirely from constants and function letters, then the value $|\tau|^{\mathcal{M}} f$ is the same for all $f \in U^\omega$, and this constant value may also be denoted $|\tau|^{\mathcal{M}}$.

A variable-assignment f will be extended to variables themselves, essentially by identifying v_n with n. Thus we write fx for fn when x is v_n, so we get $|x|^{\mathcal{M}} f = fx$. To symbolise changes in variable-assignment, we use the notation $f[a/x]$ for the function that *updates* f by assigning the value a to x and otherwise acting identically to f:

$$f[a/x]y = \begin{cases} a & \text{if } x = y, \\ fy & \text{otherwise.} \end{cases}$$

A premodel gives an interpretation $|\varphi|^{\mathcal{M}} : U^\omega \to \wp W$ to each $\mathcal{L}$-formula. This interpretation is a *propositional function*, i.e. a function whose values are propositions (not necessarily admissible ones). For each assignment f, $|\varphi|^{\mathcal{M}} f$ is to be the truth set of all worlds at which φ is true under f. This is defined by induction on the formation of φ:

- $|P\tau_1 \cdots \tau_n|^{\mathcal{M}} f = |P|^{\mathcal{M}}(|\tau_1|^{\mathcal{M}} f, \ldots, |\tau_n|^{\mathcal{M}} f)$.
- $|\mathsf{F}|^{\mathcal{M}} f = \emptyset$.
- $|\varphi \wedge \psi|^{\mathcal{M}} f = |\varphi|^{\mathcal{M}} f \cap |\psi|^{\mathcal{M}} f$.
- $|\neg\varphi|^{\mathcal{M}} f = W - |\varphi|^{\mathcal{M}} f$.
- $|\Box\varphi|^{\mathcal{M}} f = [R]|\varphi|^{\mathcal{M}} f$.
- $|\forall x\varphi|^{\mathcal{M}} f = \bigsqcap_{a\in U} \left(Ea \Rightarrow |\varphi|^{\mathcal{M}} f[a/x]\right)$.

Thus if $X \in Prop$, then $X \subseteq |\forall x\varphi|^{\mathcal{M}} f$ iff $X \subseteq Ea \Rightarrow |\varphi|^{\mathcal{M}} f[a/x]$ for all $a \in U$. We have

$$|\forall x\varphi|^{\mathcal{M}} f = \Big[\bigcap_{a\in U} Ea \Rightarrow |\varphi|^{\mathcal{M}} f[a/x]\Big]\downarrow$$
$$= \Big[\bigcap_{a\in U} (-Ea) \cup |\varphi|^{\mathcal{M}} f[a/x]\Big]\downarrow.$$

Identifying $\exists x$ with $\neg\forall x\neg$, and using one of the De Morgan laws of Lemma 1.5.3(5), leads to

$$|\exists x\varphi|^{\mathcal{M}} f = \bigsqcup_{a\in U} Ea \cap |\varphi|^{\mathcal{M}} f[a/x]$$
$$= \Big[\bigcup_{a\in U} Ea \cap |\varphi|^{\mathcal{M}} f[a/x]\Big]\uparrow.$$

Writing $\mathcal{M}, w, f \models \varphi$ to mean that $w \in |\varphi|^{\mathcal{M}} f$, we get the following clauses for this truth/satisfaction relation $\models$, with all except that for $\forall$ being standard:

- $\mathcal{M}, w, f \models P\tau_1 \cdots \tau_n$ iff $w \in |P|^{\mathcal{M}}(|\tau_1|^{\mathcal{M}} f, \ldots, |\tau_n|^{\mathcal{M}} f)$.
- $\mathcal{M}, w, f \not\models \mathsf{F}$.
- $\mathcal{M}, w, f \models \varphi \wedge \psi$ iff $\mathcal{M}, w, f \models \varphi$ and $\mathcal{M}, w, f \models \psi$.
- $\mathcal{M}, w, f \models \neg\varphi$ iff $\mathcal{M}, w, f \not\models \varphi$.
- $\mathcal{M}, w, f \models \Box\varphi$ iff for all $v \in W$ (wRv implies $\mathcal{M}, v, f \models \varphi$).
- $\mathcal{M}, w, f \models \forall x\varphi$ iff there is an $X \in Prop$ such that $w \in X$ and $X \subseteq \bigcap_{a\in U} \left(Ea \Rightarrow |\varphi|^{\mathcal{M}} f[a/x]\right)$.

From the clause for $\forall$ we see that

$$\mathcal{M}, w, f \models \forall x\varphi \ \textit{ only if } \ \text{for all } a \in Dw,\ \mathcal{M}, w, f[a/x] \models \varphi. \tag{1.6.1}$$

The converse need not hold, as we show in Example 1.6.6 below.

Reading off the semantics for the defined existential quantifier, we get

$\mathcal{M}, w, f \models \exists x\varphi$ iff for all $X \in Prop$ such that $w \in X$, there exists $u \in X$ and $a \in Du$ with $\mathcal{M}, u, f[a/x] \models \varphi$.

Consequently:

$$\text{If some } a \in Dw \text{ has } \mathcal{M}, w, f[a/x] \models \varphi, \text{ then } \mathcal{M}, w, f \models \exists x\varphi. \tag{1.6.2}$$

Again, the converse can fail.

The defined connectives have the following truth sets:

- $|\mathsf{T}|^{\mathcal{M}} f = W$.
- $|\varphi \vee \psi|^{\mathcal{M}} f = |\varphi|^{\mathcal{M}} f \cup |\psi|^{\mathcal{M}} f$.
- $|\varphi \to \psi|^{\mathcal{M}} f = |\varphi|^{\mathcal{M}} f \Rightarrow |\psi|^{\mathcal{M}} f$.
- $|\varphi \leftrightarrow \psi|^{\mathcal{M}} f = (|\varphi|^{\mathcal{M}} f \Rightarrow |\psi|^{\mathcal{M}} f) \cap (|\psi|^{\mathcal{M}} f \Rightarrow |\varphi|^{\mathcal{M}} f)$.
- $|\Diamond\varphi|^{\mathcal{M}} f = \langle R \rangle |\varphi|^{\mathcal{M}} f$.

In terms of the truth relation, these state that

- $\mathcal{M}, w, f \models \mathsf{T}$.
- $\mathcal{M}, w, f \models \varphi \vee \psi$ iff $\mathcal{M}, w, f \models \varphi$ or $\mathcal{M}, w, f \models \psi$.
- $\mathcal{M}, w, f \models \varphi \to \psi$ iff $\mathcal{M}, w, f \models \varphi$ implies $\mathcal{M}, w, f \models \psi$.
- $\mathcal{M}, w, f \models \varphi \leftrightarrow \psi$ iff $[\, \mathcal{M}, w, f \models \varphi$ iff $\mathcal{M}, w, f \models \psi \,]$.
- $\mathcal{M}, w, f \models \Diamond\varphi$ iff for some $v \in W(wRv$ and $\mathcal{M}, v, f \models \varphi)$.

As with standard semantics, truth of a formula depends only on value-assignment to *free* variables:

LEMMA 1.6.1 (Free Assignment). *In any premodel $\mathcal{M}$, for any formula φ, if assignments $f, g \in U^{\omega}$ agree on all free variables of φ, then $|\varphi|^{\mathcal{M}} f = |\varphi|^{\mathcal{M}} g$.*

PROOF. This is by induction on the formation of φ, with the only departure from the standard argument being for formulas beginning with $\forall$.

Note first that for the case of atomic φ, we need the fact that if f and g agree on the variables that occur in a term τ, then $|\tau|^{\mathcal{M}} f = |\tau|^{\mathcal{M}} g$. That itself is shown by a standard induction on the formation of τ.

For the inductive case that φ is $\forall x\psi$, suppose that the Lemma holds for ψ. Then if f and g agree on all free variables of φ, then they agree on all free variables of ψ except perhaps for x. Hence for each $a \in U$, $f[a/x]$ and $g[a/x]$ agree on all free variables of ψ, so $|\psi|^{\mathcal{M}} f[a/x] = |\psi|^{\mathcal{M}} g[a/x]$ by induction hypothesis. Thus

$$|\varphi|^{\mathcal{M}} f = \prod_{a \in U} (Ea \Rightarrow |\psi|^{\mathcal{M}} f[a/x]) = \prod_{a \in U} (Ea \Rightarrow |\psi|^{\mathcal{M}} g[a/x]) = |\varphi|^{\mathcal{M}} g,$$

so the Lemma holds for φ. ⊣

This result has the familiar consequence that if φ is a *sentence*, i.e. has no free variables, then $|\varphi|^{\mathcal{M}} f = |\varphi|^{\mathcal{M}} g$ for all f and g, i.e. the propositional function $|\varphi|^{\mathcal{M}}$ is constant. So for each world w we have $\mathcal{M}, w, f \models \varphi$ either for every assignment f, or for no f. In the first case we can just write $\mathcal{M}, w \models \varphi$ (φ is *true* at w in $\mathcal{M}$), and in the second $\mathcal{M}, w \not\models \varphi$ (φ is *false* at w in $\mathcal{M}$). Thus the constant value taken by the function $|\varphi|^{\mathcal{M}}$ is the set $\{w : \mathcal{M}, w \models \varphi\}$ of worlds at which the sentence φ is true, and we can identify $|\varphi|^{\mathcal{M}}$ with this truth set.

The Free Assignment Lemma 1.6.1 can also be used to establish the usual relationship between syntactic substitution of terms for variables and updating of evaluations. Substituting τ for x in φ and evaluating the result at f gives the same result as first updating the value of x to that of τ, and then evaluating φ at the updated $f[\,|\tau|^{\mathcal{M}}f/x]$:

LEMMA 1.6.2 (Substitution). *Let φ be any formula, and τ a term that is free for x in φ. Then in any premodel $\mathcal{M}$, for any $f \in U^{\omega}$,*

$$|\varphi(\tau/x)|^{\mathcal{M}}f = |\varphi|^{\mathcal{M}}f[\,|\tau|^{\mathcal{M}}f/x].$$

Hence for any $w \in W$ and $f \in U^{\omega}$,

$$\mathcal{M}, w, f \models \varphi(\tau/x) \text{ iff } \mathcal{M}, w, f[\,|\tau|^{\mathcal{M}}f/x] \models \varphi.$$

PROOF. Again the only nonstandard case is when φ is of the form $\forall y\psi$. First, when x is not free in φ then f and $f[\,|\tau|^{\mathcal{M}}f/x]$ agree on all free variables of φ, and $\varphi(\tau/x)$ is just φ, so the result is given by Lemma 1.6.1.

Otherwise, x occurs free in φ, so $x \neq y$ and $\varphi(\tau/x) = \forall y(\psi(\tau/x))$. Since τ is free for x in $\forall y\psi$, it must also be free for x in ψ, and y does not occur in τ. Then

$$\begin{aligned} |\varphi(\tau/x)|^{\mathcal{M}}f &= \bigsqcap_{a \in U} Ea \Rightarrow |\psi(\tau/x)|^{\mathcal{M}}f[a/y], \quad \text{and} \\ |\varphi|^{\mathcal{M}}f[\,|\tau|^{\mathcal{M}}f/x] &= \bigsqcap_{a \in U} Ea \Rightarrow |\psi|^{\mathcal{M}}f[\,|\tau|^{\mathcal{M}}f/x][a/y]. \end{aligned} \tag{1.6.3}$$

But for any $a \in U$, the induction hypothesis on ψ gives

$$|\psi(\tau/x)|^{\mathcal{M}}f[a/y] = |\psi|^{\mathcal{M}}f[a/y][\,|\tau|^{\mathcal{M}}f[a/y]/x],$$

and $|\tau|^{\mathcal{M}}f[a/y] = |\tau|^{\mathcal{M}}f$ because y is not in τ, while

$$f[a/y][\,|\tau|^{\mathcal{M}}f/x] = f[\,|\tau|^{\mathcal{M}}f/x][a/y]$$

as $y \neq x$. So altogether

$$|\psi(\tau/x)|^{\mathcal{M}}f[a/y] = |\psi|^{\mathcal{M}}f[\,|\tau|^{\mathcal{M}}f/x][a/y].$$

Hence $|\varphi(\tau/x)|^{\mathcal{M}}f = |\varphi|^{\mathcal{M}}f[\,|\tau|^{\mathcal{M}}f/x]$ in this case, by (1.6.3). ⊣

Since the set-theoretic operations $\cap$, $-$, $[R]$ are defined for all arguments from $\wp W$, admissible or not, and $\bigsqcap \mathcal{Z}$ is defined for all $\mathcal{Z} \subseteq \wp W$, the truth sets $|\varphi|^{\mathcal{M}}f$ are all defined regardless of whether they, or any component of their construction, is admissible. So these truth sets may or may not be admissible. We will say that a formula φ is *admissible in* $\mathcal{M}$ if the function $|\varphi|^{\mathcal{M}}$ has the form $U^{\omega} \to \mathit{Prop}$, i.e. $|\varphi|^{\mathcal{M}}f \in \mathit{Prop}$ for all $f \in U^{\omega}$.

Given the closure properties of *Prop* it is evident that the set of admissible formulas is closed under the Boolean connectives and $\Box$. In particular, if every atomic formula is admissible, then every *quantifier-free* formula is admissible.

A *model* for $\mathcal{L}$ is, by definition, a premodel in which every $\mathcal{L}$-formula is admissible. Note that in a model, for n-ary P the function $|P|^{\mathcal{M}}$ has the form $U^n \to \mathit{Prop}$. This holds because for any $a_1, \ldots, a_n \in U$, choosing distinct variables $x_1, \ldots, x_n$ and an $f \in U^{\omega}$ with $fx_i = a_i$ for $1 \leq i \leq n$ we get

$$|P|^{\mathcal{M}}(a_1, \ldots, a_n) = |Px_1 \cdots x_n|^{\mathcal{M}}f \in \mathit{Prop} \tag{1.6.4}$$

by admissibility of the atomic formula $Px_1 \cdots x_n$.

LEMMA 1.6.3. *In any model* $\mathcal{M}$, $|\forall x\varphi|^{\mathcal{M}} f = \bigsqcap_{a \in U} \big(Ea\!\uparrow \Rightarrow |\varphi|^{\mathcal{M}} f[a/x]\big)$.

PROOF. As φ is admissible in $\mathcal{M}$, $\{|\varphi|^{\mathcal{M}} f[a/x] : a \in U\} \subseteq \mathit{Prop}$. Hence by Lemma 1.5.3(4),

$$\bigsqcap_{a \in U} \big(Ea \Rightarrow |\varphi|^{\mathcal{M}} f[a/x]\big) = \bigsqcap_{a \in U} \big(Ea\!\uparrow \Rightarrow |\varphi|^{\mathcal{M}} f[a/x]\big). \qquad \dashv$$

A premodel $\mathcal{M}$ will be called *Kripkean* if it always has

$$|\forall x\varphi|^{\mathcal{M}} f = \bigcap_{a \in U} \big(Ea \Rightarrow |\varphi|^{\mathcal{M}} f[a/x]\big). \tag{1.6.5}$$

This means that $\forall$ gets the varying-domain semantics of [Kripke 1963b]:

$$\mathcal{M}, w, f \models \forall x\varphi \;\text{ iff }\; \text{for all } a \in Dw,\; \mathcal{M}, w, f[a/x] \models \varphi. \tag{1.6.6}$$

Correspondingly, the existential quantifier get the semantics

$$\mathcal{M}, w, f \models \exists x\varphi \;\text{ iff }\; \text{for some } a \in Dw,\; \mathcal{M}, w, f[a/x] \models \varphi \tag{1.6.7}$$

(see (1.6.2)).

Another way of describing Kripkean models is to use a premodel $\mathcal{M}$ to inductively assign to each formula φ a function $\|\varphi\|^{\mathcal{M}}$ on U^{ω} as follows:

- For atomic φ, including the case $\varphi = \mathsf{F}$, $\|\varphi\|^{\mathcal{M}} f = |\varphi|^{\mathcal{M}} f$ by definition.
- For the inductive case of the propositional connectives, $\|\varphi \wedge \psi\|^{\mathcal{M}} f = \|\varphi\||^{\mathcal{M}} f \cap \|\psi\|^{\mathcal{M}} f$, $\|\neg\varphi\|^{\mathcal{M}} f = W - \|\varphi\|^{\mathcal{M}} f$, $\|\Box\varphi\|^{\mathcal{M}} f = [R]\|\varphi\|^{\mathcal{M}} f$.
- For the quantifier case, $\|\forall x\varphi\|^{\mathcal{M}} f = \bigcap_{a \in U} \big(Ea \Rightarrow \|\varphi\|^{\mathcal{M}} f[a/x]\big)$.

$\|\varphi\|^{\mathcal{M}}$ is the *Kripkean interpretation of* φ *in* $\mathcal{M}$. We have

LEMMA 1.6.4. *A premodel* $\mathcal{M}$ *is Kripkean iff* $\|\varphi\|^{\mathcal{M}} = |\varphi|^{\mathcal{M}}$ *for all formulas* φ.

PROOF. Assuming $\mathcal{M}$ is Kripkean, we prove the result $\|\varphi\|^{\mathcal{M}} = |\varphi|^{\mathcal{M}}$ by induction on φ. For φ atomic, this holds by definition. The inductive cases of the propositional connectives are straightforward. For the case of $\forall$, assuming inductively that $\|\varphi\|^{\mathcal{M}} = |\varphi|^{\mathcal{M}}$ gives

$$\|\forall x\varphi\|^{\mathcal{M}} f = \bigcap_{a \in U} \big(Ea \Rightarrow \|\varphi\|^{\mathcal{M}} f[a/x]\big) = \bigcap_{a \in U} \big(Ea \Rightarrow |\varphi|^{\mathcal{M}} f[a/x]\big),$$

which is $|\forall x\varphi|^{\mathcal{M}} f$ by (1.6.5), so the result holds for $\forall x\varphi$. Hence it holds for all formulas.

Conversely, if $\|\varphi\|^{\mathcal{M}} = |\varphi|^{\mathcal{M}}$ and $\|\forall x\varphi\|^{\mathcal{M}} = |\forall x\varphi|^{\mathcal{M}}$, then the definition of $\|\forall x\varphi\|^{\mathcal{M}}$ immediately gives (1.6.5), so $\mathcal{M}$ is Kripkean. $\dashv$

LEMMA 1.6.5. *A* ***model*** $\mathcal{M}$ *is Kripkean iff any set of the form*

$$\bigcap_{a \in U} Ea \Rightarrow |\varphi|^{\mathcal{M}} f[a/x]$$

is admissible.

PROOF. Let $\mathcal{Z} = \{Ea \Rightarrow |\varphi|^{\mathcal{M}} f[a/x] : a \in U\}$. If $\mathcal{M}$ is Kripkean, then $|\forall x\varphi|^{\mathcal{M}} f = \bigcap \mathcal{Z}$, so $\bigcap \mathcal{Z} \in \mathit{Prop}$ by admissibility of the formula $\forall x\varphi$ in the model $\mathcal{M}$. The converse holds in any premodel: if $\bigcap \mathcal{Z} \in \mathit{Prop}$, then $\bigcap \mathcal{Z} = \bigsqcap \mathcal{Z} = |\forall x\varphi|^{\mathcal{M}} f$, so $\mathcal{M}$ is Kripkean. ⊣

If a model structure $\mathcal{S}$ is *full* in the sense that *all* subsets of W are admissible, i.e. $\mathit{Prop} = \wp W$, then in general $\bigcap \mathcal{Z}$ is admissible, so $\bigsqcap \mathcal{Z} = \bigcap \mathcal{Z}$. In that case all (pre)models on $\mathcal{S}$ are Kripkean.

EXAMPLE 1.6.6 (A Non-Kripkean Model). This is a continuation of Example 1.5.2, in which $W = \omega$, the set of natural numbers, and *Prop* consists of the finite sets of *positive* numbers and their complements in W. Make this into a model structure $\mathcal{S}$ by putting $R = \emptyset$, $U = \omega$, and specifying domains by $D(0) = \emptyset$ and $Dn = \{n\}$ for $n > 0$. Then the existence sets are given by $E(0) = \emptyset$ and $En = \{n\}$ for $n > 0$, so every existence set is admissible.

If $\mathcal{M}$ were Kripkean, then since $\mathcal{M}, w, g \not\models \mathsf{F}$, by (1.6.6) we would have

$$\mathcal{M}, w \models \forall x\mathsf{F} \text{ iff } Dw = \emptyset. \tag{1.6.8}$$

Now $\mathcal{M}, w \models \forall x\mathsf{F}$ implies $Dw = \emptyset$ in any premodel by (1.6.1). But the converse fails in all premodels on the present model structure $\mathcal{S}$, so in fact *no* premodel on $\mathcal{S}$ can be Kripkean, and in particular no *model* on $\mathcal{S}$ can be Kripkean. To see this, note that in any premodel $\mathcal{M}$ on $\mathcal{S}$, for all $n \in U$,

$$En \Rightarrow |\mathsf{F}|^{\mathcal{M}} f[n/x] = En \Rightarrow \emptyset = \begin{cases} -\{n\} & \text{if } n > 0 \\ \omega & \text{if } n = 0, \end{cases}$$

so $\bigcap_{n \in U} En \Rightarrow |\mathsf{F}|^{\mathcal{M}} f[n/x] = \{0\} \notin \mathit{Prop}$. Hence

$$|\forall x\mathsf{F}|^{\mathcal{M}} f = \bigsqcap_{n \in U} En \Rightarrow |\mathsf{F}|^{\mathcal{M}} f[n/x] = \emptyset,$$

since $\emptyset$ is the only, hence largest, admissible subset of $\{0\}$. In other words, $\forall x\mathsf{F}$ is false everywhere in any premodel $\mathcal{M}$ on $\mathcal{S}$. But then $D(0) = \emptyset$ while $\mathcal{M}, 0 \not\models \forall x\mathsf{F}$, showing that (1.6.8) fails, so $\mathcal{M}$ cannot be Kripkean.

This provides the promised counterexample to the converse of (1.6.1), since it is *vacuously* true that for all $a \in D(0)$, $\mathcal{M}, 0, f[a/x] \models \mathsf{F}$, while $\mathcal{M}, 0, f \not\models \forall x\mathsf{F}$ in general. We also see that $\|\forall x\mathsf{F}\|^{\mathcal{M}} f = \{0\} \neq |\forall x\mathsf{F}|^{\mathcal{M}} f$, showing that $\|\forall x\mathsf{F}\|^{\mathcal{M}} f$ is not admissible, while $|\forall x\mathsf{F}|^{\mathcal{M}} f$ is admissible.

Now since $\exists x\mathsf{T}$ is equivalent to $\neg\forall x\mathsf{F}$, we have $\mathcal{M}, 0 \models \exists x\mathsf{T}$ while $D(0) = \emptyset$. A Kripkean premodel would have

$$\mathcal{M}, w \models \exists x\mathsf{T} \text{ iff } Dw \neq \emptyset.$$

This also shows that the converse of (1.6.2) fails.

To exhibit a non-Kripkean *model*, it remains only to show that there is at least one premodel on $\mathcal{S}$ for which every formula is admissible. For any signature $\mathcal{L}$, define $|P|^{\mathcal{M}}$ to be the *empty function* for all predicate letters, i.e. $|P|^{\mathcal{M}}(m_1, \ldots, m_n) = \emptyset$ for all arguments, and interpret individual constants and function letters arbitrarily. Then an induction on formation shows that for any formula φ, $|\varphi|^{\mathcal{M}}$ is either the empty function, or the constantly W-valued function. Hence every $\mathcal{L}$-formula is admissible in $\mathcal{M}$, making $\mathcal{M}$ a model. ⊣

A formula φ is *valid in* premodel $\mathcal{M}$, written $\mathcal{M} \models \varphi$, if $|\varphi|^{\mathcal{M}} f = W$ for all f, i.e. if $\mathcal{M}, w, f \models \varphi$ for all $w \in W$ and $f \in U^{\omega}$. Concerning the terminology, "valid" here refers to truth for all values of the (free) variables of φ.[17]

A formula φ will be called *valid in* a model structure $\mathcal{S}$, written $\mathcal{S} \models \varphi$, if φ is valid in all ***models*** on $\mathcal{S}$.

If $\mathscr{C}$ is a class of model structures, or a class of models, then φ will be called *valid in* $\mathscr{C}$, written $\mathscr{C} \models \varphi$, if it is valid in every member of $\mathscr{C}$. A logic L is *sound for* $\mathscr{C}$ if all theorems of L are valid in $\mathscr{C}$, i.e. $\vdash_{\mathrm{L}} \varphi$ implies $\mathscr{C} \models \varphi$. Conversely, L is *complete for* $\mathscr{C}$ if every formula that is valid in $\mathscr{C}$ is an L-theorem, i.e. $\mathscr{C} \models \varphi$ implies $\vdash_{\mathrm{L}} \varphi$. L is *characterised by* $\mathscr{C}$ if it is both sound and complete for $\mathscr{C}$, i.e. $\mathrm{L} = \{\varphi : \mathscr{C} \models \varphi\}$.

A structure $\mathcal{S}$ is an L-*structure* if L is sound for $\mathcal{S}$, i.e. all theorems of L are valid in $\mathcal{S}$. Then we may say that $\mathcal{S}$ *validates* L. Likewise, an L-*model* is a model in which all theorems of L are valid, i.e. the model validates L.

In a Kripkean model, the interpretation of the quantifier $\forall$ uses the set intersection operation $\bigcap$ and does not depend on *Prop*. From that point of view we could assume that every set of worlds was admissible: it would not change the truth relation. To formalise that claim, we associate with each model structure $\mathcal{S}$ the structure $\mathcal{S}^+$ that is identical to $\mathcal{S}$ except in having the powerset $\wp W$ in place of the *Prop* of $\mathcal{S}$. Then $\mathcal{S}^+$ is full, so every premodel on $\mathcal{S}^+$ is Kripkean. Given a premodel $\mathcal{M} = (\mathcal{S}, |-|^{\mathcal{M}})$ on $\mathcal{S}$, define $\mathcal{M}^+ = (\mathcal{S}^+, |-|^{\mathcal{M}})$. The idea here is that $\mathcal{M}$ assigns to each n-ary predicate symbol P a function $|P|^{\mathcal{M}} : U^n \to \mathit{Prop}$, which we then view as being a function $U^n \to \wp W$ in $\mathcal{M}^+$. In other words, $\mathcal{M}^+$ is just $\mathcal{M}$ viewed as a premodel on $\mathcal{S}^+$. The sense in which a Kripkean model is equivalent to a full one is given by the following result.

THEOREM 1.6.7. *If $\mathcal{M}$ is Kripkean, then for any formula φ we have $|\varphi|^{\mathcal{M}} = |\varphi|^{\mathcal{M}^+}$. Hence for all $w \in W$ and $f \in U^{\omega}$,*

$$\mathcal{M}, w, f \models \varphi \quad \textit{iff} \quad \mathcal{M}^+, w, f \models \varphi.$$

Therefore φ is valid in $\mathcal{M}$ iff it is valid in the full model $\mathcal{M}^+$.

PROOF. We prove $|\varphi|^{\mathcal{M}} = |\varphi|^{\mathcal{M}^+}$ by induction on φ. We have $|\mathsf{F}|^{\mathcal{M}} = \emptyset = |\mathsf{F}|^{\mathcal{M}^+}$, and $|P\tau_1 \cdots \tau_n|^{\mathcal{M}} f = |P\tau_1 \cdots \tau_n|^{\mathcal{M}^+} f$ by the definition of $\mathcal{M}^+$. The

[17] In a model $\mathcal{M}$ for propositional modal logic we referred to $\mathcal{M} \models A$ as a relation of *truth* rather than validity, since the values of the (propositional) variables of A are fixed in such a propositional model.

inductive cases for $\wedge$ and $\neg$ are straightforward. For the case of $\Box$, assuming the result for φ, we get

$$|\Box\varphi|^{\mathcal{M}} f = [R]|\varphi|^{\mathcal{M}} f = [R]|\varphi|^{\mathcal{M}^+} f = |\Box\varphi|^{\mathcal{M}^+} f.$$

For the quantifier case,

$$\begin{aligned}
|\forall x\varphi|^{\mathcal{M}} f &= \bigcap_{a\in U} \big(Ea \Rightarrow |\varphi|^{\mathcal{M}} f[a/x]\big) && \text{as } \mathcal{M} \text{ is Kripkean,}\\
&= \bigcap_{a\in U} \big(Ea \Rightarrow |\varphi|^{\mathcal{M}^+} f[a/x]\big) && \text{by induction hypothesis on } \varphi,\\
&= |\forall x\varphi|^{\mathcal{M}^+} f && \text{as } \mathcal{M}^+ \text{ is Kripkean, being full.}
\end{aligned}$$

This completes the proof of the first statement of the Theorem, from which the rest follows directly. ⊣

COROLLARY 1.6.8. *If φ is valid in $\mathcal{S}^+$, then φ is valid in every Kripkean premodel on $\mathcal{S}$.*

PROOF. Let $\mathcal{S}^+ \models \varphi$. Then for any Kripkean $\mathcal{M}$ on $\mathcal{S}$ we have $\mathcal{M}^+ \models \varphi$, hence $\mathcal{M} \models \varphi$ by the Theorem. ⊣

We will see that are cases where a particular φ is valid in all Kripkean models on some structure $\mathcal{S}$, but is not valid in $\mathcal{S}$, because it is falsifiable in some non-Kripkean model on $\mathcal{S}$ (see Theorem 1.11.3). Also, although we now see that each individual Kripkean model $\mathcal{M}$ can be identified with one ($\mathcal{M}^+$) on a full structure, when it comes to dealing with classes of Kripkean models it does not suffice to deal with full structures. There are logics that are characterised by the set of all Kripkean models on some structure, but is not characterised by any class of full structures at all (see Example 2.8.2).

1.7. Soundness

This section shows that the quantifier axioms AI, UD and VQ from Section 1.2 are valid in all models, and that the rules UG, TI and GC preserve this validity. This implies soundness of the minimal quantified modal logic QK for validity in all models, hence in all model structures. An indication is given of just where the *admissibility* of formulas is needed. We also discuss conditions under which the Commuting Quantifiers axiom is valid.

The section is self-contained, and the reader most interested in the completeness of QK for our semantics may skip ahead to the next section with impunity.

Let $\mathcal{M}$ be a premodel that will remain fixed until further notice.

THEOREM 1.7.1. *The Actual Instantiation and Universal Distribution axioms are valid in $\mathcal{M}$, and the Universal Generalisation and Term Instantiation rules are sound for validity in $\mathcal{M}$.*

PROOF. We deal with the two rules first, as they are simplest.

For the rule UG, if $\mathcal{M} \models \varphi$, then for any variable-assignment $f \in U^\omega$ and individual $a \in U$,

$$Ea \Rightarrow |\varphi|^{\mathcal{M}} f[a/x] = Ea \Rightarrow W = W,$$

so $|\forall x\varphi|^{\mathcal{M}} f = \bigsqcap\{W\} = W$. Hence $\mathcal{M} \models \forall x\varphi$.

For the rule TI, if $\mathcal{M} \models \varphi$ and τ is free for x in φ, then for any $f \in U^\omega$, by the Substitution Lemma 1.6.2, $|\varphi(\tau/x)|^{\mathcal{M}} f = |\varphi|^{\mathcal{M}} f[\,|\tau|^{\mathcal{M}} f/x] = W$. Hence $\mathcal{M} \models \varphi(\tau/x)$.

For the axiom AI, let y be free for x in φ. It suffices to show that for any f and a,

$$Ea \subseteq |\forall x\varphi \to \varphi(y/x)|^{\mathcal{M}} f[a/y]. \tag{1.7.1}$$

For then $Ea \Rightarrow |\forall x\varphi \to \varphi(y/x)|^{\mathcal{M}} f[a/y] = W$ for all $a \in U$, so

$$|\forall y(\forall x\varphi \to \varphi(y/x))|^{\mathcal{M}} f = \bigsqcap\{W\} = W,$$

and hence $\mathcal{M} \models \forall y(\forall x\varphi \to \varphi(y/x))$.

To prove (1.7.1), let $w \in Ea$. Then if $w \in |\forall x\varphi|^{\mathcal{M}} f[a/y]$, there exists $X \in \mathit{Prop}$ with

$$w \in X \subseteq \bigcap_{b \in U} Eb \Rightarrow |\varphi|^{\mathcal{M}} f[a/y][b/x].$$

In particular, when $b = a$, since $w \in Ea$ we get $w \in |\varphi|^{\mathcal{M}} f[a/y][a/x]$. But by the Substitution Lemma 1.6.2, $|\varphi|^{\mathcal{M}} f[a/y][a/x] = |\varphi(y/x)|^{\mathcal{M}} f[a/y]$ because $|y|^{\mathcal{M}} f[a/y] = a$. Thus

$$w \in |\forall x\varphi|^{\mathcal{M}} f[a/y] \Rightarrow |\varphi(y/x)|^{\mathcal{M}} f[a/y] = |\forall x\varphi \to \varphi(y/x)|^{\mathcal{M}} f[a/y].$$

For the axiom UD, suppose that $\mathcal{M}, w, f \models \forall x(\varphi \to \psi)$ and $\mathcal{M}, w, f \models \forall x\varphi$. Then there exist $X, Y \in \mathit{Prop}$ such that

$$w \in X \subseteq \bigcap_{a \in U} Ea \Rightarrow |\varphi \to \psi|^{\mathcal{M}} f[a/x], \quad \text{and}$$

$$w \in Y \subseteq \bigcap_{a \in U} Ea \Rightarrow |\varphi|^{\mathcal{M}} f[a/x].$$

Then $w \in X \cap Y \in \mathit{Prop}$, and for all a,

$$X \cap Y \cap Ea \subseteq |\varphi \to \psi|^{\mathcal{M}} f[a/x] \cap |\varphi|^{\mathcal{M}} f[a/x] \subseteq |\psi|^{\mathcal{M}} f[a/x],$$

hence $X \cap Y \subseteq Ea \Rightarrow |\psi|^{\mathcal{M}} f[a/x]$. This shows $\mathcal{M}, w, f \models \forall x\psi$. ⊣

Next we consider the validity of the Vacuous Quantification axiom, and see that this is where admissibility becomes essential.

THEOREM 1.7.2. *Suppose that x has no free occurrence in φ. If φ is admissible in $\mathcal{M}$, then $\mathcal{M} \models \varphi \to \forall x\varphi$.*

PROOF. For any $f \in U^\omega$ and $a \in U$, the assignments f and $f[a/x]$ agree on all free variables of φ, so by the Free Assignment Lemma 1.6.1,

$$|\varphi|^{\mathcal{M}} f = |\varphi|^{\mathcal{M}} f[a/x] \subseteq Ea \Rightarrow |\varphi|^{\mathcal{M}} f[a/x].$$

But $|\varphi|^{\mathcal{M}} f \in Prop$ *by* $\mathcal{M}$*-admissibility of* φ, so then

$$|\varphi|^{\mathcal{M}} f \subseteq \prod_{a \in U} \left(Ea \Rightarrow |\varphi|^{\mathcal{M}} f[a/x]\right) = |\forall x \varphi|^{\mathcal{M}} f.$$

Hence $|\varphi \to \forall x \varphi|^{\mathcal{M}} f = |\varphi|^{\mathcal{M}} f \Rightarrow |\forall x \varphi|^{\mathcal{M}} f = W$ for all f. ⊣

COROLLARY 1.7.3. *Every model validates* VQ.

PROOF. In a model, every φ is admissible. ⊣

Next we consider the Generalisation on Constants rule

$$\text{GC:} \quad \frac{\varphi(\mathsf{c}/x)}{\varphi}, \quad \text{if c is not in } \varphi.$$

This differs from the other rules in that it is not sound for validity in a single model $\mathcal{M}$: it may be that $\mathcal{M} \models \varphi(\mathsf{c}/x)$ but $\mathcal{M} \not\models \varphi$. For instance, suppose φ is the atomic formula Px with P a monadic predicate letter, and we take an $\mathcal{M}$ with $|P|^{\mathcal{M}}(|\mathsf{c}|^{\mathcal{M}}) = W$, while $|P|^{\mathcal{M}}(a) \neq W$ for some $a \in W$. Then $\varphi(\mathsf{c}/x)$ is the sentence $P\mathsf{c}$, which is valid in $\mathcal{M}$, but any $f \in U^\omega$ with $fx = a$ has $|Px|^{\mathcal{M}} f = |P|^{\mathcal{M}}(a) \neq W$, so φ is not valid in $\mathcal{M}$.

However, GC is sound for validity in a model structure $\mathcal{S}$. If $\mathcal{S} \models \varphi(\mathsf{c}/x)$, then $\varphi(\mathsf{c}/x)$ is valid in all models on $\mathcal{S}$, which means that $\varphi(\mathsf{c}/x)$ is valid no matter what interpretation is given to c by models on $\mathcal{S}$. But that should mean that φ is valid no matter how x interpreted, that is $\mathcal{S} \models \varphi$. This sketch of an argument can be formalised and shown to be correct. The details, somewhat intricate, are as follows.

The constant c is fixed throughout. Given any premodel $\mathcal{M}$ on $\mathcal{S}$, and any individual $b \in U^\omega$, a new premodel $\mathcal{M}[b]$ on $\mathcal{S}$ is defined by declaring it to be identical to $\mathcal{M}$ except in its interpretation of the constant c, which has $|\mathsf{c}|^{\mathcal{M}[b]} = b$. (A more explicit notation for the new model would be $\mathcal{M}[b/\mathsf{c}]$, but we are taking c as understood.)

Now it can be shown that if c is not in φ, then in general $|\varphi|^{\mathcal{M}} f = |\varphi|^{\mathcal{M}[b]} f$. Taking $b = fx$ here and using the Substitution Lemma 1.6.2 and the fact that $f = f[fx/x] = f[\,|\mathsf{c}|^{\mathcal{M}[fx]}/x]$, we derive

$$|\varphi|^{\mathcal{M}} f = |\varphi(\mathsf{c}/x)|^{\mathcal{M}[fx]} f.$$

This is enough to ensure that if $\varphi(\mathsf{c}/x)$ is valid in all premodels on $\mathcal{S}$, then so is φ. But we are after a stronger result, since we want to work with validity in models rather than premodels, so we also need $\mathcal{M}[fx]$ to be a model whenever $\mathcal{M}$ is. This requires that any set of the form $|\varphi|^{\mathcal{M}[fx]} g$ can be expressed as $|\varphi'|^{\mathcal{M}} g'$ for some φ' and g'. The strategy for achieving all this is reminiscent of the way we showed, at the end of Section 1.2, that GC is derivable in logics of the form QS. We show how to replace the constant c by a new variable, and later use TI to replace that variable by the variable x we are really interested in (see the proof of Theorem 1.7.7).

Now in $\mathcal{M}^{[b]}$, the constant c is assigned the value b. To evaluate a formula φ in $\mathcal{M}^{[b]}$ at g, i.e. to form $|\varphi|^{\mathcal{M}[b]}g$, we can equivalently replace c in φ by some new variable x, and then evaluate the result in $\mathcal{M}$ with x assigned b, forming $|\varphi(x/\mathsf{c})|^{\mathcal{M}}g[b/x]$. To prove this we first show that the same equivalence holds when evaluating terms:

LEMMA 1.7.4. *For any term τ that does not contain x, for all $b \in U$ and $g \in U^{\omega}$,*

$$|\tau|^{\mathcal{M}[b]}g = |\tau(x/\mathsf{c})|^{\mathcal{M}}g[b/x].$$

PROOF. By induction on the formation of τ. If τ is c, then the left side of this equation is b, the value of c in $\mathcal{M}[b]$. But now $\tau(x/\mathsf{c})$ is x, so the right side is $g[b/x]x = b$, so the equation holds.

If τ is a constant c′ other than c, then $\tau(x/\mathsf{c})$ is also c′, and the equation holds because $\mathcal{M}[b]$ and $\mathcal{M}$ interpret c′ identically.

If τ is a variable y, then $\tau(x/\mathsf{c})$ is also y, and the equation asserts that $gy = g[b/x]y$. This is true, because g and $g[b/x]$ differ only at x, and $x \neq y$ by the hypothesis that x is not in τ.

The inductive case involving a function symbol F follows readily because $\mathcal{M}[b]$ and $\mathcal{M}$ interpret F identically. For instance if τ is $F\tau'$ with F unary, and x is not in τ, then it is not in τ' so we can assume the result inductively for τ' and calculate

$$|F\tau'|^{\mathcal{M}[b]}g = |F|^{\mathcal{M}[b]}(|\tau'|^{\mathcal{M}[b]}g) = |F|^{\mathcal{M}}(|\tau'(x/\mathsf{c})|^{\mathcal{M}}g[b/x]),$$

which is $|F(\tau'(x/\mathsf{c}))|^{\mathcal{M}}g[b/x]$. But $F(\tau'(x/\mathsf{c})) = (F\tau')(x/\mathsf{c})) = \tau(x/\mathsf{c})$, so the result holds for τ in this case. ⊣

LEMMA 1.7.5. *For any formula φ that does not contain x, for all $b \in U$ and $g \in U^{\omega}$,*

$$|\varphi|^{\mathcal{M}[b]}g = |\varphi(x/\mathsf{c})|^{\mathcal{M}}g[b/x].$$

PROOF. By induction on the formation of φ. If φ is an atomic formula $P\tau_1 \cdots \tau_n$, then $\varphi(x/\mathsf{c})$ is $P\tau_1(x/\mathsf{c}) \cdots \tau_n(x/\mathsf{c})$. But the τ_i's do not contain x, so Lemma 1.7.4 holds for them. Using that and $|P|^{\mathcal{M}[b]} = |P|^{\mathcal{M}}$, the desired result then follows readily in this case.

The inductive cases for the connectives $\wedge$, $\neg$ and $\Box$ are straightforward and left to the reader. For the quantifier case, let φ be $\forall y\psi$. Then $\varphi(x/\mathsf{c})$ is $\forall y\psi(x/\mathsf{c})$. As x is not in φ, $y \neq x$ and x is not in ψ, so we can inductively assume the result for ψ. Now $|\varphi|^{\mathcal{M}[b]}g$ is, by definition,

$$\bigsqcap_{a \in U} Ea \Rightarrow |\psi|^{\mathcal{M}[b]}g[a/y],$$

and for each $a \in U$ the induction hypothesis on ψ gives

$$|\psi|^{\mathcal{M}[b]}g[a/y] = |\psi(x/\mathsf{c})|^{\mathcal{M}}g[a/y][b/x].$$

But $g[a/y][b/x] = g[b/x][a/y]$ as $y \neq x$, so altogether we can rewrite our expression for $|\varphi|^{\mathcal{M}[b]}g$ as

$$\prod_{a \in U} Ea \Rightarrow |\psi(x/\mathsf{c})|^{\mathcal{M}}g[b/x][a/y].$$

By definition this is $|\forall y\psi(x/\mathsf{c})|^{\mathcal{M}}g[b/x]$, i.e. $|\varphi(x/\mathsf{c})|^{\mathcal{M}}g[b/x]$, so the result holds for φ. ⊣

COROLLARY 1.7.6. *If $\mathcal{M}$ is a model, then so is $\mathcal{M}[b]$. If $\mathcal{M}$ is Kripkean, then so is $\mathcal{M}[b]$.*

PROOF. Let $\mathcal{M}$ be a model. We need to show that any given formula φ is admissible in $\mathcal{M}[b]$. So, take a variable x that is not in φ. Then for any $g \in U^{\omega}$, the set $|\varphi(x/\mathsf{c})|^{\mathcal{M}}g[b/x]$ is admissible as $\mathcal{M}$ is a model, hence $|\varphi|^{\mathcal{M}[b]}g$ is admissible by the Lemma.

Now suppose $\mathcal{M}$ is Kripkean. Given any formula of the form $\forall y\psi$, take a variable x that is not in this formula. Then for any $g \in U^{\omega}$,

$$\begin{aligned} |\forall y\psi|^{\mathcal{M}[b]}g &= |\forall y\psi(x/\mathsf{c})|^{\mathcal{M}}g[b/x] && \text{by Lemma 1.7.5} \\ &= \bigcap_{a \in U} Ea \Rightarrow |\psi(x/\mathsf{c})|^{\mathcal{M}}g[b/x][a/y] && \text{as } \mathcal{M} \text{ is Kripkean} \\ &= \bigcap_{a \in U} Ea \Rightarrow |\psi(x/\mathsf{c})|^{\mathcal{M}}g[a/y][b/x] && \text{as } x \neq y \\ &= \bigcap_{a \in U} Ea \Rightarrow |\psi|^{\mathcal{M}[b]}g[a/y] && \text{by Lemma 1.7.5.} \end{aligned}$$

This proves the Kripkean condition for $\mathcal{M}[b]$. ⊣

THEOREM 1.7.7. *The Generalisation on Constants rule is sound for validity in $\mathcal{S}$, where $\mathcal{S}$ is any model structure. This rule is also sound for the property of being valid in all Kripkean models on $\mathcal{S}$.*

PROOF. Let $\mathcal{S} \models \varphi(\mathsf{c}/x)$, with c not in φ. We need to show $\mathcal{S} \models \varphi$. Let y be a new variable not occurring in φ. We will show that $\varphi(y/x)$ is valid in $\mathcal{S}$, and from this that φ is valid in $\mathcal{S}$. Note that $\varphi(\mathsf{c}/x)(y/\mathsf{c}) = \varphi(y/x)$, as y is not in φ, hence not in $\varphi(\mathsf{c}/x)$.

Let $\mathcal{M}$ be any model on $\mathcal{S}$. Then for each $f \in U^{\omega}$, $\mathcal{M}[fy]$ is a model on $\mathcal{S}$ by the Corollary just proved, so $|\varphi(\mathsf{c}/x)|^{\mathcal{M}[fy]}f = W$ as $\varphi(\mathsf{c}/x)$ is valid in $\mathcal{S}$. But by Lemma 1.7.5,

$$|\varphi(\mathsf{c}/x)|^{\mathcal{M}[fy]}f = |\varphi(\mathsf{c}/x)(y/\mathsf{c})|^{\mathcal{M}}f[fy/y].$$

But $f[fy/y]$ is just f, and $\varphi(\mathsf{c}/x)(y/\mathsf{c})$ is $\varphi(y/x)$ as just noted. So this shows that $|\varphi(\mathsf{c}/x)|^{\mathcal{M}[fy]}f = |\varphi(y/x)|^{\mathcal{M}}f$, and therefore $|\varphi(y/x)|^{\mathcal{M}}f = W$. Since this holds for all $f \in U^{\omega}$, we have shown that $\mathcal{M} \models \varphi(y/x)$.

The Term Instantiation rule is sound for validity in $\mathcal{M}$ (Theorem 1.7.1), so we can now infer $\mathcal{M} \models \varphi(y/x)(x/y)$ because x is freely substitutable for y in $\varphi(y/x)$. But $\varphi(y/x)(x/y) = \varphi$, so we have shown that $\mathcal{M} \models \varphi$. Since $\mathcal{M}$ was an arbitrary model on $\mathcal{S}$, this proves that φ is valid in $\mathcal{S}$.

For the second statement we use the fact that $\mathcal{M}[fy]$ is Kripkean whenever $\mathcal{M}$ is to show that if $\varphi(\mathsf{c}/x)$ is valid in all Kripkean models on $\mathcal{S}$, then $\varphi(y/x)$ is valid in all Kripkean models on $\mathcal{S}$, and hence so is φ by soundness of Term Instantiation in any model. ⊣

We collect up our results so far:

THEOREM 1.7.8. *For any model structure* $\mathcal{S}$:

(1) *The set* $\mathrm{L}_{\mathcal{S}} = \{\varphi : \mathcal{S} \models \varphi\}$ *of formulas valid in all models on* $\mathcal{S}$ *is a quantified modal logic.*
(2) *The set* $\mathrm{L}^{K}_{\mathcal{S}}$ *of formulas valid in all Kripkean models on* $\mathcal{S}$ *is a quantified modal logic.*

PROOF. For each model $\mathcal{M}$ on $\mathcal{S}$, the set $\{\varphi : \mathcal{M} \models \varphi\}$ of formulas valid in $\mathcal{M}$ is closed under all the axioms and rules defining a quantified modal logic except the rule GC. This shown by Theorem 1.7.1 and Corollary 1.7.3 together with the readily-shown validity of Boolean tautologies and axiom K, and soundness of Modus Ponens and the Necessitation rule, in all models.

Hence for any set $\mathscr{C}$ of models on $\mathcal{S}$, the set $\mathrm{L}_{\mathscr{C}} = \{\varphi : \mathscr{C} \models \varphi\}$ of formulas valid in all members of $\mathscr{C}$ has all the closure properties of a logic except GC. Thus by Theorem 1.7.7, if $\mathscr{C}$ is either the set of all models on $\mathcal{S}$, or the set of all Kripkean models on $\mathcal{S}$, then $\mathrm{L}_{\mathscr{C}}$ $(= \mathrm{L}_{\mathcal{S}}$ or $\mathrm{L}^{K}_{\mathcal{S}})$ is closed under GC and therefore is a logic. ⊣

REMARK 1.7.9. By part (1) of this Theorem, $\mathrm{L}_{\mathcal{S}}$ is closed under the rule

$$\frac{\varphi}{\varphi(\mathsf{c}'/\mathsf{c})},$$

since this is just an instance of the rule Sub of Lemma 1.2.3. Thus if φ is valid in all models on $\mathcal{S}$, it remains valid in all such models when any of its constants c is replaced by any other constant c'. This reflects the fact, already noted in relation to soundness of GC, that validity of φ in $\mathcal{S}$ means validity no matter what interpretation is given to c by models on $\mathcal{S}$.

Similarly, validity of φ in $\mathcal{S}$ means validity no matter what interpretation is given by models on $\mathcal{S}$ to the predicate symbols occurring in φ. This leads to the soundness, with respect to validity in $\mathcal{S}$, of inference rules that permit replacement of atomic formulas by more complex ones. We make an extensive study of this in Chapter 4. ⊣

THEOREM 1.7.10 (Soundness for QK). *If* $\vdash_{\mathrm{QK}} \varphi$, *then* φ *is valid in all model structures.*

PROOF. QK is the intersection of all quantified modal logics, so for any model structure $\mathcal{S}$, QK is included in the logic $\mathrm{L}_{\mathcal{S}}$ of the previous Theorem. Hence $\mathcal{S}$ validates QK. ⊣

This soundness argument can be refined to logics of the form QS, as follows.

THEOREM 1.7.11. *Let $\mathcal{G} = (W, R, Prop)$ be a general frame that validates a set* S *of propositional modal formulas. Then the quantified logic* QS *is validated by every model structure based on $\mathcal{G}$.*

PROOF. Let $\mathcal{S} = (W, R, Prop, U, D)$ be based on $\mathcal{G}$. We will show that every $\mathcal{L}$-substitution instance of a member of S is valid in $\mathcal{S}$. Then all such substitution instances belong to the logic $\mathrm{L}_{\mathcal{S}}$ characterised by $\mathcal{S}$ (Theorem 1.7.8). But QS is defined as the smallest quantified logic containing all such instances, so $\mathrm{QS} \subseteq \mathrm{L}_{\mathcal{S}}$, which is the claimed result.

So let $\varphi = A(\varphi_1/p_1, \ldots, \varphi_n/p_n)$ be an $\mathcal{L}$-formula that is a substitution instance of member A of S. Let $\mathcal{M}$ be any $\mathcal{L}$-model on $\mathcal{S}$. Then we must show that $\mathcal{M} \models \varphi$.

Fix any $f \in U^\omega$, and define a propositional model $\mathcal{M}[f]$ on $\mathcal{F}$ by putting $|p_i|^{\mathcal{M}[f]} = |\varphi_i|^{\mathcal{M}} f$ for all $i \leq n$. Since $\mathcal{M}$ is a model, φ_i is admissible in $\mathcal{M}$, so indeed $|p_i|^{\mathcal{M}[f]} \in Prop$ as required (for propositional variables q other than the p_i's, let $|q|^{\mathcal{M}[f]}$ be $\emptyset$, or any other member of $Prop$).

An induction on *propositional* modal formulas B whose variables are among $p_1, \ldots, p_n$ then shows that in general

$$|B(\varphi_1/p_1, \ldots, \varphi_n/p_n)|^{\mathcal{M}} f = |B|^{\mathcal{M}[f]}.$$

When $B = p_i$, this holds by definition of $\mathcal{M}[f]$. When $B = \mathsf{F}$, it holds because $|\mathsf{F}|^{\mathcal{M}} f = \emptyset = |\mathsf{F}|^{\mathcal{M}[f]}$. For the inductive case of $\Box$, assuming the result holds for B, we get

$$\begin{aligned}
&|(\Box B)(\varphi_1/p_1, \ldots, \varphi_n/p_n)|^{\mathcal{M}} f \\
&= |\Box(B(\varphi_1/p_1, \ldots, \varphi_n/p_n))|^{\mathcal{M}} f && \text{by definition of substitution} \\
&= [R]|B(\varphi_1/p_1, \ldots, \varphi_n/p_n)|^{\mathcal{M}} f && \text{by the semantics of } \Box \\
&= [R]|B|^{\mathcal{M}[f]} && \text{by induction hypothesis on } B \\
&= |\Box B|^{\mathcal{M}[f]} && \text{by the semantics of } \Box,
\end{aligned}$$

so the result holds for $\Box B$. The inductive cases for $\wedge$ and $\neg$ are similar, and left to the reader.

Now taking $B = A$ in this result, we get

$$|\varphi|^{\mathcal{M}} f = |A(\varphi_1/p_1, \ldots, \varphi_n/p_n)|^{\mathcal{M}} f = |A|^{\mathcal{M}[f]} = W,$$

because $A \in \mathrm{S}$, so A is valid in all models on $\mathcal{F}$ by assumption. Since this holds for all $f \in U^\omega$, φ is valid in $\mathcal{M}$ as required. ⊣

COROLLARY 1.7.12. *Let (W, R) be a Kripke frame that validates a set* S *of propositional modal formulas. Then the quantified logic* QS *is validated by every model structure based on (W, R).*

PROOF. If S is validated by (W, R), then it is validated by any general frame $(W, R, Prop)$ based on (W, R). ⊣

Thus if the Kripke frame underlying a particular structure $\mathcal{S}$ validates S, then $\mathcal{S}$ itself validates QS. The converse of this can also be shown in certain circumstances, for instance when $\mathcal{S}$ is full, provided that we have infinitely many monadic predicate symbols available. We give a proof of this now, as we will need to apply the result later (in Section 2.8).

THEOREM 1.7.13. *Suppose that the signature $\mathcal{L}$ has infinitely many monadic predicate symbols. If a full structure $\mathcal{S}$ validates* QS, *then its underlying Kripke frame validates* S.

PROOF. Assign to each propositional variable p a monadic predicate symbol P_p, with $P_p \neq P_q$ whenever $p \neq q$. Fixing an individual variable x, we associate with each $p \in \mathsf{PropVar}$ the atomic $\mathcal{L}$-formula $P_p x$.

By uniform replacement of each p by $P_p x$, we then obtain a mapping $A \mapsto A^*$ of *propositional* modal formulas to quantifier-free $\mathcal{L}$-formulas. Here

$$A^* = A(P_{p_1}x/p_1, \dots, P_{p_n}x/p_n),$$

where $p_1, \dots, p_n$ are all the propositional variables of A.

Now let $\mathcal{M}$ be a propositional model on the Kripke frame $\mathcal{F}$ underlying $\mathcal{S}$. Then we define an $\mathcal{L}$-premodel $\mathcal{M}'$ on $\mathcal{S}$ by putting $|P_p|^{\mathcal{M}'} a = |p|^{\mathcal{M}}$ for all $a \in U$. The interpretation by $\mathcal{M}'$ can be arbitrary for constants, function symbols, and predicate symbols other than the P_p's. Since $\mathcal{S}$ is full, it is automatic that any truth set $|\varphi|^{\mathcal{M}'} f$ of $\mathcal{M}'$ is admissible, so $\mathcal{M}'$ is a model.

For each $f \in U^\omega$ we get $|P_p x|^{\mathcal{M}'} f = |p|^{\mathcal{M}}$, i.e. $|p^*|^{\mathcal{M}'} f = |p|^{\mathcal{M}}$. A routine induction on the formation of propositional modal formulas A then shows that in general

$$|A^*|^{\mathcal{M}'} f = |A|^{\mathcal{M}}.$$

But if $A \in$ S, then A^* is an $\mathcal{L}$-substitution-instance of a member of S, hence $\vdash_{\mathrm{QS}} A^*$, so $|A^*|^{\mathcal{M}'} f = W$ for any f because $\mathcal{S}$ validates QS. Hence $|A|^{\mathcal{M}} = W$. This shows that $\mathcal{M} \models A$.

So every member of S is true in every model on $\mathcal{F}$, i.e. $\mathcal{F}$ validates S. ⊣

We now discuss some conditions under which the axiom

CQ: $\forall x \forall y \varphi \to \forall y \forall x \varphi$

is valid in a model.[18] Of course we can assume $x \neq y$ here, for otherwise there is no work to do. Then assignments $f[a/x][b/y]$ and $f[b/y][a/x]$ are identical, and may be written $f[a/x, b/y]$ or $f[b/y, a/x]$.

We make use of the notion of an atomic Boolean algebra $\mathcal{B}$. An *atom* of $\mathcal{B}$ is a minimal non-zero element, i.e. a non-zero element that has no other non-zero element below it. $\mathcal{B}$ is *atomic* if each of its non-zero members has an atom below it. If $\mathcal{B}$ is atomic, then each member of $\mathcal{B}$ is the join of all the atoms below it. Every finite Boolean algebra is atomic. Every powerset

[18]The remainder of this Section comes from Goldblatt and Hodkinson [2009], although the result of Theorem 1.7.16 was given there for models rather than premodels.

Boolean algebra $\wp W$ is atomic, with its atoms being the singleton subsets $\{w\}$ of W.

LEMMA 1.7.14. *In a premodel $\mathcal{M}$, let $f \in U^\omega$ and let $\mathcal{B}$ be any Boolean subalgebra of Prop that contains the sets $|\varphi|^{\mathcal{M}} f[a/x, b/y]$, $|\forall x\varphi|^{\mathcal{M}} f[b/y]$, and $|\forall y\varphi|^{\mathcal{M}} f[a/x]$ for all $a, b \in U$. Then exactly the same atoms of $\mathcal{B}$ are included in $|\forall x\forall y\varphi|^{\mathcal{M}} f$ and $|\forall y\forall x\varphi|^{\mathcal{M}} f$.*

PROOF. Let X be an atom of $\mathcal{B}$ with $X \not\subseteq |\forall x\forall y\varphi|^{\mathcal{M}} f$. Then as $X \in Prop$, there exists $a_0 \in U$ such that

$$X \not\subseteq Ea_0 \Rightarrow |\forall y\varphi|^{\mathcal{M}} f[a_0/x]. \tag{1.7.2}$$

Hence $X \not\subseteq |\forall y\varphi|^{\mathcal{M}} f[a_0/x]$, so again as $X \in Prop$ there exists $b_0 \in U$ such that

$$X \not\subseteq Eb_0 \Rightarrow |\varphi|^{\mathcal{M}} f[a_0/x, b_0/y]. \tag{1.7.3}$$

Hence $X \not\subseteq |\varphi|^{\mathcal{M}} f[a_0/x, b_0/y]$. But X is a $\mathcal{B}$-atom and $|\varphi|^{\mathcal{M}} f[a_0/x, b_0/y] \in \mathcal{B}$, so X must be *disjoint* from $|\varphi|^{\mathcal{M}} f[a_0/x, b_0/y] = |\varphi|^{\mathcal{M}} f[b_0/y, a_0/x]$. Since $X \cap Ea_0 \neq \emptyset$ by (1.7.2), this implies

$$X \not\subseteq Ea_0 \Rightarrow |\varphi|^{\mathcal{M}} f[b_0/y, a_0/x].$$

Hence

$$X \not\subseteq \prod_{a \in U} Ea \Rightarrow |\varphi|^{\mathcal{M}} f[b_0/y, a/x] = |\forall x\varphi|^{\mathcal{M}} f[b_0/y].$$

Again the atomicity of X then makes X disjoint from $|\forall x\varphi|^{\mathcal{M}} f[b_0/y] \in \mathcal{B}$. Since $X \cap Eb_0 \neq \emptyset$ by (1.7.3),

$$X \not\subseteq Eb_0 \Rightarrow |\forall x\varphi|^{\mathcal{M}} f[b_0/y].$$

Hence

$$X \not\subseteq \prod_{b \in U} Eb \Rightarrow |\varphi|^{\mathcal{M}} f[b/y] = |\forall y\forall x\varphi|^{\mathcal{M}} f.$$

Conversely, interchanging x and y in this argument shows that if $X \not\subseteq |\forall y\forall x\varphi|^{\mathcal{M}} f$, then $X \not\subseteq |\forall x\forall y\varphi|^{\mathcal{M}} f$. ⊣

THEOREM 1.7.15. *A **model** validates* CQ *if any of the following hold of its model structure*:

(1) *Prop is an atomic Boolean algebra.*
(2) *The structure is full, i.e. Prop* $= \wp W$.
(3) *Prop is finite.*
(4) *The universe U is finite.*

PROOF. In a given model $\mathcal{M}$, put $\mathcal{B} = Prop$. Fix a formula φ.

(1) For any f, all sets of the form $|\varphi|^{\mathcal{M}} f[a/x, b/y]$, $|\forall x\varphi|^{\mathcal{M}} f[b/y]$, and $|\forall y\varphi|^{\mathcal{M}} f[a/x]$ are in $\mathcal{B}$ by admissibility. Likewise the sets $|\forall x\forall y\varphi|^{\mathcal{M}} f$ and $|\forall y\forall x\varphi|^{\mathcal{M}} f$ are in $\mathcal{B}$, and include the same atoms of $\mathcal{B}$ by Lemma 1.7.14. If $\mathcal{B}$ is atomic, this makes $|\forall x\forall y\varphi|^{\mathcal{M}} f = |\forall y\forall x\varphi|^{\mathcal{M}} f$.
(2) By (1), as $\wp W$ is atomic.

(3) By (1), as any finite Boolean algebra is atomic.
(4) If U is finite, then for any f,

$$\{|\forall x\forall y\varphi|^{\mathcal{M}}f, |\forall y\forall x\varphi|^{\mathcal{M}}f\} \\ \cup \{|\varphi|^{\mathcal{M}}f[a/x,b/y], |\forall x\varphi|^{\mathcal{M}}f[b/y], |\forall y\varphi|^{\mathcal{M}}f[a/x] : a,b \in U\}$$

is a finite subset of *Prop*, so it generates a Boolean subalgebra $\mathcal{B}$ of *Prop* that is finite, hence atomic. The proof that $|\forall x\forall y\varphi|^{\mathcal{M}}f = |\forall y\forall x\varphi|^{\mathcal{M}}f$ in $\mathcal{B}$ then follows by the argument of (1). ⊣

Next we consider consequences of admissibility of the "existence sets" Ea and $Ea\uparrow$.

THEOREM 1.7.16. *If a premodel has $Ea \in Prop$ for all $a \in U$, then it validates* CQ.

PROOF. $|\forall x\forall y\varphi|^{\mathcal{M}}f$

$$= \bigsqcap_{a\in U} \Big(Ea \Rightarrow \bigsqcap_{b\in U} (Eb \Rightarrow |\varphi|^{\mathcal{M}}f[a/x,b/y])\Big)$$

$$= \bigsqcap_{a\in U}\bigsqcap_{b\in U} \big(Ea \Rightarrow (Eb \Rightarrow |\varphi|^{\mathcal{M}}f[a/x,b/y])\big) \quad \text{by Lemma 1.5.3(3) as } Ea \in Prop,$$

$$= \bigsqcap_{a\in U}\bigsqcap_{b\in U} \big(Ea \cap Eb \Rightarrow |\varphi|^{\mathcal{M}}f[a/x,b/y]\big) \quad \text{by set theory.}$$

Similarly, $|\forall y\forall x\varphi|^{\mathcal{M}}f = \bigsqcap_{b\in U}\bigsqcap_{a\in U}\big(Eb \cap Ea \Rightarrow |\varphi|^{\mathcal{M}}f[b/y,a/x]\big)$.

But $Eb \cap Ea \Rightarrow |\varphi|^{\mathcal{M}}f[b/y,a/x] = Ea \cap Eb \Rightarrow |\varphi|^{\mathcal{M}}f[a/x,b/y]$, so the $\bigsqcap$-commutation result of Lemma 1.5.3(2) applies to give $|\forall x\forall y\varphi|^{\mathcal{M}}f = |\forall y\forall x\varphi|^{\mathcal{M}}f$. ⊣

COROLLARY 1.7.17. *If a model structure has $Ea\uparrow \in Prop$ for all $a \in U$, then it validates* CQ.

PROOF. In a model $\mathcal{M}$, we can use Lemma 1.6.3 to replace Ea by $Ea\uparrow$ in the definition of $|\forall x\varphi|^{\mathcal{M}}$, and carry through the argument of the Theorem with $Ea\uparrow$ and $Eb\uparrow$ in place of Ea and Eb. ⊣

THEOREM 1.7.18. *Every Kripkean premodel validates* CQ.

PROOF. Repeat the proof of Theorem 1.7.16 with $\bigcap$ in place of $\bigsqcap$, using the Kripkean definition of $|\forall x\varphi|^{\mathcal{M}}f$ of (1.6.5). Instead of parts (2) and (3) of Lemma 1.5.3, use the results

$$\bigcap_{i\in I}\bigcap_{j\in J} X_{ij} = \bigcap_{j\in J}\bigcap_{i\in I} X_{ij}, \qquad X \Rightarrow \bigcap \mathcal{Z} = \bigcap_{Y\in\mathcal{Z}} (X \Rightarrow Y).$$

These are laws of set theory that hold independently of any admissibility constraints. ⊣

We see from these results that a falsifying model for CQ cannot be full, i.e. must have non-admissible sets, and must have infinite sets for U and *Prop*, and hence for W. Also *Prop* cannot be atomic, and cannot contain every

Ea, or every $Ea\uparrow$, and the model cannot be Kripkean. Models of this kind are constructed in [Goldblatt and Hodkinson 2009], where it is shown that for every consistent propositional modal logic S there is a model of our kind that validates QS together with both the Barcan Formula and its converse, but falsifies CQ.

1.8. Infinitely Many Constants

From now on we will assume, unless otherwise stated, that the background signature $\mathcal{L}$ contains infinitely many individual constants. This assumption is needed for the construction of characteristic canonical models for various logics, starting in the next section. For some logics, $\mathcal{L}$ is required to be not too large, in fact countable (see Section 2.6), but that leaves room for a countable infinity of constants.

This assumption is justified by a standard argument to the effect that any logic L over $\mathcal{L}$ can be *conservatively* extended by the addition of a set of new constants, of any size. Let $\mathcal{L}' = \mathcal{L} \cup \mathrm{C}$ where C is a set of individual constants disjoint from $\mathcal{L}$, and let L$'$ be the smallest set of $\mathcal{L}'$-formulas that forms a logic including L. Then the conservativity consists in the fact that no new $\mathcal{L}$-formulas become theorems in passing from L to L$'$: if an $\mathcal{L}$-formula is an L$'$-theorem, then it must already have been an L-theorem.

The conservativity proof is standard for traditional systems of quantificational logic (e.g. Enderton 1972, pp. 116, 129). But since we are dealing with proof theory for actualist quantification, and using the non-traditional rules TI and GC, we explain how the argument works here.

First note that for any $\mathcal{L}'$-formula φ we have $\vdash_{\mathrm{L}'} \varphi$ iff there is a L-*proof sequence over* $\mathcal{L}'$ *for* φ, by which we mean a finite sequence $\varphi_0, \dots, \varphi_n$ of $\mathcal{L}'$-formulas with $\varphi_n = \varphi$, such that for all $i \leq n$, either

- $\varphi_i \in \mathrm{L}$; or
- φ_i is an instance of one of the axiom schemes K, AI, UD and VQ; or
- φ_i follows from previous members of the sequence by one of the rules MP, N, UG, TI and GC.

This holds because the set of all $\mathcal{L}'$-formulas φ for which there exists such a proof-sequence forms a logic over $\mathcal{L}'$ that includes L, and is included in any other logic over $\mathcal{L}'$ that includes L.

Now fix an L-proof sequence $\varphi_0, \dots, \varphi_n$ over $\mathcal{L}'$. Let $\mathsf{c}_0, \dots, \mathsf{c}_{k-1}$ be a list of all the constants from C that occur in any formula φ_i of the sequence. Take a fresh list $x_0, \dots, x_{k-1}$ of distinct individual variables that do not occur in any of the formulas in the proof sequence and, for each $i \leq n$, define

$$\varphi_i' = \varphi_i(x_0/\mathsf{c}_0, \dots, x_{k-1}/\mathsf{c}_{k-1}). \tag{1.8.1}$$

Then φ_i' is an $\mathcal{L}$-formula, since any C-constants occurring in φ_i have been replaced by new variables in forming φ_i'. If φ_i itself is an $\mathcal{L}$-formula, then $\varphi_i' = \varphi_i$.

LEMMA 1.8.1. *For all* $i \leq n$, $\vdash_{\mathrm{L}} \varphi_i'$.

PROOF. By induction on the number i, using the induction hypothesis that $\vdash_{\mathrm{L}} \varphi_j'$ for all $j < i$.

- If $\varphi_i \in \mathrm{L}$, then φ_i is an $\mathcal{L}$-formula, hence $\varphi_i' = \varphi_i$ and so $\vdash_{\mathrm{L}} \varphi_i'$.
- If φ_i is an instance of one of the axiom schemes, then it is readily checked that replacing constants by fresh variables not occurring in φ_i turns φ_i into another instance of the same scheme. Hence $\vdash_{\mathrm{L}} \varphi_i'$ since L includes all $\mathcal{L}$-formulas that are axioms.
- If φ_i follows from some formulas φ_j, φ_h with $j, h < i$ by Modus Ponens, because $\varphi_h = (\varphi_j \to \varphi_i)$, then $\varphi_h' = (\varphi_j' \to \varphi_i')$. But by induction hypothesis $\vdash_{\mathrm{L}} \varphi_j'$ and $\vdash_{\mathrm{L}} \varphi_h'$, so $\vdash_{\mathrm{L}} \varphi_i'$ by MP in L.

  If φ_i follows from some φ_j with $j < i$ by Necessitation, because $\varphi_i = \Box\varphi_j$, then $\varphi_i' = \Box\varphi_j'$ and $\vdash_{\mathrm{L}} \varphi_j'$, so $\vdash_{\mathrm{L}} \varphi_i'$ by rule N in L.

  Similarly, if φ_i is obtained by Universal Generalisation, i.e. $\varphi_i = \forall y \varphi_j$ with $j < i$, then $\varphi_i' = \forall y(\varphi_j')$ as none of the variables x_l is equal to y, hence from $\vdash_{\mathrm{L}} \varphi_j'$ we get $\vdash_{\mathrm{L}} \forall y(\varphi_j')$ by UG in L, so again $\vdash_{\mathrm{L}} \varphi_i'$.

  Next, for the case of Term Instantiation, if $\varphi_i = \varphi_j(\tau/y)$ with $j < i$ and τ free for y in φ_j, then τ' is free for y in φ_j', where τ' results from τ by the substitution $(x_0/\mathsf{c}_0, \ldots, x_{k-1}/\mathsf{c}_{k-1})$. Moreover, $\varphi_i' = \varphi_j'(\tau'/y)$, so $\vdash_{\mathrm{L}} \varphi_i'$ by rule TI in L.

  Finally, there is the case of Generalisation on Constants. This is a little different, as it uses TI instead of GC in L. Suppose there is some $j < i$ with $\varphi_j = \varphi_i(\mathsf{c}/y)$ and c not in φ_i. Since c does occur in φ_j, we have $\mathsf{c} = \mathsf{c}_h$ for some $h < k$. Then

  $$\varphi_j' = \varphi_i(\mathsf{c}_h/y)(x_0/\mathsf{c}_0, \ldots, x_{k-1}/\mathsf{c}_{k-1}).$$

  Now since neither c_h nor x_h is in φ_i, applying (c_h/y) and then (x_h/c_h) to φ_i has the same effect as applying (x_h/y). Indeed we could just as well construct φ_j' by first applying $(x_0/\mathsf{c}_0, \ldots, x_{k-1}/\mathsf{c}_{k-1})$ to φ_i, giving φ_i', and then apply (x_h/y). Either way, the result is φ_i with free y replaced by x_h, and every c_l replaced by x_l.

  In other words, $\varphi_j' = \varphi_i'(x_h/y)$. Hence $\vdash_{\mathrm{L}} \varphi_i'(x_h/y)$ by induction hypothesis. Now y is free for x_h in $\varphi_i'(x_h/y)$. The reason is that, since c_h does not occur in φ_i, x_h is not present in φ_i', so x_h occurs in $\varphi_i'(x_h/y)$ exactly in the places where φ_i' has free y, therefore occurs in places that are not in the scope of $\forall y$. So from $\vdash_{\mathrm{L}} \varphi_i'(x_h/y)$ we can infer $\vdash_{\mathrm{L}} \varphi_i'(x_h/y)(y/x_h)$ by the rule TI. But given that x_h is not in φ_i', $\varphi_i'(x_h/y)(y/x_h) = \varphi_i'$, so we finally get $\vdash_{\mathrm{L}} \varphi_i'$ in this case as well.

  That completes the inductive argument. ⊣

THEOREM 1.8.2. *L' is a conservative extension of* L.

PROOF. Let φ be an $\mathcal{L}$-formula such that $\vdash_{\mathrm{L}'} \varphi$. We have to show $\vdash_{\mathrm{L}} \varphi$. Now there exists an L-proof sequence over $\mathcal{L}'$ of the form

$$\varphi_0, \ldots, \varphi_n = \varphi.$$

By the above construction, taking $i = n$ in the Lemma just proved, we have $\vdash_{\mathrm{L}} \varphi'$. But $\varphi' = \varphi$ as φ is an $\mathcal{L}$-formula. ⊣

This proof of conservativity of adding constants can be adapted to many logics defined by specified axiom schemes. Thus if L is the logic QS over $\mathcal{L}$ for some set S of propositional formulas, we can take L' to be the smallest logic over $\mathcal{L}'$ that contains every $\mathcal{L}'$-formula that is a substitution-instance of a member of S. Then a proof that L' is a conservative extension of L proceeds as above, because if an $\mathcal{L}'$-formula φ_i is an instance of an S-member, then the $\mathcal{L}$-formula φ_i' as in (1.8.1) is an instance of the same member of S. Similarly for L of the form $\mathrm{QS} + \Sigma_1 + \cdots + \Sigma_n$, where the Σ_i are various schemes. For instance, if φ_i is an instance of any of CQ, IU, BF, or CBF, then φ_i' is an instance of the same scheme.

1.9. Canonical Models and Completeness

To obtain a completeness theorem for a quantified modal logic L, we build a model $\mathcal{M}_{\mathrm{L}}$ that is *characteristic* for L in the sense that it validates precisely the L-theorems:

$$\vdash_{\mathrm{L}} \varphi \quad \text{iff} \quad \mathcal{M}_{\mathrm{L}} \models \varphi.$$

For this we adapt a standard methodology in which the worlds of $\mathcal{M}_{\mathrm{L}}$ are maximally L-consistent sets of formulas. For propositional modal logics this produces a model called *the canonical* model of the logic, but for some quantified logics more that one such model will be constructed along these lines. So we follow the usage of Chellas [1980, p. 60], who takes the term "canonical" not to be a definite description of a unique model, but rather an adjective describing any structure in which

(i) worlds are sets of formulas (typically maximally consistent), and
(ii) truth at world w means *membership*: $w \models \varphi$ iff $\varphi \in w$.

Fix a quantified logic L for a signature $\mathcal{L}$ whose set $\mathsf{Con}_{\mathcal{L}}$ of individual constants is assumed to be infinite, an assumption just justified in the previous section. A formula φ is L-*deducible from* a set Γ of formulas, written $\Gamma \vdash_{\mathrm{L}} \varphi$, if for some $n \in \omega$ there are formulas $\psi_0, \ldots \psi_{n-1} \in \Gamma$ such that

$$\vdash_{\mathrm{L}} \psi_0 \to (\cdots \to (\psi_{n-1} \to \varphi) \cdots).$$

In the case that $n = 0$, this just means that $\vdash_{\mathrm{L}} \varphi$. One of the more useful facts about this deducibility relation is that

$$\Gamma \cup \{\psi\} \vdash_{\mathrm{L}} \varphi \quad \text{iff} \quad \Gamma \vdash_{\mathrm{L}} \psi \to \varphi.$$

Γ is L-*consistent* if $\Gamma \nvdash_{\mathrm{L}} \mathsf{F}$, and L-*maximal* if it is maximally L-consistent, i.e. L-consistent but has no proper L-consistent extension. The set of all L-maximal sets is denoted W_{L}, and for each formula φ, the set of L-maximal sets containing φ is denoted $|\varphi|_{\mathrm{L}}$.

We use many standard facts about maximal sets (Chellas [1980], Hughes and Cresswell [1996], Blackburn, de Rijke, and Venema [2001]), including

- Γ is L-maximal iff it is L-consistent and *negation complete*, which means that it contains either φ or $\neg\varphi$, for all $\mathcal{L}$-formulas φ.
- Every L-consistent set has an L-maximal extension (Lindenbaum's Lemma).
- $\Gamma \vdash_{\mathrm{L}} \varphi$ iff φ belongs to every L-maximal extension of Γ.
- φ is an L-theorem iff it belongs to every L-maximal set, i.e
$$\vdash_{\mathrm{L}} \varphi \quad \text{iff} \quad \varphi \in \bigcap W_{\mathrm{L}} \quad \text{iff} \quad |\varphi|_{\mathrm{L}} = W_{\mathrm{L}}.$$
- If Γ is L-maximal then
  $\mathsf{F} \notin \Gamma$,
  $\varphi \wedge \psi \in \Gamma$ iff $\varphi \in \Gamma$ and $\psi \in \Gamma$,
  $\neg\varphi \in \Gamma$ iff $\varphi \notin \Gamma$,
  $\varphi \vee \psi \in \Gamma$ iff $\varphi \in \Gamma$ or $\psi \in \Gamma$,
  $\varphi \to \psi \in \Gamma$ iff $\varphi \notin \Gamma$ or $\psi \in \Gamma$.
- $\vdash_{\mathrm{L}} \varphi \to \psi$ iff $|\varphi|_{\mathrm{L}} \subseteq |\psi|_{\mathrm{L}}$.

These facts show that $|\mathsf{F}|_{\mathrm{L}} = \emptyset$, $|\varphi|_{\mathrm{L}} \cap |\psi|_{\mathrm{L}} = |\varphi \wedge \psi|_{\mathrm{L}}$, and $W_{\mathrm{L}} - |\varphi|_{\mathrm{L}} = |\neg\varphi|_{\mathrm{L}}$. It follows that if we define

$$Prop_{\mathrm{L}} = \{\, |\varphi|_{\mathrm{L}} : \varphi \text{ is an } \mathcal{L}\text{-sentence}\},$$

then $Prop_{\mathrm{L}}$ is a Boolean subalgebra of the powerset algebra on $\wp W_{\mathrm{L}}$.[19]

A binary relation R_{L} is defined on W_{L} by

$$\Gamma R_{\mathrm{L}} \Delta \quad \text{iff} \quad \{\varphi : \Box\varphi \in \Gamma\} \subseteq \Delta.$$

Thus $\Gamma R_{\mathrm{L}} \Delta$ iff $\Box^{-}\Gamma \subseteq \Delta$, where $\Box^{-}\Gamma = \{\varphi : \Box\varphi \in \Gamma\}$. Using only the basic propositional modal axiom K, the Necessitation rule, and PC, it can be shown that, for any L-maximal Γ,

$$\Box\varphi \in \Gamma \quad \text{iff} \quad \Box^{-}\Gamma \vdash_{\mathrm{L}} \varphi \tag{1.9.1}$$

(see Chellas 1980, p. 159). It follows that R_{L} has the property that

$$\Box\varphi \in \Gamma \quad \text{iff} \quad \text{for all } \Delta \in W_{\mathrm{L}},\ \Gamma R_{\mathrm{L}} \Delta \text{ implies } \varphi \in \Delta.$$

[19]We could allow $Prop_{\mathrm{L}}$ to consist of the sets $|\varphi|_{\mathrm{L}}$ for arbitrary $\mathcal{L}$-formulas φ, but we show throughout that it suffices to admit only those such sets that can be "named" by a sentence.

This means that $[R_L]|\varphi|_L = |\Box\varphi|_L$, and hence that $Prop_L$ is closed under the operator $[R_L]$ induced on $\wp W_L$ by R_L.

Thus $(W_L, R_L, Prop_L)$ is a general frame for propositional modal logic.

To facilitate our definition of a model structure, associate with each $\mathcal{L}$-term τ the set of formulas

$$\mathrm{UI}(\tau) = \{\forall x\varphi \to \varphi(\tau/x) : x \in \mathsf{InVar} \text{ and } \varphi \text{ is any formula}\}$$

(the notation recalls the connection with Universal Instantiation). Now put

$$\mathcal{S}_L = (W_L, R_L, Prop_L, U_{\mathcal{L}}, D_L),$$

where:

- $(W_L, R_L, Prop_L)$ is the general propositional frame defined as above.
- $U_{\mathcal{L}}$ is the set of all *closed* $\mathcal{L}$-terms, i.e. terms without variables.
- For each $w \in W_L$, $D_L(w) = \{\tau \in U_{\mathcal{L}} : \mathrm{UI}(\tau) \subseteq w\}$.

$\mathcal{S}_L$ is the *canonical varying-domains model structure for* L. The idea behind the definition of $D_L(w)$, bearing in mind the "truth means membership" notion, is that for τ to exist at w, every universally quantified formula $\forall x\varphi$ should imply its instantiation by τ at w. The existence sets defined by these domains $D_L(w)$ satisfy

$$E\tau = \{w \in W_L : \tau \in D_L(w)\} = \{w \in W_L : \mathrm{UI}(\tau) \subseteq w\}. \tag{1.9.2}$$

The *canonical premodel* $\mathcal{M}_L = (\mathcal{S}_L, |-|^{\mathcal{M}_L})$ on $\mathcal{S}_L$ is defined by:

- $|\mathsf{c}|^{\mathcal{M}_L} = \mathsf{c} \in U_{\mathcal{L}}$, for all constants $\mathsf{c} \in \mathsf{Con}_{\mathcal{L}}$.
- $|F|^{\mathcal{M}_L}(\tau_1, \dots, \tau_n) = F\tau_1 \cdots \tau_n \in U_{\mathcal{L}}$, for all $\tau_1, \dots, \tau_n \in U_{\mathcal{L}}$.
- $|P|^{\mathcal{M}_L}(\tau_1, \dots, \tau_n) = |P\tau_1 \cdots \tau_n|_L \in Prop_L$, for all $\tau_1, \dots, \tau_n \in U_{\mathcal{L}}$.

We now begin a detailed proof-theoretic analysis leading to the conclusion that $\mathcal{M}_L$ is a characteristic model for L. First, a technical fact about universal quantification that depends only on Actual Instantiation.

LEMMA 1.9.1. *Let ψ be a member $\forall z\varphi \to \varphi(\mathsf{c}/z)$ of* $\mathrm{UI}(\mathsf{c})$ *for some constant* c. *Let y be a variable not occurring in φ. Then* $\vdash_L \forall y(\psi(y/\mathsf{c}))$.

PROOF. First note that

$$\varphi(\mathsf{c}/z)(y/\mathsf{c}) = \varphi(y/\mathsf{c})(y/z), \tag{1.9.3}$$

because the formulas on each side of the equation both differ from φ only in having free y exactly where φ has c or free z.

Now $\forall y(\psi(y/\mathsf{c}))$ is $\forall y([\forall z\varphi \to \varphi(\mathsf{c}/z)](y/\mathsf{c}))$, which, using (1.9.3), is

$$\forall y((\forall z\varphi)(y/\mathsf{c}) \to \varphi(y/\mathsf{c})(y/z)).$$

But $(\forall z\varphi)(y/\mathsf{c}) = \forall z(\varphi(y/\mathsf{c}))$, so this is just the instance

$$\forall y(\forall z\varphi' \to \varphi'(y/z))$$

of axiom AI, where $\varphi' = \varphi(y/\mathsf{c})$ and y is free for z in φ' because the quantifier $\forall y$ does not occur in φ, hence does not occur in φ'. ⊣

The next result is the heart of the matter. It is particularly interesting to see how the proof uses all of the quantificational principles AI, UD, VQ and UG involved in the definition of a quantified modal logic. The result shows that our semantics for $\forall$, based on the "conjunction" operation $\sqcap$ induced by *Prop*, exactly captures the meaning of the universal quantifier in Kripke-style logics.

LEMMA 1.9.2. *If* $\forall x\varphi$ *is a sentence, then in* $\mathcal{S}_{\mathrm{L}}$,

$$|\forall x\varphi|_{\mathrm{L}} = \bigsqcap_{\tau \in U_{\mathcal{L}}} E\tau \Rightarrow |\varphi(\tau/x)|_{\mathrm{L}}.$$

PROOF. Here $\sqcap$ is determined by $Prop_{\mathrm{L}}$, and $E\tau$ is as in (1.9.2).

Consider the condition

$$X \subseteq \bigcap_{\tau \in U_{\mathcal{L}}} E\tau \Rightarrow |\varphi(\tau/x)|_{\mathrm{L}}. \tag{1.9.4}$$

Now $\bigsqcap_{\tau \in U_{\mathcal{L}}} E\tau \Rightarrow |\varphi(\tau/x)|_{\mathrm{L}}$ is, by definition, the union of all admissible subsets of the intersection $\bigcap_{\tau \in U_{\mathcal{L}}} E\tau \Rightarrow |\varphi(\tau/x)|_{\mathrm{L}}$. But $|\forall x\varphi|_{\mathrm{L}}$ is admissible, since $\forall x\varphi$ being a sentence ensures $|\forall x\varphi|_{\mathrm{L}} \in Prop_{\mathrm{L}}$. So it suffices to show that

(i) (1.9.4) holds when $X = |\forall x\varphi|_{\mathrm{L}}$, and

(ii) if X is any admissible set satisfying (1.9.4), then $X \subseteq |\forall x\varphi|_{\mathrm{L}}$.

That makes $|\forall x\varphi|_{\mathrm{L}}$ the largest admissible set satisfying (1.9.4), which proves the Lemma.

For (i), let $\tau \in U_{\mathcal{L}}$ and suppose $w \in |\forall x\varphi|_{\mathrm{L}}$. Then if $w \in E\tau$, we have $\mathrm{UI}(\tau) \subseteq w$, so $\forall x\varphi \to \varphi(\tau/x) \in w$, hence as $\forall x\varphi \in w$ we get $\varphi(\tau/x) \in w$ and thus $w \in |\varphi(\tau/x)|_{\mathrm{L}}$. This shows that $w \in E\tau \Rightarrow |\varphi(\tau/x)|_{\mathrm{L}}$. So altogether,

$$|\forall x\varphi|_{\mathrm{L}} \subseteq E\tau \Rightarrow |\varphi(\tau/x)|_{\mathrm{L}}$$

for any $\tau \in U_{\mathcal{L}}$, proving (i).

For (ii), let X be admissible and satisfy (1.9.4). Then $X = |\psi|_{\mathrm{L}}$ for some sentence ψ, so we have to show that $|\psi|_{\mathrm{L}} \subseteq |\forall x\varphi|_{\mathrm{L}}$. For this it suffices that $\vdash_{\mathrm{L}} \psi \to \forall x\varphi$.[20]

Take a constant $\mathsf{c} \in U_{\mathcal{L}}$ that does not occur in φ or ψ—*this is where we need the assumption that* $\mathsf{Con}_{\mathcal{L}}$ *is infinite*. By assumption,

$$|\psi|_{\mathrm{L}} \subseteq E\mathsf{c} \Rightarrow |\varphi(\mathsf{c}/x)|_{\mathrm{L}}.$$

Now for any $w \in W_{\mathrm{L}}$, if $\mathrm{UI}(\mathsf{c}) \cup \{\psi\} \subseteq w$, then $w \in |\psi|_{\mathrm{L}}$ as $\psi \in w$, and $w \in E\mathsf{c}$ as $\mathrm{UI}(\mathsf{c}) \subseteq w$, so it follows that $w \in |\varphi(\mathsf{c}/x)|_{\mathrm{L}}$, i.e. $\varphi(\mathsf{c}/x) \in w$. This proves that $\varphi(\mathsf{c}/x)$ belongs to every L-maximal extension of $\mathrm{UI}(\mathsf{c}) \cup \{\psi\}$, which implies

$$\mathrm{UI}(\mathsf{c}) \cup \{\psi\} \vdash_{\mathrm{L}} \varphi(\mathsf{c}/x).$$

[20] The proof that follows does not require ψ itself to be a *sentence*. Thus if we allowed $Prop_{\mathrm{L}}$ to contain $|\psi|_{\mathrm{L}}$ for arbitrary formulas ψ, then this Lemma would be true for any formula of the form $\forall x\varphi$, not just those that are sentences.

Therefore there is an L-theorem of the form

$$\psi_0 \to (\cdots \to (\psi_{n-1} \to (\psi \to \varphi(\mathsf{c}/x))\cdots)$$

for some n and some formulas $\psi_i \in \mathrm{UI}(\mathsf{c})$. Call this L-theorem ψ', and take a new variable y that is distinct from x and does not occur in ψ'. By the rule Sub of Lemma 1.2.3, $\vdash_{\mathrm{L}} \psi'(y/\mathsf{c})$. Hence by the rule UG, $\vdash_{\mathrm{L}} \forall y\, \psi'(y/\mathsf{c})$. Applying the Universal Distribution axiom to this repeatedly, with PC we deduce

$$\forall y\, \psi_0(y/\mathsf{c}) \to (\cdots \to (\forall y\, \psi_{n-1}(y/\mathsf{c}) \to (\forall y\, \psi(y/\mathsf{c}) \to \forall y\, \varphi(\mathsf{c}/x)(y/\mathsf{c}))\cdots).$$

Now y is not in any ψ_i, so by Lemma 1.9.1 all of the formulas $\forall y\, \psi_i(y/\mathsf{c})$ are L-theorems, so can be successively detached by Modus Ponens. Moreover, $\psi(y/\mathsf{c}) = \psi$ because c is not in ψ, and $\varphi(\mathsf{c}/x)(y/\mathsf{c}) = \varphi(y/x)$ because c is not in φ. The upshot is that we obtain

$$\vdash_{\mathrm{L}} \forall y\, \psi \to \forall y\, \varphi(y/x).$$

But by Vacuous Quantification, $\vdash_{\mathrm{L}} \psi \to \forall y\, \psi$ as y is not in ψ, and by Lemma 1.2.1(4) $\vdash_{\mathrm{L}} \forall y\, \varphi(y/x) \to \forall x\varphi$ as y is not in φ. These facts combine by PC to derive $\vdash_{\mathrm{L}} \psi \to \forall x\varphi$, which proves (ii) as explained above, and completes the proof of the Lemma. ⊣

We want to show that the canonical premodel $\mathcal{M}_{\mathrm{L}}$ is actually a model, i.e. that its truth sets $|\varphi|^{\mathcal{M}_{\mathrm{L}}} f$ are all admissible in $Prop_{\mathrm{L}}$. To do this we observe that the value-assignment f assigns to each variable a member of the set $U_{\mathcal{L}}$ of closed terms, so f can function as a substitution operator, acting on any formula by replacing each of its free variables x by the term fx. Since fx is closed it is free for x in any formula. The result of applying this substitution induced by f to a formula φ will be denoted φ^f. In the formal notation of Section 1.1,

$$\varphi^f = \varphi(fv_0/v_0, \ldots, fv_n/v_n, \ldots).$$

In addition, for each variable x we write $\varphi^{f\setminus x}$ for the formula obtained from φ by leaving x alone and otherwise applying the substitution f. Formally, if $x = v_n$, then

$$\varphi^{f\setminus x} = \varphi(fv_0/v_0, \ldots, , fv_{n-1}/v_{n-1}, v_n/v_n, fv_{n+1}/v_{n+1}, \ldots).$$

Note that φ^f is a *sentence*, as the free variables of φ have all been replaced by closed terms in φ^f. $\varphi^{f\setminus x}$ however may have x free.

For terms, we likewise write τ^f for the term resulting from applying f as a substitution to τ. Here is a collection of basic facts about this kind of substitution.

LEMMA 1.9.3. (1) *The substitution induced by f commutes with formation of function terms, atomic formulas, and the propositional connectives*:

$(F\tau_1\cdots\tau_n)^f = F(\tau_1^f)\cdots(\tau_n^f)$, $(P\tau_1\cdots\tau_n)^f = P(\tau_1^f)\cdots(\tau_n^f)$, $(\varphi\wedge\psi)^f = \varphi^f\wedge\psi^f$, $(\neg\varphi)^f = \neg(\varphi^f)$, $(\Box\varphi)^f = \Box(\varphi^f)$.

(2) *For all formulas φ, $f \in U_{\mathcal{L}}{}^{\omega}$ and $\tau \in U_{\mathcal{L}}$,*

$$(\forall x\varphi)^f = \forall x(\varphi^{f\backslash x}).$$

(3) *For all formulas φ, $f \in U_{\mathcal{L}}{}^{\omega}$ and $\tau \in U_{\mathcal{L}}$,*

$$\varphi(\tau/x)^f = \varphi(\tau/x)^{f\backslash x} = \varphi^{f\backslash x}(\tau/x) = \varphi^{f[\tau/x]}.$$

(4) *For all terms τ and $f \in U_{\mathcal{L}}{}^{\omega}$, $\tau^f = |\tau|^{\mathcal{M}_{\mathrm{L}}}f$. If τ is closed, then $\tau = |\tau|^{\mathcal{M}_{\mathrm{L}}}$.*

(5) *For any logic* L, *formula φ and $f \in U_{\mathcal{L}}{}^{\omega}$,*

$$|(\forall x\varphi)^f|_{\mathrm{L}} = \prod_{\tau\in U_{\mathcal{L}}} E\tau \Rightarrow |\varphi^{f[\tau/x]}|_{\mathrm{L}}.$$

PROOF. (1) is routine. For (2), observe that $(\forall x\varphi)^f = (\forall x\varphi)^{f\backslash x}$, as x is not free in $\forall x\varphi$. But $(\forall x\varphi)^{f\backslash x} = \forall x(\varphi^{f\backslash x})$, as $f\backslash x$ leaves free x alone, and does not introduce any new variables as each fy is closed, so the operations of prefixing $\forall x$ and applying the substitution $f\backslash x$ can be done to φ in either order, with the same overall result.

For (3), since τ is closed it does not contain x, and $\varphi(\tau/x)$ does not contain free x. Hence applying f as a substitution to $\varphi(\tau/x)$ has the same effect as applying $f\backslash x$, i.e. $\varphi(\tau/x)^f = \varphi(\tau/x)^{f\backslash x}$. Indeed, in both cases we get the formula which arises from φ by substitution of the closed term τ for free x and then the closed term fy for free y whenever $y \neq x$. Moreover, it is evident from this description that the substitutions (τ/x) and $f\backslash x$ can be applied to φ in either order, and that the overall effect is the same as applying the substitution $f[\tau/x]$, so $\varphi(\tau/x)^{f\backslash x} = \varphi^{f\backslash x}(\tau/x) = \varphi^{f[\tau/x]}$.

For (4), we use induction on the formation of the term τ, which may contain variables. If τ is the variable x, then $x^f = fx = |x|^{\mathcal{M}_{\mathrm{L}}}f$ by definition of $|x|^{\mathcal{M}}$ for any premodel $\mathcal{M}$. If τ is the constant c, then $\mathsf{c}^f = \mathsf{c} = |\mathsf{c}|^{\mathcal{M}_{\mathrm{L}}}$ by definition of $\mathcal{M}_{\mathrm{L}}$. For the inductive case that τ is $F\tau_1\cdots\tau_n$, assume the result for the τ_i and observe

$$\begin{aligned}
\tau^f &= F(\tau_1^f)\cdots(\tau_n^f) && \text{by (1)}\\
&= |F|^{\mathcal{M}_{\mathrm{L}}}(\tau_1^f,\ldots,\tau_n^f) && \text{by definition of } \mathcal{M}_{\mathrm{L}}\\
&= |F|^{\mathcal{M}_{\mathrm{L}}}(|\tau_1|^{\mathcal{M}_{\mathrm{L}}}f,\ldots,|\tau_n|^{\mathcal{M}_{\mathrm{L}}}f) && \text{by induction hypothesis}\\
&= |\tau|^{\mathcal{M}_{\mathrm{L}}}f && \text{by definition of } |\tau|^{\mathcal{M}},
\end{aligned}$$

so the result holds for τ.

If τ is closed, then the value $|\tau|^{\mathcal{M}_{\mathrm{L}}}f$ is the same for all f, and is what we mean here by $|\tau|^{\mathcal{M}_{\mathrm{L}}}$. But when τ is closed, $\tau^f = \tau$ for all f.

For (5), we apply (2) and (3) to Lemma 1.9.2, noting that $\forall x(\varphi^{f\backslash x})$ is a *sentence*, to reason that

$$\begin{aligned} |(\forall x\varphi)^f|_{\mathrm{L}} &= |\forall x(\varphi^{f\backslash x})|_{\mathrm{L}} && \text{by part (2)} \\ &= \prod_{\tau\in U_{\mathcal{L}}} E\tau \Rightarrow |\varphi^{f\backslash x}(\tau/x)|_{\mathrm{L}} && \text{by Lemma 1.9.2} \\ &= \prod_{\tau\in U_{\mathcal{L}}} E\tau \Rightarrow |\varphi^{f[\tau/x]}|_{\mathrm{L}} && \text{by part (3).} \end{aligned}$$

⊣

We can now show that the premodel $\mathcal{M}_{\mathrm{L}}$ satisfies a version of what is often called the "Truth Lemma". In our version, this involves the substitutions φ^f.

THEOREM 1.9.4 (Truth Is Membership). *Let φ be any formula. Then for all $f \in U_{\mathcal{L}}{}^{\omega}$,*

$$|\varphi|^{\mathcal{M}_{\mathrm{L}}} f = |\varphi^f|_{\mathrm{L}},$$

and hence for all $w \in W_{\mathrm{L}}$,

$$\mathcal{M}_{\mathrm{L}}, w, f \models \varphi \quad \textit{iff} \quad \varphi^f \in w.$$

PROOF. The second part of the statement is an immediate consequence of the first. The first is proved by induction on the formation of φ.

If φ is F, then $|\mathsf{F}|^{\mathcal{M}_{\mathrm{L}}} f = \emptyset = |\mathsf{F}|_{\mathrm{L}} = |\mathsf{F}^f|_{\mathrm{L}}$. If φ is the atomic $P\tau_1\cdots\tau_n$, then

$$\begin{aligned} |\varphi|^{\mathcal{M}_{\mathrm{L}}} f &= |P|^{\mathcal{M}_{\mathrm{L}}}(|\tau_1|^{\mathcal{M}_{\mathrm{L}}} f, \dots, |\tau_n|^{\mathcal{M}_{\mathrm{L}}} f) && \text{definition of } |\varphi|^{\mathcal{M}} \\ &= |P|^{\mathcal{M}_{\mathrm{L}}}(\tau_1^f, \dots, \tau_n^f) && \text{Lemma 1.9.3(4)} \\ &= |P(\tau_1^f)\cdots(\tau_n^f)|_{\mathrm{L}} && \text{definition of } |P|^{\mathcal{M}_{\mathrm{L}}} \\ &= |\varphi^f|_{\mathrm{L}} && \text{Lemma 1.9.3(1).} \end{aligned}$$

For the inductive case of conjunction, assume the result for φ and for ψ. Then using this and previous observations,

$$|\varphi \wedge \psi|^{\mathcal{M}_{\mathrm{L}}} f = |\varphi|^{\mathcal{M}_{\mathrm{L}}} f \cap |\psi|^{\mathcal{M}_{\mathrm{L}}} f = |\varphi^f|_{\mathrm{L}} \cap |\psi^f|_{\mathrm{L}} = |\varphi^f \wedge \psi^f|_{\mathrm{L}},$$

which is $|(\varphi \wedge \psi)^f|_{\mathrm{L}}$, so the result holds for $\varphi \wedge \psi$.

Now assuming the result for φ we have

$$|\neg\varphi|^{\mathcal{M}_{\mathrm{L}}} f = -|\varphi|^{\mathcal{M}_{\mathrm{L}}} f = -|\varphi^f|_{\mathrm{L}} = |\neg(\varphi^f)|_{\mathrm{L}} = |(\neg\varphi)^f|_{\mathrm{L}},$$

and

$$|\Box\varphi|^{\mathcal{M}_{\mathrm{L}}} f = [R_{\mathrm{L}}]|\varphi|^{\mathcal{M}_{\mathrm{L}}} f = [R_{\mathrm{L}}]|\varphi^f|_{\mathrm{L}} = |\Box(\varphi^f)|_{\mathrm{L}} = |(\Box\varphi)^f|_{\mathrm{L}},$$

so the result holds for $\neg\varphi$ and $\Box\varphi$.

The really new case is for the quantifier $\forall$. If the result holds for φ, then

$$\begin{aligned} |\forall x\varphi|^{\mathcal{M}_{\mathrm{L}}} f &= \prod_{\tau\in U_{\mathcal{L}}} E\tau \Rightarrow |\varphi|^{\mathcal{M}_{\mathrm{L}}} f[\tau/x] && \text{semantics of } \forall \\ &= \prod_{\tau\in U_{\mathcal{L}}} E\tau \Rightarrow |\varphi^{f[\tau/x]}|_{\mathrm{L}} && \text{induction hypothesis on } \varphi \\ &= |(\forall x\varphi)^f|_{\mathrm{L}} && \text{Lemma 1.9.3(5),} \end{aligned}$$

so it holds for $\forall x\varphi$. ⊣

COROLLARY 1.9.5. *$\mathcal{M}_{\mathrm{L}}$ is a model.*

PROOF. Let φ be any formula and $f \in U_{\mathcal{L}}{}^{\omega}$. Then φ^f is a sentence, and so $|\varphi^f|_{\mathrm{L}} \in Prop_{\mathrm{L}}$, which means by the Theorem that $|\varphi|^{\mathcal{M}_{\mathrm{L}}} f$ is admissible. ⊣

THEOREM 1.9.6. *For any quantified modal logic* L, *the canonical model* $\mathcal{M}_{\mathrm{L}}$ *characterises* L:

$$\vdash_{\mathrm{L}} \varphi \quad \textit{iff} \quad \mathcal{M}_{\mathrm{L}} \models \varphi.$$

PROOF. Let the free variables of φ be $x_1, \ldots, x_n$. Then for any $f \in U_{\mathcal{L}}{}^{\omega}$,

$$\varphi^f = \varphi(fx_1/x_1, \ldots fx_n/x_n).$$

Suppose $\vdash_{\mathrm{L}} \varphi$. Then $\vdash_{\mathrm{L}} \varphi^f$ by the extended Term Instantiation rule TI* of Lemma 1.2.3. Hence $|\varphi^f|_{\mathrm{L}} = W_{\mathrm{L}}$, i.e. φ^f belongs to every L-maximal set. So $|\varphi|^{\mathcal{M}_{\mathrm{L}}} f = W_{\mathrm{L}}$ by Theorem 1.9.4. Since f is arbitrary here, this shows that φ is valid in $\mathcal{M}_{\mathrm{L}}$.

For the converse, suppose $\nvdash_{\mathrm{L}} \varphi$. Take constants $\mathsf{c}_1, \ldots, \mathsf{c}_n$ that do not occur in φ (for a second time we use the assumption that $\mathsf{Con}_{\mathcal{L}}$ is infinite), and let f be any member of $U_{\mathcal{L}}{}^{\omega}$ that has $fx_i = \mathsf{c}_i$ for $1 \leq i \leq n$. Then

$$\varphi^f = \varphi(\mathsf{c}_1/x_1, \ldots \mathsf{c}_n/x_n),$$

and so $\nvdash_{\mathrm{L}} \varphi^f$ by the extended Generalisation on Constants rule GC* of Lemma 1.2.3. Hence there exists some L-maximal $w \in W_{\mathrm{L}}$ such that $\varphi^f \notin w$. Therefore $\mathcal{M}_{\mathrm{L}}, w, f \not\models \varphi$ by Theorem 1.9.4, so φ is not valid in $\mathcal{M}_{\mathrm{L}}$.

Note that if φ is a sentence, without free variables, then every $f \in U_{\mathcal{L}}{}^{\omega}$ has $\varphi^f = \varphi$, hence $\vdash_{\mathrm{L}} \varphi$ iff $\vdash_{\mathrm{L}} \varphi^f$, and these arguments go through more directly, without needing to use TI* or GC*. ⊣

COROLLARY 1.9.7. L *is complete for* $\mathcal{S}_{\mathrm{L}}$, *i.e.* $\mathcal{S}_{\mathrm{L}} \models \varphi$ *implies* $\vdash_{\mathrm{L}} \varphi$.

PROOF. If φ is valid in the canonical model structure $\mathcal{S}_{\mathrm{L}}$ for L, then it valid in the model $\mathcal{M}_{\mathrm{L}}$ on $\mathcal{S}_{\mathrm{L}}$, so $\vdash_{\mathrm{L}} \varphi$ by the Theorem. ⊣

THEOREM 1.9.8 (Completeness for QK). *The canonical model structure* $\mathcal{S}_{\mathrm{QK}}$ *characterises* QK :

$$\vdash_{\mathrm{QK}} \varphi \quad \textit{iff} \quad \mathcal{S}_{\mathrm{QK}} \models \varphi.$$

Hence $\vdash_{\mathrm{QK}} \varphi$ *iff* φ *is valid in all model structures.*

PROOF. $\vdash_{QK} \varphi$ implies $\mathcal{S}_{QK} \models \varphi$ by Soundness (Theorem 1.7.10), and the converse holds by the Corollary just proved.

The second statement of the Theorem follows from the first and Soundness. ⊣

1.10. Completeness and Canonicity for QS

We can now demonstrate the full generality of our admissible model theory, by showing that every quantified modal logic of the form QS is characterised by validity in its model structures. In fact, more strongly, QS is characterised by validity in model structures whose underlying general frame validates the propositional formulas in S. The key to this is the behaviour of the general frame

$$\mathcal{G}_L = (W_L, R_L, Prop_L)$$

underlying the canonical model structure $\mathcal{S}_L$ of a logic:

LEMMA 1.10.1. *If* L *is any quantified modal logic that extends* QS, *then*

(1) $\mathcal{G}_L$ *validates the set* S *of propositional modal formulas*; *and*

(2) $\mathcal{S}_L$ *validates* QS.

PROOF. Let $\mathcal{M}$ be any model on $\mathcal{G}_L$ for the propositional language. Thus $\mathcal{M}$ assigns to each variable $p \in \mathsf{PropVar}$ a set $|p|^{\mathcal{M}} \in Prop_L$. Hence $|p|^{\mathcal{M}} = |\varphi_p|_L$ for some $\mathcal{L}$-sentence φ_p, by definition of $Prop_L$.

By uniform replacement of each p by φ_p, we then obtain a mapping $A \mapsto A^*$ of *propositional* modal formulas to $\mathcal{L}$-formulas. Here

$$A^* = A(\varphi_{p_1}/p_1, \dots, \varphi_{p_n}/p_n),$$

where $p_1, \dots, p_n$ are all the propositional variables of A. An induction on the formation of propositional formulas then shows that in general

$$|A|^{\mathcal{M}} = |A^*|_L.$$

When A is a variable p, then $A^* = \varphi_p$ so this result holds by definition. When A is F, then $A^* = \mathsf{F}$ so the result holds as $|\mathsf{F}|^{\mathcal{M}} = \emptyset = |\mathsf{F}|_L$. For the inductive case of $\Box$, assuming the result holds for A, we get

$$|\Box A|^{\mathcal{M}} = [R_L]|A|^{\mathcal{M}} = [R_L]|A^*|_L = |\Box(A^*)|_L = |(\Box A)^*|_L,$$

so the result holds for $\Box A$. The inductive cases for $\wedge$ and $\neg$ are similar, and left to the reader.

Now if $A \in$ S, then A^* is a substitution-instance of a member of S, hence $\vdash_{QS} A^*$, so $\vdash_L A^*$ and hence A^* belongs to every L-maximal set. Therefore

$$|A|^{\mathcal{M}} = |A^*|_L = W_L,$$

showing that $\mathcal{M} \models A$.

Thus every member of S is true in every model on $\mathcal{G}_L$, which proves (1).

(2) then follows from (1) by by Theorem 1.7.11. ⊣

Taking L in this Lemma to be QS itself, we obtain our desired characterisation:

THEOREM 1.10.2 (Completeness for QS). *Let* S *be any set of propositional modal formulas.*

(1) QS *is characterised by validity in* $\mathcal{S}_{\mathrm{QS}}$, *i.e.* $\vdash_{\mathrm{QS}} \varphi$ *iff* $\mathcal{S}_{\mathrm{QS}} \models \varphi$.
(2) QS *is characterised by validity in all model structures whose underlying general frame validates* S.
(3) QS *is characterised by validity in all* QS-*structures.*

PROOF. (1): For soundness, the Lemma just proved shows $\mathcal{S}_{\mathrm{QS}}$ validates QS. For completeness, we saw already in Corollary 1.9.7 that $\mathcal{S}_{\mathrm{QS}} \models \varphi$ implies $\vdash_{\mathrm{QS}} \varphi$.

(2): Soundness is again Theorem 1.7.11, which states that QS is valid in all model structures whose underlying general frame validates S. For completeness, if φ is valid in all model structures whose underlying general frame validates S, then in particular it is valid in $\mathcal{S}_{\mathrm{QS}}$ (Lemma 1.10.1), so $\vdash_{\mathrm{QS}} \varphi$ by (1).

(3): A QS-structure is one validating QS, so QS is sound for validity in such structures by definition. Completeness follows from (1), as $\mathcal{S}_{\mathrm{QS}}$ is a QS-structure. ⊣

The characterisations in this Theorem raise the question of whether they can be refined to show that QS is characterised by model structures whose underlying *Kripke frame* validates S. The answer is negative in general, as will be seen in the next section, but now we show that there is a positive answer when S is a propositional modal logic that is *canonical*. This intensely-studied property means that S is validated by its *canonical Kripke frame* $\mathcal{F}_{\mathrm{S}} = (W_{\mathrm{S}}, R_{\mathrm{S}})$. Here W_{S} is the collection of all maximally S-consistent sets of formulas of the propositional modal language, with

$$wR_{\mathrm{S}}u \quad \text{iff} \quad \{A : \Box A \in w\} \subseteq u.$$

Putting $|A|_{\mathrm{S}} = \{w \in W_{\mathrm{S}} : A \in w\}$ for any propositional modal formula A, the *canonical* S-*model* $\mathcal{M}_{\mathrm{S}}$ *on* $\mathcal{F}_{\mathrm{S}}$ is defined by putting $|p|^{\mathcal{M}_{\mathrm{S}}} = |p|_{\mathrm{S}}$ for each propositional variable p. An induction on formulas shows that $|A|^{\mathcal{M}_{\mathrm{S}}} = |A|_{\mathrm{S}}$ in general, giving the *Truth Lemma*

$$\mathcal{M}_{\mathrm{S}}, w \models A \quad \text{iff} \quad A \in w.$$

From this it follows that S is characterised by truth in $\mathcal{M}_{\mathrm{S}}$:

$$\vdash_{\mathrm{S}} A \quad \text{iff} \quad \mathcal{M}_{\mathrm{S}} \models A,$$

and hence that S is complete for validity in $\mathcal{F}_{\mathrm{S}}$:

$$\mathcal{F}_{\mathrm{S}} \models A \quad \text{implies} \quad \vdash_{\mathrm{S}} A.$$

But the converse of this last implication can fail, so $\mathcal{F}_S$ may not characterise S. When it does do so, we say that S is *canonical*. Thus a canonical propositional logic S is one that is validated by $\mathcal{F}_S$, i.e. $\vdash_S A$ implies $\mathcal{F}_S \models A$, or more briefly, $\mathcal{F}_S \models S$.

The family of canonical propositional logics is wide: there is a celebrated theorem of Fine [1975] which states that if $\mathscr{C}$ is any class of Kripke frames that is defined by some first-order conditions on a binary relation R, then the logic $S_{\mathscr{C}} = \{A : \mathscr{C} \models A\}$ characterised by $\mathscr{C}$ is canonical. There are many other canonical logics besides these [Goldblatt, Hodkinson, and Venema 2003, 2004].

We now establish a tight relationship between the canonical Kripke frame $\mathcal{F}_S$ of any propositional logic S and the Kripke frame (W_{QS}, R_{QS}) underlying the canonical model structure $\mathcal{S}_{QS}$ of QS. For this we need a fragment of the theory of inner subframes and bounded morphisms. Given two frames $\mathcal{F} = (W, R)$ and $\mathcal{F}' = (W', R')$, we say that $\mathcal{F}$ is an *inner subframe*[21] of $\mathcal{F}'$ if

(i) W is a subset of W';
(ii) R is the restriction of R' to W, i.e. for all $w, u \in W$, wRu iff $wR'u$; and
(iii) W is *R'-closed in W'*, meaning that if $w \in W$, and $wR'u$, then $u \in W$.

Conditions (i) and (ii) here define the notion of $\mathcal{F}$ being a *subframe* of $\mathcal{F}'$. It is condition (iii) that makes the subframe *inner*. The importance of this notion is that it preserves validity:

LEMMA 1.10.3. *If $\mathcal{F}$ is an inner subframe of $\mathcal{F}'$, then $\mathcal{F}' \models A$ implies $\mathcal{F} \models A$.*

PROOF. If $\mathcal{M}$ is a model on $\mathcal{F}$, we can view it as a model $\mathcal{M}'$ on $\mathcal{F}'$ by putting $|p|^{\mathcal{M}'} = |p|^{\mathcal{M}}$ for all $p \in \mathsf{PropVar}$. An induction on propositional formulas A then shows that for all $w \in W$,

$$\mathcal{M}, w \models A \quad \text{iff} \quad \mathcal{M}', w \models A.$$

Hence $\mathcal{M}' \models A$ implies $\mathcal{M} \models A$. Thus if A is true in all models on $\mathcal{F}'$, it is true in all models on $\mathcal{F}$. ⊣

A *bounded morphism*[22] from frame $\mathcal{F}$ to frame $\mathcal{F}'$ is a function $h : W \to W'$ satisfying

(i) wRu implies $h(w)R'h(u)$;
(ii) if $h(w)R'v$, then there exists $u \in W$ with wRu and $h(u) = v$.

The *h-image* of $\mathcal{F}$, i.e. the frame $(h(W), R' \restriction h(W))$, is then an inner subframe of $\mathcal{F}'$. If h is injective, this image is isomorphic to $\mathcal{F}$ itself. But isomorphic frames validate the same formulas, so in combination with the Lemma just proved we get the following useful fact:

LEMMA 1.10.4. *If there exists an injective bounded morphism from $\mathcal{F}$ to $\mathcal{F}'$, then $\mathcal{F}' \models A$ implies $\mathcal{F} \models A$.* ⊣

[21] Also known as a "generated" subframe.

[22] Also known as a "p-morphism".

Now let $\mathcal{F}_{\mathrm{L}} = (W_{\mathrm{L}}, R_{\mathrm{L}})$ be the Kripke frame underlying the canonical model structure $\mathcal{S}_{\mathrm{L}}$ of a quantified logic L. The relationship between such frames, built from the quantified language, and the canonical frames of propositional logics is as follows.

THEOREM 1.10.5. *If* L *is any quantified modal logic that extends* QS, *then there is an injective bounded morphism*

$$h : \mathcal{F}_{\mathrm{L}} \longrightarrow \mathcal{F}_{\mathrm{S}}$$

from the Kripke frame underlying the canonical model structure $\mathcal{S}_{\mathrm{L}}$ *of* L *into the canonical Kripke frame of* S. *Hence* $\mathcal{F}_{\mathrm{L}}$ *is isomorphic to an inner subframe of* $\mathcal{F}_{\mathrm{S}}$.

PROOF. [23] For a given signature $\mathcal{L}$, we need to assume that our infinite set PropVar of propositional variables is at least as big as $\mathcal{L}$, hence at least as big as the set of all $\mathcal{L}$-formulas.[24] Then there exists a mapping $p \mapsto p^*$ of PropVar *onto* the set of $\mathcal{L}$-formulas. Under this mapping every $\mathcal{L}$-formula is equal to p^* for some $p \in$ PropVar.

This mapping lifts to all propositional modal formulas A: let A^* be the result of uniformly substituting p^* for p in A, for all p that occur in A. As usual, this operation commutes with the propositional connectives: $\mathsf{F}^* = \mathsf{F}$, $(A \wedge B)^* = A^* \wedge B^*$, $(\Box A)^* = \Box(A^*)$ etc. Moreover, if A is an S-theorem, then its substitution-instance A^* is a QS-theorem, and hence $\vdash_{\mathrm{L}} A^*$.

Define $h : W_{\mathrm{L}} \to W_{\mathrm{S}}$ by putting $h(w) = \{A : A^* \in w\}$ for all maximally L-consistent sets w. Of course it has to be checked that this $h(w)$ is a maximally S-consistent set of propositional formulas. First, $h(w)$ is S-consistent, for otherwise there would be $A_1, \dots, A_n \in h(w)$ such that $(A_1 \wedge \cdots \wedge A_n) \to \mathsf{F}$ is an S-theorem, hence $(A_1^* \wedge \cdots \wedge A_n^*) \to \mathsf{F}$ is an L-theorem with $A_1^*, \dots, A_n^* \in w$, contrary to the L-consistency of w. Second, w contains one of A^* and $\neg(A^*) = (\neg A)^*$ so $h(w)$ contains one of A and $\neg A$, for all A, so $h(w)$ is negation complete, as required.

f is injective: if $w \neq u$ in W_{L}, then there is some formula $A = p^* \in w$ with $\neg A = (\neg p)^* \in u$, hence $p \in h(w)$ and $\neg p \in h(u)$, so $p \notin h(u)$ and $h(w) \neq h(u)$.

If $wR_{\mathrm{L}}u$, then $\Box A \in h(w)$ implies $\Box(A^*) = (\Box A)^* \in w$, hence $A^* \in u$ and thus $A \in h(u)$; so $h(w)R_{\mathrm{S}}h(u)$.

Finally, to complete the proof that h is a bounded morphism, suppose $h(w)R_{\mathrm{S}}v$ in W_{S}. We have to show that $wR_{\mathrm{L}}u$ and $h(u) = v$ for some $u \in W_{\mathrm{L}}$. Put

$$u_0 = \{A^* : \Box A^* \in w\} \cup \{B^* : B \in v\}.$$

[23] A similar construction is given by Schurz [1997, A11, p. 295], between frames underlying canonical models for certain quantified modal logics.

[24] See page 3 for reference to the size of PropVar.

If u_0 were not L-consistent, then since the two sets that make up u_0 are each closed under finite conjunctions, there would be formulas A, B with $\Box A^* \in w$ and $B \in v$, such that $A \to \neg B^*$ is an L-theorem. But then $\Box A^* \to \Box\neg B^*$ is an L-theorem, so belongs to w, implying that $\Box\neg B^* \in w$, hence $\Box\neg B \in h(w)$, and so $\neg B \in v$ as $h(w)R_S v$, contradicting the S-consistency of v.

Thus u_0 is L-consistent, and hence is included in some $u \in W_L$. Since $\{A^* : \Box A^* \in w\} \subseteq u$ and every $\mathcal{L}$-formula is of the form A^*, we get $wR_L u$. Since $\{B^* : B \in v\} \subseteq u$ we get $v \subseteq h(u)$, and so $v = h(u)$ as required, by *maximal* S-consistency of v.

Altogether then, h is an injective bounded morphism from (W_L, R_L) into $\mathcal{F}_S$. Its image is an inner subframe of $\mathcal{F}_S$ that is isomorphic to (W_L, R_L). ⊣

As something of a side remark, the bounded morphism h of this proof can be used to give a more structural explanation of why the general frame $\mathcal{G}_L$ underlying $\mathcal{S}_L$ validates S, as shown in Lemma 1.10.1. As well as its canonical Kripke frame $\mathcal{F}_S$, any propositional logic S has the canonical *general* frame

$$\mathcal{G}_S = (W_S, R_S, Prop_S), \tag{1.10.1}$$

where $Prop_S = \{|A|_S : A \text{ is a propositional formula}\}$. $\mathcal{G}_S$ always validates S, regardless of whether the Kripke frame $\mathcal{F}_S$ does. This can be shown by an argument quite similar to that used in Lemma 1.10.1. Now the image of the map $h : W_L \to W_S$ can be made into a general frame whose admissible propositions are the restrictions of the admissible propositions of $\mathcal{G}_S$ to $h(W)$, i.e. the sets $|A|_S \cap h(W)$. The following relationships are readily shown:

$$|A|_S \cap h(W) = h|A|_L.$$
$$h^{-1}(|A|_S \cap h(W)) = |A|_L.$$

This correspondence between $|A|_L$ and $|A|_S \cap h(W)$ is in fact a general-frame isomorphism between $\mathcal{G}_L$ and the general frame based on the image set $h(W)$. Moreover this h-image of $\mathcal{G}_L$ is an inner subframe of $\mathcal{G}_S$, in the sense that applies to general frames [Goldblatt 1993, Section 1.4]. So validity is preserved in passing from $\mathcal{G}_S$ to the image frame, and hence to $\mathcal{G}_L$. Since $\mathcal{G}_S \models$ S, it follows that $\mathcal{G}_L \models$ S, as stated by Lemma 1.10.1.

Another useful observation about the proof of Theorem 1.10.5 is that we can show directly that $h(w)R_S h(u)$ implies $wR_L u$, without invoking any results about the existence of maximal sets. For if $h(w)R_S h(u)$, and $\Box\varphi \in w$ with $\varphi = p^*$, then as $(\Box p)^* = \Box(p^*) \in w$ we have $\Box p \in h(w)$, hence $p \in h(u)$ as $h(w)R_S h(u)$, so $\varphi = p^* \in u$. This shows that $wR_L u$.

The upshot is that in general $wR_L u$ iff $h(w)R_S h(u)$, so for any subset X of W_L, the restriction of R_L to X gives a frame $(X, R_L \restriction X)$ isomorphic to the h-image $(h(X), R_S \restriction h(X))$ of X with the restriction of R_S to $h(X)$. This h-image is a subframe of the Kripke frame $\mathcal{F}_S$, but for some X it will not be an *inner* subframe. This has consequences that will be discussed in Section 2.8 (see the discussion prior to Theorem 2.8.3).

Using Theorem 1.10.5, we can now prove our main result on canonicity:

THEOREM 1.10.6. *Let* S *be a* ***canonical*** *propositional logic. Then*:

(1) *If* L *is any quantified modal logic that extends* QS, *then* S *is validated by the Kripke frame* $\mathcal{F}_L = (W_L, R_L)$ *underlying the model structure* $\mathcal{S}_L$.

(2) QS *is characterised by the class of all model structures whose underlying Kripke frame validates* S.

PROOF. Since S is canonical, the Kripke frame $\mathcal{F}_S$ validates S.

(1): Since validity of propositional formulas is preserved by inner subframes (Lemma 1.10.3) and by isomorphism, it then follows from Theorem 1.10.5 that (W_L, R_L) validates S.

(2): In particular the Kripke frame (W_{QS}, R_{QS}) underlying the canonical model structure $\mathcal{S}_{QS}$ validates S. But $\mathcal{S}_{QS}$ characterises QS (Theorem 1.10.2), so QS is *complete* for the class of all model structures whose underlying Kripke frame validates S. For soundness relative to this class, if $\mathcal{S}$ is any model structure whose underlying Kripke frame validates S, then by Corollary 1.7.12, $\mathcal{S}$ validates QS. ⊣

EXAMPLE 1.10.7 (The Brouwerian Axiom). KB is the smallest propositional modal logic that includes the scheme

$$\text{B}: \quad A \to \Box\Diamond A,$$

or equivalently $\Diamond\Box A \to A$, sometimes called the "Brouwerian axiom".[25] A Kripke frame validates scheme B iff its relation R is symmetric. The canonical frame $\mathcal{F}_S$ of any logic S including KB is symmetric. In particular $\mathcal{F}_{KB}$ is symmetric and validates B, so KB is canonical.

Thus from Theorem 1.10.6 we conclude that if a quantified logic L includes QKB, then the canonical model structure $\mathcal{S}_L$ is based on a KB-frame, so its relation R_{QKB} is symmetric. In particular,

> QKB *is characterised by the class of all model structures based on symmetric Kripke frames.*

This would appear to be the only structural semantic characterisation of QKB available. Currently there are no known general canonical model constructions giving characteristic *Kripkean* models for logics that include QKB [Hughes and Cresswell 1996, p. 309].

Actually it is not necessary to first prove Theorem 1.10.6, and hence Theorem 1.10.5, to obtain our characterisation of QKB. The simple proof of symmetry for $\mathcal{F}_{KB}$ works directly for $\mathcal{S}_L$ when L is any quantified logic including QKB. The condition $\{\varphi : \Box\varphi \in w\} \subseteq u$ defining $wR_L u$ in $\mathcal{S}_L$ is equivalent

[25] See [Goldblatt 2006b, p. 7] or [Hughes and Cresswell 1996, p. 70] for the historical background.

to $\{\Diamond\varphi : \varphi \in u\} \subseteq w$. Using this, observe that if $wR_{\mathrm{L}}u$, then $\Box\varphi \in u$ implies $\Diamond\Box\varphi \in w$, hence $\varphi \in w$ by the Brouwerian axiom, showing that $uR_{\mathrm{L}}w$.

This is just one amongst many examples. There are numerous canonical propositional logics S characterised by elementary frame conditions that hold for the canonical frame, and for which the proof that $\mathcal{F}_{\mathrm{S}}$ satisfies these conditions works in exactly the same way for $\mathcal{S}_{\mathrm{QS}}$.

1.11. Kinds of Incompleteness

The two distinctive features of our model theory are

(i) admissibility: the use of structures that are not full, i.e. $Prop \neq \wp W$; and
(ii) models that are not Kripkean, interpreting $\forall$ by the infinitary conjunction operation $\bigsqcap$ that may not be given by the set-theoretic intersection $\bigcap$.

Without these features the general characterisation result of Theorem 1.10.2 would not be possible. We confirm this in the present section by exhibiting failures of completeness with respect to Kripkean models. This includes showing that in Theorem 1.10.6, both the hypothesis of canonicity of S, and the inclusion of non-Kripkean models, are essential.

1.11.1. Incompleteness for Kripkean Models. To confirm that the use of admissible propositions and non-Kripkean models is necessary, we observe that, for an arbitrary consistent propositional modal logic S:

- QS is incomplete for any class of Kripkean models.
- QS is incomplete for any class of models or model structures that are full, or have all existence sets *Ea* admissible.
- The canonical model $\mathcal{M}_{\mathrm{QS}}$ is not Kripkean.
- The canonical model structure $\mathcal{S}_{\mathrm{QS}}$ is not full, and does not have all its existence sets admissible.

These results are all consequences of the fact that the Commuting Quantifiers axiom CQ is not derivable in QS (as already mentioned at the end of Section 1.7). The first two results then follow because all Kripkean models validate CQ (Theorem 1.7.18), and all full structures are Kripkean (Theorem 1.7.15), as are structures with admissible existence sets (Theorem 1.7.16). The last two results follow from the first two, because QS is complete for validity in $\mathcal{M}_{\mathrm{QS}}$ (Theorem 1.9.6) and in $\mathcal{S}_{\mathrm{QS}}$ (Theorem 1.10.2).

It was shown by Fine [1983] that CQ cannot be derived from AI, UD and VQ by using UG and Boolean reasoning. In [Goldblatt and Hodkinson 2009], models of our present kind were used to give a proof that CQ is not derivable in QS for any consistent S. We now given an account of what is involved in this. One ingredient is the result of Makinson [1971] that every consistent propositional modal logic is included either in the *Trivial system*, known as Triv, or in the *Verum system*, known as Ver.

Triv is the smallest propositional modal logic containing the axiom scheme $\Box A \leftrightarrow A$, and is characterised by the one-element reflexive Kripke frame. Ver is the smallest propositional modal logic containing $\Box$F (and hence every formula $\Box A$), and is characterised by the one-element irreflexive Kripke frame [Hughes and Cresswell 1996, p. 121]. Makinson's proof was algebraic. A more proof-theoretic version can be found in [Hughes and Cresswell 1996, pp. 67, 108]. Here is a model-theoretic argument:

THEOREM 1.11.1. *If* S *is any consistent propositional modal logic, then either* $\mathrm{S} \subseteq \mathsf{Triv}$ *or* $\mathrm{S} \subseteq \mathsf{Ver}$, *and hence either* $\mathrm{QS} \subseteq \mathsf{QTriv}$ *or* $\mathrm{QS} \subseteq \mathsf{QVer}$.

PROOF. This involves two cases concerning the canonical general S-frame $\mathcal{G}_\mathrm{S}$ of (1.10.1). In the first case, suppose there exists some $w \in [R_\mathrm{S}]\emptyset$. Then there is no $u \in W_\mathrm{S}$ with $wR_\mathrm{S}u$, so every propositional formula of the form $\Box A$ is true at w in every model on $\mathcal{G}_\mathrm{S}$. Let $\mathcal{F}_\bullet = (\{0\}, \emptyset)$ be the irreflexive frame on the singleton set $\{0\}$. Since $\mathcal{F}_\bullet$ characterises Ver, it is enough to show that $\mathcal{F}_\bullet \models \mathrm{S}$ to conclude that $\mathrm{S} \subseteq \mathsf{Ver}$.

For any propositional model $\mathcal{M}$ on $\mathcal{F}_\bullet$, define a model $\mathcal{M}'$ on $\mathcal{G}_\mathrm{S}$ by putting

$$|p|^{\mathcal{M}'} = \begin{cases} W_\mathrm{S} & \text{if } \mathcal{M}, 0 \models p, \\ \emptyset & \text{otherwise.} \end{cases} \tag{1.11.1}$$

Then $\mathcal{M}, 0 \models p$ iff $\mathcal{M}', w \models p$. An induction on formulas then shows that $\mathcal{M}, 0 \models A$ iff $\mathcal{M}', w \models A$, for all A. The inductive case of $\Box$ uses the fact that all formulas beginning with $\Box$ are true at 0 in $\mathcal{M}$ and at w in $\mathcal{M}'$. Now if $\vdash_\mathrm{S} A$, then A is valid in $\mathcal{G}_\mathrm{S}$, hence $\mathcal{M}', w \models A$, and so $\mathcal{M}, 0 \models A$. This shows that every S-theorem is true in every model on $\mathcal{F}_\bullet$, as required to show $\mathrm{S} \subseteq \mathsf{Ver}$.

The alternative case is that $[R_\mathrm{S}]\emptyset = \emptyset$. Then let $\mathcal{F}_\circ = (\{0\}, \{(0,0)\})$ be the reflexive frame on $\{0\}$. Since $\mathcal{F}_\circ$ characterises Triv, it is enough to show that $\mathcal{F}_\circ \models \mathrm{S}$ to conclude that $\mathrm{S} \subseteq \mathsf{Triv}$. Given a model $\mathcal{M}$ on $\mathcal{F}_\circ$, again define a model $\mathcal{M}'$ on $\mathcal{G}_\mathrm{S}$ by (1.11.1). This time we get

$$|A|^{\mathcal{M}'} = \begin{cases} W_\mathrm{S} & \text{if } \mathcal{M}, 0 \models A, \\ \emptyset & \text{otherwise,} \end{cases} \tag{1.11.2}$$

for all A. The inductive case of $\Box$ uses that $\mathcal{M}, 0 \models \Box A$ iff $\mathcal{M}, 0 \models A$; $|A|^{\mathcal{M}'} = W_\mathrm{S}$ implies $|\Box A|^{\mathcal{M}'} = W_\mathrm{S}$; and $|A|^{\mathcal{M}'} = \emptyset$ implies $|\Box A|^{\mathcal{M}'} = [R_\mathrm{S}]\emptyset = \emptyset$. But now if $\vdash_\mathrm{S} A$, then A is valid in $\mathcal{G}_\mathrm{S}$, hence $|A|^{\mathcal{M}'} = W_\mathrm{S}$, and so $\mathcal{M}, 0 \models A$. We conclude that $\mathcal{F}_\circ \models A$ as required to show $\mathrm{S} \subseteq \mathsf{Triv}$. ⊣

Countermodels to CQ can be built on the model structure $\mathcal{S}$ for which:

- $W = U = \mathbb{Q}$, the set of rational numbers.
- R is the identity relation on $\mathbb{Q}$.

- *Prop* is the Boolean subalgebra of $\wp(\mathbb{Q})$ generated by the set of all half-open intervals $[a,b) = \{x \in \mathbb{Q} : a \leq x < b\}$, where $a,b \in \mathbb{Q}$ and $a < b$.
- $Da = \{a\}$ for each $a \in \mathbb{Q}$.

Prop is an atomless Boolean algebra. The existence sets of $\mathcal{S}$ are given by $Ea = \{a\}$ and are not admissible.

The signature $\mathcal{L}$ is taken to consist of two binary relation symbols, P and $\sim$, with $\sim$ being identified with a fixed equivalence relation on $\mathbb{Q}$ having infinitely many equivalence classes, each of which is dense in $\mathbb{Q}$. A premodel $\mathcal{M}$ on $\mathcal{S}$ is defined by:

- $|P|^{\mathcal{M}}(a,b) = \begin{cases} \mathbb{Q}, & \text{if } a \sim b, \\ \text{some non-empty interval} & \\ [b,c) \text{ not containing } a, & \text{otherwise.} \end{cases}$
- $|{\sim}|^{\mathcal{M}}(a,b) = \begin{cases} \mathbb{Q}, & \text{if } a \sim b, \\ \emptyset, & \text{otherwise.} \end{cases}$

It turns out that every $\mathcal{L}$-formula is equivalent in $\mathcal{M}$ to a quantifier free formula, and hence is admissible in $\mathcal{M}$, so $\mathcal{M}$ is a model [Goldblatt and Hodkinson 2009, 5.3]. Also, for any $f \in U^{\omega}$, $|\forall x \forall y Pxy| f = \mathbb{Q}$ while $|\forall y \forall x Pxy| f = \emptyset$, so $\mathcal{M}$ does not validate $\forall x \forall y Pxy \to \forall y \forall x Pxy$, an instance of CQ [Goldblatt and Hodkinson 2009, 5.2].

Now the fact that R is the identity relation ensures that the Triv axiom $\Box\varphi \leftrightarrow \varphi$ is valid in $\mathcal{M}$. The same holds for the axioms K, UD and VQ, while the rules MP, N and UG preserve this validity. Therefore QTriv is validated by $\mathcal{M}$, implying that CQ is not derivable in QTriv.

If we change R to the empty relation, then $\mathcal{M}$ validates the Ver axiom $\Box$F, and so QVer is validated by $\mathcal{M}$, hence CQ is not derivable in QVer. Combining these facts with Theorem 1.11.1 then yields that CQ is not derivable in QS for any consistent S.

It is notable that the Barcan Formula $\forall x \Box\varphi \to \Box\forall x\varphi$ and its converse $\Box\forall x\varphi \to \forall x\Box\varphi$ are valid in the two models we have just described, so CQ is not derivable in QS+CBF+BF.

1.11.2. Kripkean S-frame Incompleteness. If a propositional logic S is canonical, then we know that QS is characterised by model structures based on frames (W,R) that validate S (Theorem 1.10.6). We also know that non-Kripkean models play an essential role in this characterisation: QS is incomplete for *Kripkean* models on structures based on S-frames, because we have just seen that QS is not characterised by any class of Kripkean models at all.

But it might be said that this incompleteness has little to do with modality, since it resulted from the independence of the purely quantificational principle CQ. So there is some interest in showing that the incompleteness cannot be overcome by adopting CQ as a further axiom. There are canonical logics S such that QS+CQ is incomplete for the class of Kripkean models on structures whose underlying Kripke frame is an S-frame.

One such example is S4M, the the extension of S4 by the McKinsey axiom

$$\text{M:} \quad \Box\Diamond A \to \Diamond\Box A.$$

A frame (W, R) validates S4M iff R is reflexive and transitive and satisfies the condition

$$\forall w\, \exists u(wRu \text{ and } \forall v(uRv \text{ implies } u = v)), \tag{1.11.3}$$

stating that every world w has an R-alternative u that is *R-final* in the sense that it is R-related at most to itself [van Benthem 1975]. The canonical S4M-frame $\mathcal{F}_{\text{S4M}}$ satisfies this condition [Lemmon 1977], so S4M is a canonical propositional logic.

To show that QS4M+CQ is incomplete for Kripkean models on structures based on S4M-frames, let the signature $\mathcal{L}$ have a monadic predicate symbol P, and consider the sentence

$$\Box\exists x Px \to \Diamond\exists x\Box Px. \tag{1.11.4}$$

LEMMA 1.11.2. *Let $\mathcal{S}$ be a model structure whose underlying Kripke frame satisfies* (1.11.3). *Then any Kripkean model on $\mathcal{S}$ validates the sentence* (1.11.4).

PROOF. This is an adaptation of [Hughes and Cresswell 1996, Lemma 14.11] to our framework. Let $\mathcal{M}, w, f \models \Box\exists x Px$, where $\mathcal{M}$ is a Kripkean model on $\mathcal{S}$. By (1.11.3), take an R-final u with wRu. Then $\mathcal{M}, u, f \models \exists x Px$, so as $\mathcal{M}$ is Kripkean, there is some $a \in Du$ with $\mathcal{M}, u, f[a/x] \models Px$ (1.6.7). Since u is R-final, $\mathcal{M}, u, f[a/x] \models \Box Px$. Hence $\mathcal{M}, u, f \models \exists x\Box Px$ (1.6.2), and so $\mathcal{M}, w, f \models \Diamond\exists x\Box Px$. ⊣

The claimed incompleteness result will thus follow by showing that (1.11.4) not a theorem of QS4M+CQ. That can be done by using a model $\mathcal{M}$, defined in [Hughes and Cresswell 1996, p. 266], that validates QS4M+CQ but falsifies (1.11.4). The key point is that $\mathcal{M}$ validates the McKinsey axiom even though its underlying frame does not satisfy (1.11.3). In our present terms, $\mathcal{M}$ is the premodel specified by:

- $W = U = \omega$, the set of natural numbers.
- R is the numerical order relation $\leq$.
- $Prop = \wp(W)$, so every set is admissible.
- $Dw = U$ for each $w \in W$.
- $|P|^{\mathcal{M}} n = \{n\}$ for each $n \in U$.

Since $\mathcal{M}$ is full, it is automatically a model, i.e. all truth sets are admissible, and is Kripkean. Hence it validates CQ. It validates the S4 axioms because $\leq$ is reflexive and transitive. To show that it validates M requires the following technical fact about an arbitrary formula φ, shown in [Hughes and Cresswell 1996, 14.13]:

If $n > fx$ for every x free in φ, and $m \geq n$, then $\mathcal{M}, n, f \models \varphi$ iff $\mathcal{M}, m, f \models \varphi$.

This implies that the truth set $|\varphi|^{\mathcal{M}} f$ is either finite or cofinite, so the truth value of φ along ω is "ultimately constant", a property that ensures the truth of M [Thomason 1972]. For if $|\varphi|^{\mathcal{M}} f$ is finite, then $\mathcal{M}, n, f \not\models \Box\Diamond\varphi$ for all n, while if $|\varphi|^{\mathcal{M}} f$ is cofinite, then $\mathcal{M}, n, f \models \Diamond\Box\varphi$ for all n. So $\mathcal{M}, n, f \models \Box\Diamond \to \Diamond\Box\varphi$ in general.

It remains to show that $\mathcal{M}$ falsifies (1.11.4). Notice that $\mathcal{M}, n, f \models Px$ iff $n = fx$; so $\mathcal{M}, n, f[n/x] \models Px$ and hence $\mathcal{M}, n, f \models \exists x Px$, for all n and f. On the other hand, if $\mathcal{M}, n, f \models \exists x \Box Px$, then as $\mathcal{M}$ is Kripkean we get $\mathcal{M}, n, f[m/x] \models \Box Px$ for some m. But then every $k \geq n$ has $\mathcal{M}, k, f[m/x] \models Px$ and so $k = m$, which is impossible. So $\mathcal{M}, n, f \not\models \exists x \Box Px$.

The upshot is that $\exists x Px$ is true everywhere in $\mathcal{M}$, hence so is $\Box\exists x Px$; while $\exists x \Box Px$ is false everywhere, hence so is $\Diamond\exists x \Box Px$; and consequently (1.11.4) is false at every world of $\mathcal{M}$.

This model delivers more than we asked of it. It validates the Universal Instantiation scheme $\forall x\varphi \to \varphi(\tau/x)$, as it has a single domain equal to the universe of the model. The single domain together with the Kripkean property means that it also validates CBF and BF. Thus it shows that (1.11.4) is not a theorem of QS4M + CQ + UI + CBF + BF. We will make use of that fact later in Section 2.8, in showing that a certain Kripkean model characterising QS4M + CQ + CBF + BF cannot be based on an S4M-frame.

We also obtain from this analysis a proof, promised at the end of Section 1.5, that a formula can be valid in all Kripkean models on a structure $\mathcal{S}$ without being valid in $\mathcal{S}$:

THEOREM 1.11.3. *Let* L *be any quantified modal logic that includes* QS4M *and is included in* QS4M + CQ + UI + CBF + BF. *The sentence* (1.11.4) *is valid in all Kripkean models on the canonical structure* $\mathcal{S}_{\mathrm{L}}$, *but is not valid in* $\mathcal{S}_{\mathrm{L}}$.

PROOF. By Theorem 1.10.6(1), the Kripke frame underlying $\mathcal{S}_{\mathrm{L}}$ is an S4M-frame. Hence Lemma 1.11.2 ensures that (1.11.4) is valid in all Kripkean models on $\mathcal{S}_{\mathrm{L}}$. But (1.11.4) is not valid in $\mathcal{S}_{\mathrm{L}}$, as it is not an L-theorem (see Corollary 1.9.7). ⊣

1.11.3. Non-Canonical S-frame Incompleteness. In Theorem 1.10.6, the hypothesis that S is canonical is essential. Without it, QS may be incomplete for any class of model structures based on Kripke frames that validate S. This occurs when S is the Gödel-Löb logic GL, the smallest propositional modal logic that includes the scheme

$$\text{W:} \quad \Box(\Box A \to A) \to \Box A.$$

A frame (W, R) validates GL iff R is transitive and *inverse well-founded*, i.e. there are no infinite R-chains $w_0 R \cdots w_n R w_{n+1} \cdots$ [Boolos 1979, p. 82]. If a frame has such an R-chain, then taking A to be a propositional variable that is false exactly at the members of the chain gives a model falsifying W at each member.

The members of an R-chain do not have to be distinct: if wRw, then $wRwRwR\cdots$ is an infinite R-chain. The canonical frame for GL contains such a reflexive point w, so does not validate W (see [Goldblatt 1992, p. 51] or [Hughes and Cresswell 1996, p. 140]). Hence GL is not canonical.

Now take a signature $\mathcal{L}$ having a monadic predicate symbol P, and a unary function symbol F, and consider the sentence

$$\forall x\Box(Px \to \Diamond PFx) \to \forall x\neg\Diamond Px. \tag{1.11.5}$$

LEMMA 1.11.4. *The sentence* (1.11.5) *is valid in any model structure whose underlying Kripke frame validates GL.*

PROOF. We work contrapositively: let $\mathcal{M}$ be an $\mathcal{L}$-model that falsifies (1.11.5). Suppose further that the underlying Kripke frame is transitive. Then we show that it contains an R-chain. Hence it cannot be a GL-frame.

Now there is some point w in $\mathcal{M}$ at which(1.11.5) is false, so $\mathcal{M}, w \models \forall x\Box(Px \to \Diamond PFx)$ and $\mathcal{M}, w \not\models \forall x\neg\Diamond Px$. By our semantics for $\forall$, there must be some admissible set X with $w \in X$ and $X \subseteq |\Box(Pa \to \Diamond PFa)|^{\mathcal{M}}$ for all $a \in U$ (treating a as an individual constant). Since $\forall x\neg\Diamond Px$ is false at w, we have $X \nsubseteq |\neg\Diamond Pb|^{\mathcal{M}}$ for some $b \in U$. Hence there is some world $w_0 \in X$ at which $w_0 \models \Diamond Pb$ in $\mathcal{M}$, so there is some w_1 such that w_0Rw_1 and $w_1 \models Pb$. Now suppose inductively that we have defined w_n $(n \geq 1)$ such that $w_n \models PF^{n-1}b$ and w_0Rw_n. Here F^n is the n-th iteration of F, with $F^0b = b$ and $F^nb = F(F^{n-1}b)$. Then

$$w_0 \in X \subseteq |\Box(PF^{n-1}b \to \Diamond PF^nb)|^{\mathcal{M}},$$

so $w_n \models PF^{n-1}b \to \Diamond PF^nb$. Hence there is some w_{n+1} with w_nRw_{n+1} and $w_{n+1} \models PF^nb$. Then w_0Rw_{n+1} by transitivity of R. This shows, by induction, that there is an infinite R-chain $w_0Rw_1R\cdots Rw_nR\cdots$ as claimed.

Thus if $\mathcal{M}$ is based on a GL-frame, it cannot falsify (1.11.5). $\dashv$

Thus to show that QGL is incomplete for validity in structures based on GL-frames, it suffices to show that the sentence (1.11.5) is not a QGL-theorem. This sentence is a variation on an example of Montagna [1984], who used one equivalent to $\forall x\exists y\Box(Px \to \Diamond Py) \to \forall x\neg\Diamond Px$. In our case, F was introduced to make the value of y that is asserted to exist into a function of x, in order to control the complexities of admissible semantics in the Lemma just proved.

Montagna constructed a counter-model to his sentence by a beautiful application of nonstandard models of arithmetic. This works also for sentence (1.11.5), as we now explain. Let $\mathcal{N} = (N, +, \cdot, \mathfrak{s}, 0)$ be a nonstandard model of first-order Peano Arithmetic, with $\mathfrak{s}$ its successor function, i.e. $\mathfrak{s}(a) = a + 1$ where $1 = \mathfrak{s}(0)$. $\mathcal{N}$ includes the standard model of arithmetic based on ω as an initial segment, but also has nonstandard members, which are all larger in $\mathcal{N}$ than the standard ones. Fix a particular nonstandard $\mu \in N$. Thus $\mathcal{N} \models \mu > n$ for all $n \in \omega$. Taking an object $\infty \notin N$, an $\mathcal{L}$-model for quantified modal logic can be specified by:

- $W = N \cup \{\infty\}$.
- $R = \{(a,b) \in N^2 : \mathcal{N} \models a > b\} \cup \{(\infty, a) : a \in N\}$.
- $Prop = \wp(W)$.
- $U = N$.
- $D\infty = \omega$ while $Dw = N$ for each $w \in N$.
- $|P|^{\mathcal{M}}a = \{w \in W : w \in N \text{ and } \mathcal{N} \models w \cdot a > \mu\}$, for each $a \in U$.
- $|F|^{\mathcal{M}} = \mathfrak{s}$, the successor function on N.

$\mathcal{M}$ is a full model, hence Kripkean. The relation R extends the linear "greater than" relation on $\mathcal{N}$ by the addition of a new greatest element ∞. The Kripke frame underlying $\mathcal{M}$ does not validate GL, because it has infinite R-chains, for instance $\mu R(\mu-1)R(\mu-2)R\cdots\cdots$.

The satisfaction relation for $\mathcal{L}$-formulas at points in N can be defined in the (non-modal) language of $\mathcal{N}$ with members of N as constants. Using the first-order induction scheme that holds in $\mathcal{N}$, it can then be shown that if a truth-set $|\varphi|^{\mathcal{M}}f$ of an $\mathcal{L}$-formula φ contains a member of N, then it contains an $<$-least such member. That is enough to ensure that $\mathcal{M}$ validates any instance of the GL-axiom W. For if $\mathcal{M}, w, f \not\models \Box\varphi$, then (suppressing f), φ is false at some u with wRu. Then $u \in N$ and $\neg\varphi$ is true at u, so there is a $<$-least $v \leq u$ at which $\neg\varphi$ is true. Hence $\Box\varphi$ is true at v, so $\Box\varphi \to \varphi$ is false at v, and wRv, therefore $\Box(\Box\varphi \to \varphi)$ is false at w. This shows that $\mathcal{M}, w, f \models \Box(\Box\varphi \to \varphi) \to \Box\varphi$ for all w and f. Hence $\mathcal{M}$ is a model of QGL.

It remains to show that (1.11.5) is not true in $\mathcal{M}$, in order to show that it is not a QGL-theorem. We show that it is false at the world ∞. For this we use the notation $w \models \varphi[n]$, where φ is a formula with one free variable, to mean that φ is true in $\mathcal{M}$ at w when its free variable is assigned the value n. Now it is shown in [Montagna 1984] by nonstandard arithmetic that

$$\text{if } u \in N \text{ and } n \in \omega, \text{ then } u \models Px[n] \text{ implies } u-1 \models Px[n+1].$$

The reason is that if $\mathcal{N} \models u \cdot n > \mu$, then $u \cdot n$ is nonstandard, so u must be nonstandard, thus $\mathcal{N} \models u > n+1$ and hence

$$\mathcal{N} \models (u-1)\cdot(n+1) = u \cdot n + u - (n+1) > u \cdot n > \mu.$$

We can use this to show that

$$\text{for all } n \in \omega, \quad \infty \models \Box(Px \to \Diamond PFx)[n]. \tag{1.11.6}$$

Indeed if ∞Ru, then $u \models Px[n]$ implies that $u-1 \models Px[n+1]$ as above, hence $u-1 \models PFx[n]$ as $|F|^{\mathcal{M}}$ is the successor function, so $u \models \Diamond PFx[n]$ as $uR(u-1)$; therefore $u \models (Px \to \Diamond PFx)[n]$. Since $\omega = D\infty$, we conclude from (1.11.6) that $\infty \models \forall x\Box(Px \to \Diamond PFx)$.

But choosing a $u \in N$ with $\mathcal{N} \models u > \mu$ (e.g. $u = \mu + 1$), we have $\mathcal{N} \models u \cdot 1 > \mu$, hence $u \models Px[1]$, so $\infty \models \Diamond Px[1]$ as $\infty R u$. This shows that $\infty \not\models \forall x \neg \Diamond Px$, completing the proof that (1.11.5) is false in $\mathcal{M}$ at ∞, and therefore is not a QGL theorem.

That concludes for now our discussion of kinds of incompleteness. We will take it up again in the next chapter, after constructing Kripkean canonical models.

Chapter 2

THE BARCAN FORMULAS

A model structure $\mathcal{S}$ has

- *contracting domains* if, for all $w, u \in W$, wRu implies $Dw \supseteq Du$;
- *expanding domains* if wRu implies $Dw \subseteq Du$;
- *constant domains* if wRu implies $Dw = Du$.

The contracting domains condition is often identified with validity of the Barcan Formula BF: $\forall x\Box\varphi \to \Box\forall x\varphi$, and expanding domains with validity of the Converse scheme CBF: $\Box\forall x\varphi \to \forall x\Box\varphi$.

Indeed a *full* model structure has contracting domains iff it validates the scheme BF; and has expanding domains iff it validates CBF (see Fitting and Mendelsohn [1998, Section 4.9]).

However, we have seen that full models are not adequate to characterise logics in general. We will now see that, for our adequate admissible models, the relationship between the above structural conditions and the schemes BF and CBF is more complicated and surprising.

CBF is valid in admissible models with expanding domains, and any logic of the form QS + CBF is characterised by model structures with expanding domains. But the surprising part is that these same logics are also characterised by model structures with *constant* domains. The class of expanding domain structures includes the constant domain ones, and these constant ones are sufficient to characterised QS + CBF, even when BF is not amongst its theorems.

The point is that on contracting domain structures, validity of BF holds in general only in *Kripkean* models. That accounts for the above observation about *full* structures, on which all models are Kripkean. It turns out that *every* logic QS is characterised by contracting domain structures, and addition of CBF allows us to make these domains constant. The main model-theoretic function performed by BF itself is to allow us to build models that are Kripkean, as we shall see in due course.

This chapter works out the details of these facts, including demonstrating that there are contracting-domain model structures, and even constant-domain ones, that do not validate BF. In the next chapter, where we introduce an existence predicate, we will be able to explain from an *axiomatic* point of view the sense in which the presence or absence of BF is independent of the property of contracting domains.

2.1. Logics with CBF

The expanding-domains condition in any model structure is equivalent to the requirement that

$$Ea \subseteq [R]Ea, \quad \text{for all } a \in U, \tag{2.1.1}$$

since this asserts that if $a \in Dw$ and wRu, then $a \in Du$. We use this in the basic soundness result for CBF, for which we also need the facts that $[R]$ is monotonic, i.e. $X \subseteq Y$ implies $[R]X \subseteq [R]Y$, and distributes over Boolean implication in the sense that $[R](X \Rightarrow Y) \subseteq [R]X \Rightarrow [R]Y$.

THEOREM 2.1.1. CBF *is valid in every model structure that has expanding domains.*

PROOF. Let $\mathcal{S}$ have expanding domains, and $\mathcal{M}$ be a model on $\mathcal{S}$. Let $f \in U^{\omega}$. Then if $b \in U$,

$$\begin{aligned}
|\Box\forall x\varphi|^{\mathcal{M}} f &= [R]\big(\textstyle\bigsqcap_{a\in U} Ea \Rightarrow |\varphi|^{\mathcal{M}} f[a/x]\big) && \\
&\subseteq [R](Eb \Rightarrow |\varphi|^{\mathcal{M}} f[b/x]) && \text{as } [R] \text{ is monotonic} \\
&\subseteq [R]Eb \Rightarrow [R]|\varphi|^{\mathcal{M}} f[b/x] && \text{as } [R] \text{ distributes over } \Rightarrow \\
&\subseteq Eb \Rightarrow [R]|\varphi|^{\mathcal{M}} f[b/x] && \text{as } Eb \subseteq [R]Eb \text{ (2.1.1)} \\
&= Eb \Rightarrow |\Box\varphi|^{\mathcal{M}} f[b/x]. &&
\end{aligned}$$

Since this holds for all $b \in U$, and $|\Box\forall x\varphi|^{\mathcal{M}} f$ is admissible,

$$|\Box\forall x\varphi|^{\mathcal{M}} f \subseteq \bigsqcap_{b\in U} Eb \Rightarrow |\Box\varphi|^{\mathcal{M}} f[b/x] = |\forall x\Box\varphi|^{\mathcal{M}} f.$$

Hence $\mathcal{M} \models \Box\forall x\varphi \to \forall x\Box\varphi$. ⊣

Now we turn to completeness for logics that include CBF. First, some deductive equivalents of this scheme:

LEMMA 2.1.2. *For any quantified modal logic* L, *the following are equivalent.*

(1) $\vdash_{\mathrm{L}}$ CBF.
(2) $\vdash_{\mathrm{L}} \forall y\Box(\forall x\varphi \to \varphi(y/x))$ *whenever* y *is free for* x *in* φ.
(3) $\vdash_{\mathrm{L}} \forall x\Box(\forall x\varphi \to \varphi)$.

PROOF. (1) implies (2): from Actual Instantiation and the Necessitation rule,

$$\vdash_{\mathrm{L}} \Box\forall y(\forall x\varphi \to \varphi(y/x)).$$

(2) follows from this by CBF.

(2) implies 3: put $y = x$ in (2).

(3) implies (1): from (3) by distribution of $\Box$ and $\forall$ we get

$$\vdash_{\mathrm{L}} \forall x \Box \forall x \varphi \to \forall x \Box \varphi.$$

But by Vacuous Quantification,

$$\vdash_{\mathrm{L}} \Box \forall x \varphi \to \forall x \Box \forall x \varphi.$$

Hence $\vdash_{\mathrm{L}} \Box \forall x \varphi \to \forall x \Box \varphi$ by PC. ⊣

THEOREM 2.1.3. *If the logic* L *includes* CBF, *then in the canonical model structure* $\mathcal{S}_{\mathrm{L}}$, *if a closed term* τ *belongs to* $D_{\mathrm{L}}(w)$, *then*

$$\mathrm{UI}(\tau) \subseteq \Box^{-} w = \{\varphi : \Box \varphi \in w\}.$$

PROOF. Recall that

$$\mathrm{UI}(\tau) = \{\forall x \varphi \to \varphi(\tau/x) : x \in \mathsf{InVar} \text{ and } \varphi \text{ is any formula}\}.$$

Take $w \in W_{\mathrm{L}}$ and $\tau \in D_{\mathrm{L}}(w)$, i.e. $\mathrm{UI}(\tau) \subseteq w$. Given any formula $\forall x \varphi$, choose a variable y that does not occur in this formula. By part (2) of the Lemma just proved, $\forall y \Box(\forall x \varphi \to \varphi(y/x)) \in w$. Hence as $\mathrm{UI}(\tau) \subseteq w$,

$$(\Box(\forall x \varphi \to \varphi(y/x)))(\tau/y) \in w.$$

But this last formula is just $\Box(\forall x \varphi \to \varphi(\tau/x))$, as y is not in $\forall x \varphi$, so we get $\Box(\forall x \varphi \to \varphi(\tau/x)) \in w$, hence $\forall x \varphi \to \varphi(\tau/x) \in \Box^{-} w$ as required. ⊣

COROLLARY 2.1.4. *If* L *includes* CBF, *then the canonical model structure* $\mathcal{S}_{\mathrm{L}}$ *has expanding domains.*

PROOF. Let $w R_{\mathrm{L}} u$. If $\tau \in D_{\mathrm{L}}(w)$, then $\mathrm{UI}(\tau) \subseteq \Box^{-} w \subseteq u$, hence $\tau \in D_{\mathrm{L}}(u)$. So $D_{\mathrm{L}}(w) \subseteq D_{\mathrm{L}}(u)$. ⊣

THEOREM 2.1.5 (Completeness for QS + CBF). *Let* S *be any propositional modal logic.*

(1) QS + CBF *is characterised by its canonical model structure.*
(2) QS + CBF *is characterised by validity in all expanding-domain model structures whose underlying general frame validates* S.
(3) *If* S *is canonical, then* QS+CBF *is characterised by validity in all expanding-domain model structures whose underlying Kripke frame validates* S.

PROOF. Let L = QS + CBF.

(1): For soundness, the canonical structure $\mathcal{S}_{\mathrm{L}}$ validates QS by Lemma 1.10.1, and has expanding domains by the Corollary just proved, so validates CBF by Theorem 2.1.1. Hence $\mathcal{S}_{\mathrm{S}}$ validates the logic QS+CBF. Completeness is given by Corollary 1.9.7, which states that every logic is complete for its canonical model structure: $\mathcal{S}_{\mathrm{L}} \models \varphi$ implies $\vdash_{\mathrm{L}} \varphi$.

(2): For soundness, if an expanding-domain model structure has a general frame validating S, then the structure validates QS (Theorem 1.7.11) and CBF. For completeness, if φ is valid in all expanding-domain model structures whose underlying general frame validates S, then in particular it is valid in $\mathcal{S}_L$, so $\vdash_L \varphi$.

(3): If S is canonical then the Kripke frame underlying $\mathcal{S}_L$ validates S by Theorem 1.10.6. The proof of (3) is then similar to (2). ⊣

2.2. Contracting Domains for All

The canonical model structure $\mathcal{S}_L$ of any logic L can be modified into a new structure that has contracting domains and still carries a characteristic model for L. The essential idea, which could in principle be applied to other structures, is to replace each world w by the set of pairs

$$\{(w, C) : C \subseteq Dw\},$$

and declare that only the members of C exist at world (w, C) in the new structure. The original accessibility relation R is replaced by the relation in which (w', C') is accessible from (w, C) when wRw' and $C \supseteq C'$, thereby building in the contracting-domains condition.

Applying this idea to $\mathcal{S}_L$ for a given signature $\mathcal{L}$ produces the structure

$$\mathcal{S}_L^\flat = (W_L^\flat, R_L^\flat, Prop_L^\flat, U_\mathcal{L}, D_L^\flat),$$

based on the set

$$W_L^\flat = \{(\Gamma, C) : \Gamma \in W_L \text{ and } C \subseteq D_L(\Gamma)\}.$$

A typical member w of $W_L^\flat$ will be denoted $w = (\Gamma_w, C_w)$. The rest of the structure is then defined by:

- $wR_L^\flat u$ iff $\Gamma_w R_L \Gamma_u$, i.e. $\{\varphi : \Box\varphi \in \Gamma_w\} \subseteq \Gamma_u$, and $C_w \supseteq C_u$.
- $Prop_L^\flat = \{\, |\varphi|_L^\flat : \varphi \text{ is an } \mathcal{L}\text{-sentence}\}$, where

$$|\varphi|_L^\flat = \{w \in W_L^\flat : \varphi \in \Gamma_w\}.$$

- $D_L^\flat(w) = C_w$.

Thus $wR_L^\flat u$ implies $C_w \supseteq C_u$ and hence $D_L^\flat(w) \supseteq D_L^\flat(u)$, so $\mathcal{S}_L^\flat$ has contracting domains. It is readily shown that

$$|\mathsf{F}|_L^\flat = \emptyset, \quad |\varphi|_L^\flat \cap |\psi|_L^\flat = |\varphi \wedge \psi|_L^\flat, \quad \text{and} \quad W_L^\flat - |\varphi|_L^\flat = |\neg\varphi|_L^\flat, \tag{2.2.1}$$

so $Prop_L^\flat$ is a Boolean set algebra.

LEMMA 2.2.1. $[R_L^\flat]|\varphi|_L^\flat = |\Box\varphi|_L^\flat$.

PROOF. Assume that $w \in [R_L^\flat]|\varphi|_L^\flat$. If $\Delta \in W_L$ and $\Gamma_w R_L \Delta$, define

$$u = (\Delta, C_w \cap D_L(\Delta)) \in W_L^\flat.$$

Then $\Gamma_w R_L \Gamma_u$ and $C_w \supseteq C_u$, so $wR_L^\flat u$. It follows from the assumption on w that $u \in |\varphi|_L^\flat$, so $\varphi \in \Gamma_u = \Delta$, and hence $\Delta \in |\varphi|_L$.

This shows that $\Gamma_w \in [R_L]|\varphi|_L = |\Box\varphi|_L$, so $\Box\varphi \in \Gamma_w$, i.e. $w \in |\Box\varphi|^\flat_L$ as required.

Conversely, let $w \in |\Box\varphi|^\flat_L$. If $wR^\flat_L u$, then $\Gamma_w R_L \Gamma_u$, so $\varphi \in \Gamma_u$ as $\Box\varphi \in \Gamma_w$, hence $u \in |\varphi|^\flat_L$. This shows that $w \in [R^\flat_L]|\varphi|^\flat_L$. ⊣

By this Lemma, $Prop^\flat_L$ is closed under the operation $[R^\flat_L]$. It follows that $\mathcal{G}^\flat_L = (W^\flat_L, R^\flat_L, Prop^\flat_L)$ is a general propositional frame.

The next result characterises the conjunction operation $\bigsqcap$ induced by $Prop^\flat_L$ in $\mathcal{S}^\flat_L$. It corresponds to Lemma 1.9.2 for $\mathcal{S}_L$.

LEMMA 2.2.2. *If* $\forall x\varphi$ *is a sentence, then in* $\mathcal{S}^\flat_L$,

$$|\forall x\varphi|^\flat_L = \bigsqcap_{\tau \in U_\mathcal{L}} E\tau \Rightarrow |\varphi(\tau/x)|^\flat_L.$$

PROOF. Here $E\tau$ is the existence set

$$\{w \in W^\flat_L : \tau \in D^\flat_L(w)\} = \{w \in W^\flat_L : \tau \in C_w\}$$

in $\mathcal{S}^\flat_L$. Now if $w \in |\forall x\varphi|^\flat_L$, then if $w \in E\tau$ we have $\forall x\varphi \in \Gamma_w$ and $\tau \in C_w \subseteq D_L(\Gamma_w)$; so $\mathrm{UI}(\tau) \subseteq \Gamma_w$ and hence $\forall x\varphi \to \varphi(\tau/x) \in \Gamma_w$; therefore $\varphi(\tau/x) \in \Gamma_w$, giving $w \in |\varphi(\tau/x)|^\flat_L$. This shows that

$$|\forall x\varphi|^\flat_L \subseteq E\tau \Rightarrow |\varphi(\tau/x)|^\flat_L$$

for every $\tau \in U_\mathcal{L}$. Since $|\forall x\varphi|^\flat_L \in Prop^\flat_L$, it follows that

$$|\forall x\varphi|^\flat_L \subseteq \bigsqcap_{\tau \in U_\mathcal{L}} E\tau \Rightarrow |\varphi(\tau/x)|^\flat_L.$$

For the converse inclusion, it suffices to take any $|\psi|^\flat_L \in Prop^\flat_L$ such that

$$|\psi|^\flat_L \subseteq E\tau \Rightarrow |\varphi(\tau/x)|^\flat_L$$

for every $\tau \in U_\mathcal{L}$, and show that $|\psi|^\flat_L \subseteq |\forall x\varphi|^\flat_L$ (see the explanation in the proof of Lemma 1.9.2).

Now given such an admissible set $|\psi|^\flat_L$, pick a constant $\mathsf{c} \in U_\mathcal{L}$ that does not occur in φ or ψ. Then we show

$$\mathrm{UI}(\mathsf{c}) \cup \{\psi\} \vdash_L \varphi(\mathsf{c}/x). \tag{2.2.2}$$

For if Γ is any L-maximal set with $\mathrm{UI}(\mathsf{c}) \cup \{\psi\} \subseteq \Gamma$, then $\mathsf{c} \in D_L(\Gamma)$ as $\mathrm{UI}(\mathsf{c}) \subseteq \Gamma$, so the pair $w = (\Gamma, \{c\})$ belongs to $W^\flat_L$. Since $\psi \in \Gamma$, we have $w \in |\psi|^\flat_L$, so $w \in E\mathsf{c} \Rightarrow |\varphi(\mathsf{c}/x)|^\flat_L$ by assumption on $|\psi|^\flat_L$. But $w \in E\mathsf{c}$ as $\mathsf{c} \in \{\mathsf{c}\} = D^\flat_L(w)$; hence $w \in |\varphi(\mathsf{c}/x)|^\flat_L$ and so $\varphi(\mathsf{c}/x) \in \Gamma_w = \Gamma$.

This proves that every L-maximal extension of $\mathrm{UI}(\mathsf{c}) \cup \{\psi\}$ contains $\varphi(\mathsf{c}/x)$, which implies (2.2.2). Now from (2.2.2) we obtain $\vdash_L \psi \to \forall x\varphi$ exactly as in the proof of Lemma 1.9.2. This means that any $w \in W^\flat_L$ has $\psi \in \Gamma_w$ only if $\forall x\varphi \in \Gamma_w$. Hence $|\psi|^\flat_L \subseteq |\forall x\varphi|^\flat_L$. ⊣

A premodel $\mathcal{M}^\flat_L = (\mathcal{S}^\flat_L, |-|^{\mathcal{M}^\flat_L})$ on $\mathcal{S}^\flat_L$ is defined by:

- $|\mathsf{c}|^{\mathcal{M}_{\mathrm{L}}^{\flat}} = \mathsf{c} \in U_{\mathcal{L}}$, for all constants $\mathsf{c} \in \mathsf{Con}_{\mathcal{L}}$.
- $|F|^{\mathcal{M}_{\mathrm{L}}^{\flat}}(\tau_1, \dots, \tau_n) = F\tau_1 \cdots \tau_n \in U_{\mathcal{L}}$, for all $\tau_1, \dots, \tau_n \in U_{\mathcal{L}}$.
- $|P|^{\mathcal{M}_{\mathrm{L}}^{\flat}}(\tau_1, \dots, \tau_n) = |P\tau_1 \cdots \tau_n|_{\mathrm{L}}^{\flat} \in Prop_{\mathrm{L}}$, for all $\tau_1, \dots, \tau_n \in U_{\mathcal{L}}$.

The interpretation of individual constants and function letters in $\mathcal{M}_{\mathrm{L}}^{\flat}$ is identical to that in the canonical model $\mathcal{M}_{\mathrm{L}}$: for all terms τ and $f \in U_{\mathcal{L}}{}^{\omega}$, we have $|\tau|^{\mathcal{M}_{\mathrm{L}}^{\flat}} f = |\tau|^{\mathcal{M}_{\mathrm{L}}} f = \tau^f$ (see Lemma 1.9.3). There is a Truth Lemma for $\mathcal{M}_{\mathrm{L}}^{\flat}$ that corresponds to Theorem 1.9.4 for $\mathcal{M}_{\mathrm{L}}$:

THEOREM 2.2.3 (Truth Is Membership). *Let φ be any formula. Then for all $f \in U_{\mathcal{L}}{}^{\omega}$,*

$$|\varphi|^{\mathcal{M}_{\mathrm{L}}^{\flat}} f = |\varphi^f|_{\mathrm{L}}^{\flat},$$

and hence for all $w \in W_{\mathrm{L}}^{\flat}$,

$$\mathcal{M}_{\mathrm{L}}^{\flat}, w, f \models \varphi \quad \textit{iff} \quad \varphi^f \in \Gamma_w.$$

PROOF. The second part of the statement is an immediate consequence of the first. The proof of the first is formally just like that for Theorem 1.9.4, but with $|\varphi|^{\mathcal{M}_{\mathrm{L}}^{\flat}}$ in place of $|\varphi|^{\mathcal{M}_{\mathrm{L}}}$ and $|\varphi^f|_{\mathrm{L}}^{\flat}$ in place of $|\varphi^f|_{\mathrm{L}}$; using the results of (2.2.1), Lemma 2.2.1 and Lemma 2.2.2. ⊣

THEOREM 2.2.4. *For any quantified modal logic* L, *$\mathcal{M}_{\mathrm{L}}^{\flat}$ is a model that characterises* L:

$$\vdash_{\mathrm{L}} \varphi \quad \textit{iff} \quad \mathcal{M}_{\mathrm{L}}^{\flat} \models \varphi.$$

PROOF. $\mathcal{M}_{\mathrm{L}}^{\flat}$ is a model: each truth set $|\varphi|^{\mathcal{M}_{\mathrm{L}}^{\flat}} f$ of $\mathcal{M}_{\mathrm{L}}^{\flat}$ is admissible, because it is equal to $|\varphi^f|_{\mathrm{L}}^{\flat} \in Prop_{\mathrm{L}}^{\flat}$.

Now combining Theorem 2.2.3 just proved with the corresponding Theorem 1.9.4 for $\mathcal{M}_{\mathrm{L}}$, we have

$$\mathcal{M}_{\mathrm{L}}^{\flat}, w, f \models \varphi \quad \text{iff} \quad \varphi^f \in \Gamma_w \quad \text{iff} \quad \mathcal{M}_{\mathrm{L}}, \Gamma_w, f \models \varphi.$$

Hence if φ is valid in $\mathcal{M}_{\mathrm{L}}$, then $\mathcal{M}_{\mathrm{L}}, \Gamma_w, f \models \varphi$ and hence $\mathcal{M}_{\mathrm{L}}^{\flat}, w, f \models \varphi$ for every $w \in W_{\mathrm{L}}^{\flat}$ and $f \in U_{\mathcal{L}}{}^{\omega}$, so φ is valid in $\mathcal{M}_{\mathrm{L}}^{\flat}$.

Conversely, assume φ is valid in $\mathcal{M}_{\mathrm{L}}^{\flat}$. For each $\Gamma \in W_{\mathrm{L}}$, let $w = (\Gamma, D_{\mathrm{L}}(\Gamma)) \in W_{\mathrm{L}}^{\flat}$. Then for any $f \in U_{\mathcal{L}}{}^{\omega}$ we have $\mathcal{M}_{\mathrm{L}}^{\flat}, w, f \models \varphi$, and therefore $\mathcal{M}_{\mathrm{L}}, \Gamma_w, f \models \varphi$, i.e. $\mathcal{M}_{\mathrm{L}}, \Gamma, f \models \varphi$. This shows that φ is valid in $\mathcal{M}_{\mathrm{L}}$.

Thus $\mathcal{M}_{\mathrm{L}}^{\flat} \models \varphi$ iff $\mathcal{M}_{\mathrm{L}} \models \varphi$. But $\mathcal{M}_{\mathrm{L}} \models \varphi$ iff $\vdash_{\mathrm{L}} \varphi$ by Theorem 1.9.6. ⊣

COROLLARY 2.2.5. L *is complete for $\mathcal{S}_{\mathrm{L}}^{\flat}$, i.e. $\mathcal{S}_{\mathrm{L}}^{\flat} \models \varphi$ implies $\vdash_{\mathrm{L}} \varphi$.* ⊣

From this result we obtain a new characterisation of the smallest quantified modal logic:

THEOREM 2.2.6. QK *is characterised by the contracting-domains model structure $\mathcal{S}_{\mathrm{QK}}^{\flat}$, and hence is characterised by the class of all contracting-domains model structures.*

PROOF. If $\vdash_{QK} \varphi$ then φ is valid in all model structures, hence valid in all contracting-domain ones, hence valid in $\mathcal{S}^{\flat}_{QK}$. But $\mathcal{S}^{\flat}_{QK} \models \varphi$ implies $\vdash_{QK} \varphi$ by the above Corollary. ⊣

We can use the following fact about the Barcan Formula to show that the model $\mathcal{M}^{\flat}_{QK}$ is not Kripkean.

LEMMA 2.2.7. *If $\mathcal{M}$ is a Kripkean model on a contracting-domains model structure, then $\mathcal{M}$ validates the Barcan Formula.*

PROOF. Let $\mathcal{M}$ be Kripkean with contracting domains, and $\mathcal{M}, w, f \models \forall x \Box \varphi$. Suppose wRv. Then if $a \in Dv$ we have $a \in Dw$ by the contracting-domains condition, so $\mathcal{M}, w, f[a/x] \models \Box\varphi$. Hence $\mathcal{M}, v, f[a/x] \models \varphi$ as wRv.

Since now $\mathcal{M}, v, f[a/x] \models \varphi$ for all $a \in Dv$, we get $\mathcal{M}, v, f \models \forall x\varphi$ as $\mathcal{M}$ is Kripkean (1.6.6). As this holds for every v such that wRv, we conclude $\mathcal{M}, w, f \models \Box\forall x\varphi$. ⊣

Now the Barcan Formula is not derivable in QK (see below), so $\mathcal{M}^{\flat}_{QK} \not\models$ BF since $\mathcal{M}^{\flat}_{QK}$ characterises QK. Hence by this Lemma, as $\mathcal{M}^{\flat}_{QK}$ has contracting domains it cannot be Kripkean. We also see by this reasoning that QK cannot be characterised by any class of Kripkean models on contracting-domains model structures.

More generally, if L is any logic for which $\nvdash_{L}$ BF, then $\mathcal{M}^{\flat}_{L}$ will be a contracting-domains model that falsifies BF and is not Kripkean. So the contraction of domains is certainly not sufficient to ensure validity of BF in our admissible semantics.

EXAMPLE 2.2.8. QK $\nvdash$ BF.

A proof of this was given in Kripke's original paper [1963b], providing a simple two-world two-individual full model that falsifies an instance of BF. A similar model can be defined by putting $W = \{w, u\}$, $R = W \times W$, $Prop = \wp W$, $U = \{a, b\}$, $Dw = \{a\}$, $Du = \{b\}$. Let $\mathcal{M}$ be a model on this structure that has $|P|^{\mathcal{M}}a = W$ and $|P|^{\mathcal{M}}b = \emptyset$, where P is a monadic predicate symbol. Then $\mathcal{M}, w \models \forall x \Box Px$ as P holds of a in both worlds, but $\mathcal{M}, w \not\models \Box\forall x Px$ as $\mathcal{M}, u \not\models \forall x Px$ because P is false of b at u.

In fact this model shows that BF is not derivable in QS5 + CQ. ⊣

We now extend the analysis to show that every logic of the form QS is characterised by contracting-domain structures. For this we need to know that $\mathcal{S}^{\flat}_{QS}$ validates QS. The result is an analogue of Lemma 1.10.1 for $\mathcal{S}_{QS}$:

LEMMA 2.2.9. *If* L *is any quantified modal logic that extends* QS, *then* S *is validated by the general frame $\mathcal{G}^{\flat}_{L}$ underlying $\mathcal{S}^{\flat}_{L}$. Hence $\mathcal{S}^{\flat}_{L}$ validates* QS.

PROOF. For the algebraically minded, one way to see this is to observe that $|\varphi|_L \mapsto |\varphi|^{\flat}_L$ is a well-defined bijection between $Prop_L$ and $Prop^{\flat}_L$ that makes them isomorphic as modal algebras. Since $Prop_{\rm L}$ validates S, so too does $Prop^\flat_{\rm L}$.

Alternatively, a model-theoretic proof proceeds similarly to Lemma 1.10.1. If $\mathcal{M}$ is any propositional model on $\mathcal{G}^\flat_{\rm L}$, then $\mathcal{M}$ assigns to each $p \in$ PropVar a set $|p|^{\mathcal{M}} \in Prop^\flat_{\rm L}$. Hence $|p|^{\mathcal{M}} = |\varphi_p|^\flat_{\rm L}$ for some $\mathcal{L}$-sentence φ_p. Let $A \mapsto A^*$ be the mapping of propositional formulas to $\mathcal{L}$-formulas induced by the uniform substitution of φ_p for p. Then an inductive proof shows that $|A|^{\mathcal{M}} = |A^*|^\flat_{\rm L}$ for all propositional A. The induction step for $\Box$ is taken care of by Lemma 2.2.1.

Now if $\vdash_{\rm S} A$, then $\vdash_{\rm QS} A^*$, so $\vdash_{\rm L} A^*$ and hence for every $w \in W^\flat_{\rm L}$, $A^* \in \Gamma_w$ and so $w \in |A^*|^\flat_{\rm L}$. Therefore $|A|^{\mathcal{M}} = |A^*|^\flat_{\rm L} = W^\flat_{\rm L}$, i.e. $\mathcal{M} \models A$. This proves that $\mathcal{G}^\flat_{\rm L}$ validates S.

The fact that $\mathcal{S}^\flat_{\rm L}$ validates QS then follows by Theorem 1.7.11. ⊣

Taking L = QS in this Lemma gives our general result about the sufficiency of contracting-domains structures:

THEOREM 2.2.10. *For any set* S *of propositional formulas, the logic* QS *is characterised by the contracting-domains model structure* $\mathcal{S}^\flat_{\rm QS}$. *Hence* QS *is characterised by the class of all contracting-domains model structures whose underlying general frame validates* S.

PROOF. The Lemma just proved shows $\mathcal{S}^\flat_{\rm QS}$ validates QS, i.e. QS is sound for $\mathcal{S}^\flat_{\rm QS}$. Completeness of QS for $\mathcal{S}^\flat_{\rm QS}$ is given by Corollary 2.2.5.

The second statement of the Theorem follows because QS is sound for all model structures whose general frame validates S (Theorem 1.7.11), hence is sound for the contracting-domains ones; and QS is complete for the latter class because it includes $\mathcal{S}^\flat_{\rm QS}$. ⊣

2.3. Constant Domains for CBF

The ideas of the previous two sections can be combined to show that any quantified modal logic L that includes the Converse Barcan Formula is characterised by models with *constant* domains, regardless of whether it includes BF. For this we replace the relation $R^\flat_{\rm L}$ in the structure $\mathcal{S}^\flat_{\rm L}$ by a new relation $R^\circ_{\rm L}$, defined, for $w, u \in W^\flat_{\rm L}$, by

$$wR^\circ_{\rm L}u \quad \text{iff} \quad \Gamma_w R_{\rm L} \Gamma_u \text{ and } C_w = C_u.$$

Then $R^\circ_{\rm L} \subseteq R^\flat_{\rm L}$, and using $R^\circ_{\rm L}$ in place of $R^\flat_{\rm L}$ would give us constant domains. But to be able to do this, $R^\circ_{\rm L}$ must satisfy the property of $R^\flat_{\rm L}$ expressed in Lemma 2.2.1. In fact this holds in the presence of CBF:

LEMMA 2.3.1. *Let* L *include* CBF. *Then the operations* $[R^\circ_{\rm L}]$ *and* $[R^\flat_{\rm L}]$ *are identical on* $Prop^\flat_{\rm L}$, *i.e.* $[R^\circ_{\rm L}]|\varphi|^\flat_{\rm L} = [R^\flat_{\rm L}]|\varphi|^\flat_{\rm L}$ *for any* $\mathcal{L}$*-formula* φ. *Hence* $[R^\circ_{\rm L}]|\varphi|^\flat_{\rm L} = |\Box\varphi|^\flat_{\rm L}$.

PROOF. Since $R^\circ_{\rm L} \subseteq R^\flat_{\rm L}$, it is immediate that $[R^\flat_{\rm L}]|\varphi|^\flat_{\rm L} \subseteq [R^\circ_{\rm L}]|\varphi|^\flat_{\rm L}$.

For the converse, let $w \in [R^\circ_{\rm L}]|\varphi|^\flat_{\rm L}$. Then if $wR^\flat_{\rm L}u$ we have $\Gamma_w R_{\rm L} \Gamma_u$, so $D_{\rm L}(\Gamma_w) \subseteq D_{\rm L}(\Gamma_u)$ because the canonical structure $\mathcal{S}_{\rm L}$ has expanding domains in the presence of CBF (Corollary 2.1.4). But $C_w \subseteq D_{\rm L}(\Gamma_w)$ as $w \in W^\flat_{\rm L}$, so $C_w \subseteq D_{\rm L}(\Gamma_u)$, and hence putting $v = (\Gamma_u, C_w)$ defines a member v of $W^\flat_{\rm L}$. Now $\Gamma_w R_{\rm L} \Gamma_v = \Gamma_u$ and $C_w = C_v$, which means that $wR^\circ_{\rm L}v$. Since $w \in [R^\circ_{\rm L}]|\varphi|^\flat_{\rm L}$ we then get $v \in |\varphi|^\flat_{\rm L}$, hence $\varphi \in \Gamma_v = \Gamma_u$, and so $u \in |\varphi|^\flat_{\rm L}$. Altogether this shows that $wR^\flat_{\rm L}u$ implies $u \in |\varphi|^\flat_{\rm L}$, i.e. $w \in [R^\flat_{\rm L}]|\varphi|^\flat_{\rm L}$ as required.

Now $[R^\flat_{\rm L}]|\varphi|^\flat_{\rm L} = |\Box\varphi|^\flat_{\rm L}$ by Lemma 2.2.1, so $[R^\circ_{\rm L}]|\varphi|^\flat_{\rm L} = |\Box\varphi|^\flat_{\rm L}$ follows. ⊣

We now define the structure

$$\mathcal{S}^\circ_{\rm L} = (W^\flat_{\rm L}, R^\circ_{\rm L}, Prop^\flat_{\rm L}, U_{\mathcal{L}}, D^\flat_{\rm L}),$$

which differs from $\mathcal{S}^\flat_{\rm L}$ only in having the relation $R^\circ_{\rm L}$ in place of $R^\flat_{\rm L}$. The fact that $[R^\circ_{\rm L}]|\varphi|^\flat_{\rm L} = |\Box\varphi|^\flat_{\rm L}$ means that $Prop^\flat_{\rm L}$ is closed under $[R^\circ_{\rm L}]$, so $\mathcal{S}^\circ_{\rm L}$ is indeed a model structure. It has *constant* domains, since

$$wR^\circ_{\rm L}u \text{ implies } D^\flat_{\rm L}(w) = C_w = C_u = D^\flat_{\rm L}(u).$$

Let $\mathcal{M}^\circ_{\rm L}$ be the premodel on $\mathcal{S}^\circ_{\rm L}$ that interprets the signature $\mathcal{L}$ exactly as in $\mathcal{M}^\flat_{\rm L}$, i.e.:

- $|\mathsf{c}|^{\mathcal{M}^\circ_{\rm L}} = \mathsf{c}$.
- $|F|^{\mathcal{M}^\circ_{\rm L}}(\tau_1, \dots, \tau_n) = F\tau_1 \cdots \tau_n$.
- $|P|^{\mathcal{M}^\circ_{\rm L}}(\tau_1, \dots, \tau_n) = |P\tau_1 \cdots \tau_n|^\flat_{\rm L}$.

$\mathcal{M}^\circ_{\rm L}$ and $\mathcal{M}^\flat_{\rm L}$ are not identical: they have different accessibility relations and $\mathcal{M}^\circ_{\rm L}$ has constant domains. But, in the presence of CBF, they do have identical truth relations, so $\mathcal{M}^\circ_{\rm L}$ is a model:

THEOREM 2.3.2. *Let* L *include* CBF. *Then for any* $\mathcal{L}$*-formula* φ *and all* $f \in U_{\mathcal{L}}{}^\omega$,

$$|\varphi|^{\mathcal{M}^\circ_{\rm L}} f = |\varphi|^{\mathcal{M}^\flat_{\rm L}} f = |\varphi^f|^\flat_{\rm L}.$$

Hence for all $w \in W^\flat_{\rm L}$,

$$\mathcal{M}^\circ_{\rm L}, w, f \models \varphi \quad \textit{iff} \quad \mathcal{M}^\flat_{\rm L}, w, f \models \varphi \quad \textit{iff} \quad \varphi^f \in \Gamma_w.$$

In particular, $\mathcal{M}^\circ_{\rm L} \models \varphi$ *iff* $\mathcal{M}^\flat_{\rm L} \models \varphi$.

PROOF. Theorem 2.2.3 showed that $|\varphi|^{\mathcal{M}^\flat_{\rm L}} f = |\varphi^f|^\flat_{\rm L}$. The proof that $|\varphi|^{\mathcal{M}^\circ_{\rm L}} f = |\varphi^f|^\flat_{\rm L}$ proceeds by exactly the same induction on φ. The only significant point of difference is in the inductive case for $\Box$, where

$$|\Box\varphi|^{\mathcal{M}^\circ_{\rm L}} f = [R^\circ_{\rm L}](|\varphi|^{\mathcal{M}^\circ_{\rm L}} f) = [R^\circ_{\rm L}]|\varphi^f|^\flat_{\rm L}$$

by induction hypothesis on φ. But we have $[R^\circ_{\mathrm{L}}]|\varphi^f|^\flat_{\mathrm{L}} = |\Box\varphi^f|^\flat_{\mathrm{L}}$ from Lemma 2.3.1. ⊣

This equivalance of truth and validity between $\mathcal{M}^\circ_{\mathrm{L}}$ and $\mathcal{M}^\flat_{\mathrm{L}}$ allows us to characterise logics that include CBF by constant-domains models and structures:

THEOREM 2.3.3. (1) *If* L *includes* CBF, *then the constant-domains model* $\mathcal{M}^\circ_{\mathrm{L}}$ *characterises* L, *and so* L *is complete for its underlying structure* $\mathcal{S}^\circ_{\mathrm{L}}$.
(2) QK + CBF, *the smallest logic including* CBF, *is characterised by the constant-domains model structure* $\mathcal{S}^\circ_{\mathrm{QK+CBF}}$. *Hence* QK + CBF *is characterised by the class of all constant-domains model structures.*

PROOF. (1): This follows from the Theorem just proved, since $\mathcal{M}^\flat_{\mathrm{L}}$ characterises L by Theorem 2.2.4.

(2): CBF is valid in the class of expanding-domains model structures (Theorem 2.1.1), and so is valid in the subclass $\mathscr{C}_{\mathrm{const}}$ of all constant-domains structures. Hence QK + CBF is sound for $\mathscr{C}_{\mathrm{const}}$, and therefore sound for the constant-domains structure $\mathcal{S}^\circ_{\mathrm{QK+CBF}}$. But QK + CBF is complete for $\mathcal{S}^\circ_{\mathrm{QK+CBF}}$ by (1), hence is complete for $\mathscr{C}_{\mathrm{const}}$. ⊣

The Barcan Formula is not derivable in QK + CBF, so $\mathcal{M}^\circ_{\mathrm{QK+CBF}} \not\models$ BF since this model characterises QK+CBF. Thus by Lemma 2.2.7, as $\mathcal{M}^\circ_{\mathrm{QK+CBF}}$ has constant domains, hence contracting domains, it cannot be Kripkean. We also see by this reasoning that QK+CBF cannot be characterised by any class of Kripkean models on constant-domains model structures.

More generally, if L is any logic that includes CBF but not BF, then $\mathcal{M}^\circ_{\mathrm{L}}$ will be a *constant*-domains model that falsifies BF and is not Kripkean. So even constancy of domains is not sufficient to ensure validity of BF.

EXAMPLE 2.3.4. QS4 + CQ + CBF $\nvdash$ BF.

Here is a simple full model validating CBF while falsifying BF. It is similar to the one in Example 2.2.8, but has $W = \{w, u\}$, $Prop = \wp W$, $U = \{a, b\}$, $Dw = \{a\}$, $Du = \{a, b\}$, and $R = \{(w, w), (w, u), (u, u)\}$. Thus R is the partial order on $\{w, u\}$ in which wRu but not uRw.

Let $\mathcal{M}$ be a model on this structure that has $|P|^{\mathcal{M}}a = W$ and $|P|^{\mathcal{M}}b = \emptyset$. Then the BF-instance $\forall x\Box Px \to \Box\forall xPx$ is false in $\mathcal{M}$ at w, for the reasons given in Example 2.2.8. On the other hand, the structure underlying $\mathcal{M}$ has expanding domains, since $Dw \subseteq Du$, so $\mathcal{M}$ validates CBF. Also, the Kripke frame of $\mathcal{M}$ is reflexive and transitive, so validates S4.

Altogether then, $\mathcal{M}$ is a model of the logic QS4 + CQ + CBF, showing that this logic does not include BF. So the model

$$\mathcal{M}^\circ_{\mathrm{QS4+CQ+CBF}}$$

has constant domains but does not validate the Barcan Formula, hence is not Kripkean.

This Example shows in fact that QS + CQ + CBF $\nvdash$ BF, where S is the propositional logic characterised by the two-world Kripke frame underlying the given model. This S is an extension of S4.3: as well as the S4.3 axiom

$$\Box(\Box A \to B) \vee \Box(\Box B \to A),$$

corresponding to linearity in S4-frames, S includes the McKinsey axiom $\Box\Diamond\varphi \to \Diamond\Box\varphi$, the Grzegorczyk axiom

$$\Box(\Box(A \to \Box A) \to A) \to A,$$

corresponding to antisymmetry in finite S4-frames, and the scheme

$$\Box A_1 \vee \Box(A_1 \to A_2) \vee \Box(A_1 \wedge A_2 \to A_3),$$

corresponding to the condition that $\{u : wRu\}$ has at most two members, for all w. ⊣

We now show that every logic of the form QS + CBF is characterised by constant-domains model structures. For this we need to know that the structure $\mathcal{S}^{\circ}_{\text{QS+CBF}}$ validates QS + CBF:

LEMMA 2.3.5. *If* L *is any quantified modal logic that extends* QS + CBF, *then* S *is validated by the general frame* $\mathcal{G}^{\circ}_{\text{L}}$ *underlying* $\mathcal{S}^{\circ}_{\text{L}}$. *Hence* $\mathcal{S}^{\circ}_{\text{L}}$ *validates* QS + CBF.

PROOF. Here $\mathcal{G}^{\circ}_{\text{L}} = (W^{\flat}_{\text{L}}, R^{\circ}_{\text{L}}, Prop^{\flat}_{\text{L}})$. We can show that $\mathcal{G}^{\circ}_{\text{L}}$ validates S in just the same way that we showed that $\mathcal{G}^{\flat}_{\text{L}} = (W^{\flat}_{\text{L}}, R^{\flat}_{\text{L}}, Prop^{\flat}_{\text{L}})$ validates S (Lemma 2.2.9).

Even more directly, if $\mathcal{M}$ is any propositional model on $\mathcal{G}^{\circ}_{\text{L}}$, we can view it as a model $\mathcal{M}'$ on $\mathcal{G}^{\flat}_{\text{L}}$ by putting $|p|^{\mathcal{M}'} = |p|^{\mathcal{M}}$ for all $p \in \mathsf{PropVar}$. An inductive proof then shows that $\mathcal{M}'$ and $\mathcal{M}$ are truth-equivalent: $|A|^{\mathcal{M}'} = |A|^{\mathcal{M}}$ for all propositional formulas A. The case of $\Box$ has

$$|\Box A|^{\mathcal{M}'} = [R^{\flat}_{\text{L}}]|A|^{\mathcal{M}'} = [R^{\circ}_{\text{L}}]|A|^{\mathcal{M}} = |\Box A|^{\mathcal{M}},$$

using the induction hypothesis on A and the fact that $[R^{\flat}_{\text{L}}]$ and $[R^{\circ}_{\text{L}}]$ agree on $Prop^{\flat}_{\text{L}}$.

Now since $\mathcal{G}^{\flat}_{\text{L}}$ validates S, if $A \in \text{S}$ then $\mathcal{M}' \models A$, hence $\mathcal{M} \models A$. Since $\mathcal{M}$ was an arbitrary model on $\mathcal{G}^{\circ}_{\text{L}}$, this shows that $\mathcal{G}^{\circ}_{\text{L}}$ validates S.

Hence $\mathcal{S}^{\circ}_{\text{L}}$ validates QS by Theorem 1.7.11. But $\mathcal{S}^{\circ}_{\text{L}}$ has constant domains and so validates CBF. Therefore it validates QS + CBF. ⊣

Taking L = QS + CBF in this Lemma gives our general result about the sufficiency of constant-domains structures for logics of this form:

THEOREM 2.3.6. *For any set* S *of propositional formulas, the logic* QS + CBF *is characterised by the constant-domains model structure* $\mathcal{S}^{\circ}_{\text{QS+CBF}}$. *Hence* QS + CBF *is characterised by the class of all constant-domains model structures whose underlying general frame validates* S. ⊣

Finally on this topic, we consider what happens when S is a *canonical* propositional logic. In that case, the structure $\mathcal{S}^{\circ}_{\text{QS+CBF}}$ has an underlying Kripke frame that validates S. To explain why, we need the notion of a *point-generated* inner subframe.

Given a frame $\mathcal{F} = (W, R)$ and a point $w \in W$, we write $\mathcal{F}(w)$ for the smallest inner subframe of $\mathcal{F}$ that contains the point w. It consists of w and all points in W reachable from w by some finite R-chain $wRw_1R\cdots Rw_n$. More formally, $\mathcal{F}(w)$ is based on the set

$$W(w) = \{u \in W : wR^*u\}, \tag{2.3.1}$$

where R^* is the *reflexive-transitive* closure of R, the smallest reflexive and transitive relation that includes R. This is defined on W by putting vR^*u iff for some $n \in \omega$ there is a sequence $v = u_0, \ldots, u_n = u$ such that u_iRu_{i+1} for all $i < n$ (if $n = 0$, this just says that $v = u$). The set $W(w)$ is R-closed, and is a subset of any R-closed set that contains w.

The importance of this notion for us is that validity in a Kripke frame is determined by validity in these pointed-generated subframes $\mathcal{F}(w)$.

LEMMA 2.3.7. *$\mathcal{F} \models A$ iff for all $w \in W$, $\mathcal{F}(w) \models A$.*

PROOF. $\mathcal{F} \models A$ implies $\mathcal{F}(w) \models A$ for any w by Lemma 1.10.3, since $\mathcal{F}(w)$ is an inner subframe of $\mathcal{F}$.

For the converse, if $\mathcal{F} \not\models A$ then there is some model $\mathcal{M}$ on $\mathcal{F}$ with $\mathcal{M}, w \not\models A$ for some w. For this w, define a model $\mathcal{M}'$ on $\mathcal{F}(w)$ by putting $|p|^{\mathcal{M}'} = |p|^{\mathcal{M}} \cap W(w)$ for all $p \in \mathsf{PropVar}$. An induction then shows that

$$\mathcal{M}', u \models B \quad \text{iff} \quad \mathcal{M}, u \models B$$

for all u in $\mathcal{F}(w)$ and all propositional formulas B. In particular $\mathcal{M}', w \not\models A$, showing that A is not valid in $\mathcal{F}(w)$. ⊣

Recall from Section 1.10 that $\mathcal{F}_{\text{L}} = (W_{\text{L}}, R_{\text{L}})$ is the Kripke frame of the canonical model structure $\mathcal{S}_{\text{L}}$ of a quantified logic L. Let $\mathcal{F}^{\circ}_{\text{L}} = (W^{\flat}_{\text{L}}, R^{\circ}_{\text{L}})$ be the Kripke frame of the structure $\mathcal{S}^{\circ}_{\text{L}}$. Since each point of $\mathcal{F}^{\circ}_{\text{L}}$ has the form $w = (\Gamma_w, C_w)$, there is a natural map $\theta_{\text{L}} : W^{\flat}_{\text{L}} \to W_{\text{L}}$ defined by $\theta_{\text{L}}(w) = \Gamma_w$. This is surjective, as each $\Delta \in W_{\text{L}}$ is equal to $\theta_{\text{L}}(\Delta, D_{\text{L}}(\Delta))$. It is far from injective, since Δ is also equal to $\theta_{\text{L}}(\Delta, C)$ for all $C \subseteq D_{\text{L}}(\Delta)$. However, because $\mathcal{S}^{\circ}_{\text{L}}$ has constant domains, θ_{L} acts isomorphically between point-generated subframes of $\mathcal{F}^{\circ}_{\text{L}}$ and $\mathcal{F}_{\text{L}}$:

THEOREM 2.3.8. *If* L *extends* QS + CBF, *then for each $w \in W^{\flat}_{\text{L}}$, the point-generated frames $\mathcal{F}^{\circ}_{\text{L}}(w)$ and $\mathcal{F}_{\text{L}}(\theta_{\text{L}}w)$ are isomorphic under θ_{L}.*

PROOF. $\mathcal{F}^{\circ}_{\text{L}}(w)$ is based on the set

$$W^{\flat}_{\text{L}}(w) = \{u \in W^{\flat}_{\text{L}} : w(R^{\circ}_{\text{L}})^*u\},$$

while $\mathcal{F}_{\text{L}}(\theta_{\text{L}}w)$ is based on

$$W_{\text{L}}(\theta_L w) = \{\Delta \in W_{\text{L}} : \Gamma_w R^*_{\text{L}} \Delta\}.$$

The essential point of the proof is that since $\mathcal{S}^{\circ}_{\mathrm{L}}$ has constant domains, all members of $\mathcal{F}^{\circ}_{\mathrm{L}}(w)$ have the same domain as w. Indeed if $u \in W^{\flat}_{\mathrm{L}}(w)$ then there is a finite sequence $w = u_0, \dots, u_n = u$ such that for all $i < n$, $u_i R^{\circ}_{\mathrm{L}} u_{i+1}$, i.e. $\Gamma_{u_i} R_{\mathrm{L}} \Gamma_{u_{i+1}}$ and $C_{u_i} = C_{u_{i+1}}$. Thus $\Gamma_w = \Gamma_0 R_{\mathrm{L}} \cdots R_{\mathrm{L}} \Gamma_n = \Gamma_u$ and $C_w = C_0 = \cdots = C_n = C_u$. Hence $\Gamma_w R^*_{\mathrm{L}} \Gamma_u$, so $\Gamma_u \in W_{\mathrm{L}}(\theta_{\mathrm{L}} w)$, and $u = (\Gamma_u, C_w)$.

Similarly, if $\Delta \in W_{\mathrm{L}}(\theta_L w)$, then since $\Gamma_w R^*_{\mathrm{L}} \Delta$ and $w \in W^{\flat}_{\mathrm{L}}$ we get $C_w \subseteq D_{\mathrm{L}}(\Gamma_w) \subseteq D_{\mathrm{L}}(\Delta)$, hence $(\Delta, C_w) \in W^{\flat}_{\mathrm{L}}$, and $w = (\Gamma_w, C_w)(R^{\circ}_{\mathrm{L}})^*(\Delta, C_w)$, implying that $(\Delta, C_w) \in W^{\flat}_{\mathrm{L}}(w)$. So we see that

$$W^{\flat}_{\mathrm{L}}(w) = \{(\Delta, C_w) : \Delta \in W_{\mathrm{L}}(\theta_L w)\} = W_{\mathrm{L}}(\theta_L w) \times \{C_w\}.$$

The map $\theta_{\mathrm{L}} : (\Delta, C_w) \mapsto \Delta$ is therefore a bijection from $W^{\flat}_{\mathrm{L}}(w)$ onto $W_{\mathrm{L}}(\theta_L w)$. Since $(\Delta, C_w) R^{\circ}_{\mathrm{L}} (\Delta', C_w)$ iff $\Delta R_{\mathrm{L}} \Delta'$, this map is an isomorphism between the frames $\mathcal{F}^{\circ}_{\mathrm{L}}(w)$ and $\mathcal{F}_{\mathrm{L}}(\theta_{\mathrm{L}} w)$. ⊣

COROLLARY 2.3.9. *If* S *is a* ***canonical*** *propositional logic, and* L *is any quantified modal logic that extends* QS + CBF, *then* S *is validated by the Kripke frame* $\mathcal{F}^{\circ}_{\mathrm{L}} = (W^{\flat}_{\mathrm{L}}, R^{\circ}_{\mathrm{L}})$ *underlying the model structure* $\mathcal{S}^{\circ}_{\mathrm{L}}$.

PROOF. By Theorem 1.10.6, $\mathcal{F}_{\mathrm{L}}$ validates S. Hence for each w in $\mathcal{F}^{\circ}_{\mathrm{L}}$, the frame $\mathcal{F}_{\mathrm{L}}(\theta_{\mathrm{L}} w)$ validates S as it is an inner subframe of $\mathcal{F}_{\mathrm{L}}$ (Lemma 1.10.3). Therefore $\mathcal{F}^{\circ}_{\mathrm{L}}(w)$ validates S as validity is preserved by isomorphism.

This shows that all the point-generated inner subframes of $\mathcal{F}^{\circ}_{\mathrm{L}}$ validate S, which is enough to ensure that $\mathcal{F}^{\circ}_{\mathrm{L}}$ itself validates S, by Lemma 2.3.7. ⊣

Using this we obtain a refinement of Theorem 1.10.6, characterising QS + CBF when S is canonical:

THEOREM 2.3.10. *Let* S *be a canonical propositional logic. Then*:

(1) S *is validated by the Kripke frame underlying the constant-domains characteristic model structure* $\mathcal{S}^{\circ}_{\mathrm{QS+CBF}}$ *for* QS + CBF.
(2) QS + CBF *is characterised by the class of all constant-domains model structures whose underlying Kripke frame validates* S. ⊣

There are numerous standard propositional logics covered by this result. For instance, QS4 + CBF is characterised by the class of all constant-domains model structures based on reflexive and transitive frames. This class does not validate BF, since BF is not a theorem of QS4 + CBF, as we saw in Example 2.3.4. In particular $\mathcal{S}^{\circ}_{\mathrm{QS4+CBF}}$ is a characteristic model structure for QS4+CBF that has constant domains, and is reflexive and transitive. BF is falsified by the non-Kripkean model $\mathcal{M}^{\circ}_{\mathrm{QS4+CBF}}$ on the structure $\mathcal{S}^{\circ}_{\mathrm{QS4+CBF}}$.

2.4. One Universal Domain

If a model structure has constant domains, it need not follow that all of its worlds have the same domain of actual individuals. All that follows is that the *R-connected* worlds have the same domain. Here, w is R-connected to u when, in graphical terms, we can pass from w to u in finitely many steps by going back and/or forth along R. Equivalently, this means that w is related to u by R^{er}, the smallest equivalence relation that includes R. We have $wR^{\text{er}}u$ iff for some $n \in \omega$ there is a sequence $w = u_0, \dots, u_n = u$ such that for each $i < n$, either $u_i R u_{i+1}$ or $u_{i+1} R u_i$. Now $u_{i+1} R u_i$ iff $u_i R^{-1} u_{i+1}$, where R^{-1} is the inverse relation to R, so using the reflexive-transitive-closure notion defined in the previous section, we have

$$R^{\text{er}} = (R \cup R^{-1})^*.$$

The equivalence classes of R^{er} are the *R-components* of a Kripke frame (W, R). Two worlds in the same component are R-connected, while two worlds in different components have no $(R \cup R^{-1})$-chain connecting them.

Thus a frame may be thought of as being the union of its disjoint components. The constant-domains condition on a model structure means that the domain function D is constant on each component. All members of the same component have the same domain, which may be different to the common domain of the members of some other component.

If a model structure $\mathcal{S}$ has a single component, then all of its worlds do have the same domain. But this domain need not be equal to the universe U. In the case that the one domain is equal to U, then the quantifier $\forall x$ ranges over all possible individuals, so the structure should validate the Universal Instantiation axiom

UI: $\forall x\varphi \to \varphi(\tau/x)$, where τ is free for x in φ.

To study this, we define $\mathcal{S}$ to have *one universal domain* if $D_w = U$ for all w in $\mathcal{S}$. If this holds, then $Ea = W$ for all $a \in U$, and so $(Ea \Rightarrow X) = X$ in general. Thus in any model $\mathcal{M}$ on $\mathcal{S}$ we have

$$|\forall x\varphi|^{\mathcal{M}} f = \bigsqcap_{a \in U} |\varphi|^{\mathcal{M}} f[a/x],$$

so the truth condition for the quantifier $\forall$ becomes

$\mathcal{M}, w, f \models \forall x\varphi$ iff there is an $X \in Prop$ such that $w \in X$ and $X \subseteq \bigcap_{a \in U} |\varphi|^{\mathcal{M}} f[a/x]$.

THEOREM 2.4.1. *If $\mathcal{S}$ has one universal domain, then it validates* UI.

PROOF. Let $\mathcal{M}$ be any model on $\mathcal{S}$. We have just observed that

$$|\forall x\varphi|^{\mathcal{M}} f = \bigsqcap_{a \in U} |\varphi|^{\mathcal{M}} f[a/x],$$

and so $|\forall x\varphi|^{\mathcal{M}} f \subseteq |\varphi|^{\mathcal{M}} f[a/x]$ for each $f \in U^\omega$ and $a \in U$. Thus for a given term τ free for x in φ, putting $a = |\tau|^{\mathcal{M}} f$ and using the Substitution Lemma 1.6.2, we get

$$|\forall x\varphi|^{\mathcal{M}} f \subseteq |\varphi|^{\mathcal{M}} f[\,|\tau|^{\mathcal{M}} f/x] = |\varphi(\tau/x)|^{\mathcal{M}} f.$$

Hence $\mathcal{M} \models \forall x\varphi \to \varphi(\tau/x)$. (Note that this argument only required $\mathcal{M}$ to be a premodel.) ⊣

Completeness theorems for logics having UI can be derived straightforwardly from our work so far. Recall from Section 1.9 that in a canonical model structure $\mathcal{S}_{\mathrm{L}}$, the domain function is given by

$$D_{\mathrm{L}}(w) = \{\tau \in U_{\mathcal{L}} : \mathrm{UI}(\tau) \subseteq w\},$$

where $U_{\mathcal{L}}$ is the set of closed $\mathcal{L}$-terms and

$$\mathrm{UI}(\tau) = \{\forall x\varphi \to \varphi(\tau/x) : x \in \mathsf{InVar} \text{ and } \varphi \text{ is any formula}\}.$$

LEMMA 2.4.2. *If* L *includes* UI, *then* $\mathcal{S}_{\mathrm{L}}$ *has one universal domain and validates* CBF.

PROOF. An arbitrary closed term τ is free for any x in any φ, and if L includes the scheme UI, then $\forall x\varphi \to \varphi(\tau/x)$ belongs to every L-maximal set. Hence $\mathrm{UI}(\tau) \subseteq w$ for all $w \in W_{\mathrm{L}}$, so $\tau \in D_{\mathrm{L}}(w)$ for all w. Thus $D_{\mathrm{L}}(w) = U_{\mathcal{L}}$ for all w.

This shows that $\mathcal{S}_{\mathrm{L}}$ has one universal domain. Thus it has constant domains, hence expanding domains, and so validates CBF by Theorem 2.1.1. ⊣

Combining the results of this section with those of Section 1.10, we obtain a number of characterisations for logics having Universal Instantiation:

THEOREM 2.4.3. *For any propositional modal logic* S,

(1) *The canonical model structure* $\mathcal{S}_{\mathrm{QS+UI}}$ *has one universal domain and characterises* QS + UI.

(2) QS + UI *is characterised by the class of all one-universal-domain model structures whose underlying general frame validates* S.

(3) *If* S *is canonical, then the Kripke frame of* $\mathcal{S}_{\mathrm{QS+UI}}$ *validates* S, *and* QS + UI *is characterised by the class of all one-universal-domain model structures whose underlying Kripke frame validates* S. ⊣

For logics including UI, the very definition of a "logic" can be simplified in its quantificational postulates. All that are required are UI itself and the ∀-*Introduction* rule

$$\forall\text{-Intro:} \quad \frac{\varphi \to \psi}{\varphi \to \forall x\psi}, \quad \text{if } x \text{ is not free in } \varphi$$

(see Lemma 1.2.1(1)). From these it is standard to derive the schemes UD and VQ and the rule UG, as well as the scheme AI.

Note that if $\vdash_{\mathrm{L}}$ UI, then by Lemma 2.4.2, the characteristic L-model $\mathcal{M}_{\mathrm{L}}$ on $\mathcal{S}_{\mathrm{L}}$ validates CBF, so CBF is derivable in L. In fact that conclusion can be obtained deductively quite simply. We only need the instance $\forall x\varphi \to \varphi$ of UI to infer $\Box\forall x\varphi \to \Box\varphi$ by rule N and axiom K, hence $\Box\forall x\varphi \to \forall x\Box\varphi$ by ∀-Intro, as x is not free in the antecedent $\Box\forall x\varphi$.

Also, the one-universal-domain condition implies that every existence set Ea is admissible, since $Ea = W \in \mathit{Prop}$, and so the structure validates CQ

(Theorem 1.7.16). So $\vdash_{\mathrm{L}}$ UI implies $\mathcal{S}_{\mathrm{L}} \models$ CQ and hence $\vdash_{\mathrm{L}}$ CQ. But again that can be simply shown deductively: from $\forall y\varphi \to \varphi$ we infer $\forall x\forall y\varphi \to \forall x\varphi$ by rule UG and axiom UD, hence $\forall x\forall y\varphi \to \forall y\forall x\varphi$ as y is not free in $\forall x\forall y\varphi$.

What we can *not* derive from UI is the Barcan Formula. BF is not a theorem of QK + UI, or of many other systems of the form QS + UI. For such systems, our canonical model structure $\mathcal{S}_{\mathrm{QS+UI}}$ shows that even having one universal domain is not enough to ensure validity of BF.

A case in point is QS4 + UI. By Theorem 2.4.3, QS4 + UI is characterised by the class of all one-universal-domain model structures with reflexive and transitive accessibility relations, and also characterised by the single structure $\mathcal{S}_{\mathrm{QS4+UI}}$, which has one-universal-domain and is reflexive and transitive. In fact QS4 + UI $\nvdash$ BF, and BF is falsified by the non-Kripkean model $\mathcal{M}_{\mathrm{QS4+UI}}$ on $\mathcal{S}_{\mathrm{QS4+UI}}$.

EXAMPLE 2.4.4. QS4 + UI $\nvdash$ BF.

A pre-Kripkean proof that BF is not derivable in QS4 + UI was given by Lemmon [1960], using an algebraic-topological semantics for QS4 + UI due to Rasiowa [1951]. Lemmon's counter-model was based on the algebra of subsets of the set $\mathbb{R}$ of real numbers, with its standard topology used to interpret $\Box$. We can turn this into a model very like the kind we have developed here, except that there is no accessibility relation R, and the topological interior operation is used in place of $[R]$. For real numbers u and v, let

$$(u, v) = \{w \in \mathbb{R} : u < w < v\}$$

be the open interval with endpoints u, v. A real w is *interior* to a set $X \subseteq \mathbb{R}$ if there exist u, v such that $w \in (u, v) \subseteq X$. Let $Int(X)$ be the set of all interior points of X. We use the fact that $Int(u, v) = (u, v)$.

Assume that the signature $\mathcal{L}$ has a monadic predicate symbol P. A pre-model-like structure $\mathcal{M}$ is defined as follows:

- $W = \mathbb{R}$.
- $Prop = \wp W$.
- $U = \{1, 2, 3, \dots\}$, the set of positive integers.
- $Dw = U$ for all $w \in W$.
- $|P|^{\mathcal{M}}n = (-\frac{1}{n}, \frac{1}{n}) \subseteq \mathbb{R}$, for all $n \in U$.

Constants, functional symbols, and any other predicate symbols, can be interpreted arbitrarily by $\mathcal{M}$. The function $|\varphi|^{\mathcal{M}} : U^{\omega} \to Prop$ is defined inductively as for our models, except for the case of $\Box$, which has

$$|\Box\varphi|^{\mathcal{M}} f = Int(|\varphi|^{\mathcal{M}} f).$$

In terms of the truth relation, this says that

$$\mathcal{M}, w, f \models \Box\varphi \quad \text{iff} \quad w \text{ belongs to some open interval } (u, v) \subseteq |\varphi|^{\mathcal{M}} f.$$

This is the "progressive tense" interpretation of $\Box$ [Scott 1970, p. 160].

Since *Prop* is the full powerset of W, the conjunction operator $\sqcap$ is just the intersection $\bigcap$, and so as $\mathcal{M}$ has one universal domain, we have

$$|\forall x\varphi|^{\mathcal{M}} f = \bigcap_{n\in U} |\varphi|^{\mathcal{M}} f[n/x]$$

(see the remarks prior to Theorem 2.4.1). Thus $\forall$ gets the "universal" interpretation

$$\mathcal{M}, w, f \models \forall x\varphi \quad \text{iff} \quad \text{for all } n \in U,\ \mathcal{M}, w, f[n/x] \models \varphi.$$

Hence $\mathcal{M}$ validates UI (as in Theorem 2.4.1). The properties of the topological interior operator *Int* ensure that $\mathcal{M}$ validates the S4 axioms, so is a model of QS4+UI. But BF is not valid in $\mathcal{M}$, and therefore is not a QS4+UI-theorem, since

$$|\forall x\Box Px|^{\mathcal{M}} f = \bigcap_{n\in U} |\Box Px|^{\mathcal{M}} f[n/x] = \bigcap_{n\in U} Int(-\tfrac{1}{n}, \tfrac{1}{n}) = \{0\},$$

while

$$|\Box\forall x Px|^{\mathcal{M}} f = Int \bigcap_{n\in U} |Px|^{\mathcal{M}} f[n/x] = Int \bigcap_{n\in U} (-\tfrac{1}{n}, \tfrac{1}{n}) = Int\{0\} = \emptyset.$$

Thus $\mathcal{M}, 0 \not\models \forall x\Box Px \to \Box\forall x Px$.

We can also use this example to show that the one-universal-domain characteristic canonical model for QS4 + UI is non-Kripkean. Of course this follows from the failure of the model to validate BF, but it is instructive to directly demonstrate the failure of the Kripkean interpretation of $\forall$.

Let L be QS4 + UI. Since $\nvdash_{\mathrm{L}} \forall x\Box Px \to \Box\forall x Px$, there is an L-maximal set $w \in W_{\mathrm{L}}$ with $\forall x\Box Px \in w$ but $\Box\forall x Px \notin w$. Hence there is some $u \in W_{\mathrm{L}}$ with $wR_{\mathrm{L}}u$ and $\forall x Px \notin u$. Since $\forall x\Box Px \in w$, for each closed term $\tau \in U_{\mathcal{L}}$ the axiom UI yields that $\Box P\tau \in w$; hence $P\tau \in u$.

Thus in $\mathcal{M}_{\mathrm{L}}$ we get $u \models Px(\tau/x)$ for all τ in the universe $U_{\mathcal{L}}$ of the model, but $u \not\models \forall x Px$. This one-universal-domain canonical model has the non-Kripkean property that

$$|\forall x Px|^{\mathcal{M}_{\mathrm{L}}} = \bigsqcap_{\tau\in U_{\mathcal{L}}} |P\tau|^{\mathcal{M}_{\mathrm{L}}} \subsetneq \bigcap_{\tau\in U_{\mathcal{L}}} |P\tau|^{\mathcal{M}_{\mathrm{L}}}.$$

⊣

REMARK 2.4.5. Non-derivability of BF from UI can also be shown by other methods. The approach of Hughes and Cresswell [1996, Chapter 15] is to use a modified notion of validity in a model. This takes φ to be valid in model $\mathcal{M}$ if for all worlds w we have $\mathcal{M}, w, f \models \varphi$ for all those f such that $fx \in D_w$ for all $x \in \mathsf{InVar}$ (i.e. such that every fx is *actual* at w). Provided that $\mathcal{M}$ has expanding domains, the set of formulas valid in $\mathcal{M}$ in this sense will be a logic including UI—at least for signatures having only predicate symbols. We could apply this to the model of Example 2.3.4 to show that BF is not derivable in QS + UI, where S is the extension of S4.3 characterised by the Kripke frame of that model.

2.5. The Deductive Role of Commuting Quantifiers

The last two sections have exhibited constant-domain model structures that do not validate the Barcan Formula. Only Kripkean models can be guaranteed to validate BF on such structures. Kripkean models also validate the scheme CQ of Commuting Quantifiers.

We are now going to develop an axiomatisation of the logic of Kripkean models on constant-domain structures. It will involve CQ in an essential way. This section develops the required proof-theoretic facts that depend on Commuting Quantifiers.

First we introduce the important concept of *C-completeness*, where C is a set of closed terms. This will form part of the Kripkean property of canonical models. Given a quantified modal logic L, a set Σ of formulas is called *C-complete in* L if, for any formula φ and individual variable x, if $\Sigma \vdash_{\mathrm{L}} \varphi(\tau/x)$ for all $\tau \in C$, then $\Sigma \vdash_{\mathrm{L}} \forall x\varphi$. We review some standard facts about C-completeness.

LEMMA 2.5.1. *If Σ is C-complete in* L, *then so is $\Sigma \cup \Gamma$ for every finite set Γ of formulas.*

PROOF. Suppose $\Sigma \cup \Gamma \vdash_{\mathrm{L}} \varphi(\tau/x)$ for all $\tau \in C$. We have to show $\Sigma \cup \Gamma \vdash_{\mathrm{L}} \forall x\varphi$.

Let ψ be the conjunction of the finitely many members of Γ. Then $\Sigma \vdash_{\mathrm{L}} \psi \to \varphi(\tau/x)$ for all $\tau \in C$. Choose a variable y that does not occur in φ or ψ. Then

$$(\psi \to \varphi(y/x))(\tau/y) = \psi \to \varphi(\tau/x),$$

so $\Sigma \vdash_{\mathrm{L}} (\psi \to \varphi(y/x))(\tau/y)$ for all $\tau \in C$. If Σ is C-complete, it follows that $\Sigma \vdash_{\mathrm{L}} \forall y(\psi \to \varphi(y/x))$. Since y does not occur in ψ, this leads to $\Sigma \vdash_{\mathrm{L}} \psi \to \forall y\varphi(y/x))$. But $\vdash_{\mathrm{L}} \forall y\varphi(y/x) \to \forall x\varphi$ by Lemma 1.2.1(4), as y does not occur in φ, so we get $\Sigma \vdash_{\mathrm{L}} \psi \to \forall x\varphi$. Hence $\Sigma \cup \Gamma \vdash_{\mathrm{L}} \forall x\varphi$ as required. ⊣

From this we can obtain the basic result on the extension of C-complete sets to maximal ones. This holds for countable languages, as shown by Henkin [1957]. The following is the relevant version of "Lindenbaum's Lemma" in this context.

THEOREM 2.5.2. *If the signature is countable, then every* L*-consistent C-complete set of formulas has an* L*-maximal C-complete extension.*

PROOF. Let Σ_0 be L-consistent and C-complete. If the signature is countable, then there are countably many formulas, so there is an enumeration $\{\psi_n : n \in \omega\}$ of the set of all formulas of the form $\forall x\varphi$, i.e. all the formulas that begin with $\forall$. We define a nested sequence $\Sigma_0 \subseteq \cdots \subseteq \Sigma_n \subseteq \cdots$ of L-consistent sets such that $\Sigma_n - \Sigma_0$ is finite for all n.

Suppose inductively that we have defined Σ_n that is L-consistent and has $\Sigma_n - \Sigma_0$ finite. Then Σ_n is C-complete by the Lemma just proved. If $\Sigma_n \vdash_{\mathrm{L}} \psi_n$, put $\Sigma_{n+1} = \Sigma_n \cup \{\psi_n\}$. If however $\Sigma_n \nvdash_{\mathrm{L}} \psi_n$, with $\psi_n = \forall x\varphi$, by the C-completeness of Σ_n there is some term $\tau \in C$ with $\Sigma_n \nvdash_{\mathrm{L}} \varphi(\tau/x)$. Put $\Sigma_{n+1} = \Sigma_n \cup \{\neg\varphi(\tau/x)\}$. In both cases we get that Σ_{n+1} is L-consistent with $\Sigma_{n+1} - \Sigma_0$ finite.

Now put $\Sigma = \bigcup_{n \in \omega} \Sigma_n$. Then Σ is L-consistent, so extends to an L-maximal set Γ in the usual way. It remains to show that Γ is C-complete. But if $\Gamma \nvdash_{\mathrm{L}} \forall x\varphi$, with $\forall x\varphi = \psi_n$, then $\Sigma_n \nvdash_{\mathrm{L}} \psi_n$ as $\Sigma_n \subseteq \Gamma$, so by our construction there is a τ with $\neg\varphi(\tau/x) \in \Sigma_{n+1} \subseteq \Gamma$, so $\Gamma \nvdash_{\mathrm{L}} \varphi(\tau/x)$ as Γ is L-consistent. ⊣

The Barcan Formula plays a role in relation to C-completeness that was first pointed out by R. H. Thomason [1970]. To explain this, define $\Box^{-\mathrm{L}}\Sigma = \{\varphi : \Sigma \vdash_{\mathrm{L}} \Box\varphi\}$. Then standard modal reasoning (see Chellas 1980, p. 159) shows that

$$\Box^{-\mathrm{L}}\Sigma \vdash_{\mathrm{L}} \psi \quad \text{iff} \quad \Sigma \vdash_{\mathrm{L}} \Box\psi. \tag{2.5.1}$$

LEMMA 2.5.3 (BF-Lemma). *Let* L *be a quantified modal logic including* BF. *Then if Σ is C-complete, so is $\Box^{-\mathrm{L}}\Sigma = \{\varphi : \Sigma \vdash_{\mathrm{L}} \Box\varphi\}$.*

PROOF. Let $\Box^{-\mathrm{L}}\Sigma \vdash_{\mathrm{L}} \varphi(\tau/x)$ for all $\tau \in C$. Then for all such τ we have $\Sigma \vdash_{\mathrm{L}} \Box(\varphi(\tau/x))$, i.e. $\Sigma \vdash_{\mathrm{L}} (\Box\varphi)(\tau/x)$. Hence $\Sigma \vdash_{\mathrm{L}} \forall x\Box\varphi$ as Σ is C-complete, and therefore $\Sigma \vdash_{\mathrm{L}} \Box\forall x\varphi$ by BF. Thus $\forall x\varphi \in \Box^{-\mathrm{L}}\Sigma$, giving $\Box^{-\mathrm{L}}\Sigma \vdash_{\mathrm{L}} \forall x\varphi$ as required. ⊣

We will also need some facts about derivability of formulas of the form $\forall x_1 \cdots \forall x_n \mathsf{F}$:

LEMMA 2.5.4. *For any quantified modal logic* L,

(1) $\vdash_{\mathrm{L}} \forall y\forall x\mathsf{F} \leftrightarrow \forall y\mathsf{F}$.
(2) $\vdash_{\mathrm{L}} \forall x\mathsf{F} \leftrightarrow \forall y\mathsf{F}$.
(3) $\vdash_{\mathrm{L}} \forall x_n \cdots \forall x_1\mathsf{F} \leftrightarrow \forall y_m \cdots \forall y_1\mathsf{F}$, *for any $n, m \geq 1$, and any x_i, y_j.*

PROOF. (1): $\forall y(\forall x\mathsf{F} \to \mathsf{F})$ is an Actual Instantiation axiom, from which $\forall y\forall x\mathsf{F} \to \forall y\mathsf{F}$ is derived by UD and PC. But from the tautology $\mathsf{F} \to \forall x\mathsf{F}$ we derive $\forall y\mathsf{F} \to \forall y\forall x\mathsf{F}$ by the $\forall$-Monotonicity rule (Lemma 1.2.1(1)).

(2): If $y = x$, then $\forall x\mathsf{F} \leftrightarrow \forall y\mathsf{F}$ is a tautologous axiom. But if $y \neq x$, then $\forall x\mathsf{F} \to \forall y\forall x\mathsf{F}$ is a Vacuous Quantification axiom, from which $\forall x\mathsf{F} \to \forall y\mathsf{F}$ is derivable by (1) and PC. Interchanging x and y derives $\forall y\mathsf{F} \to \forall x\mathsf{F}$ likewise.

(3): First we show $\vdash_{\mathrm{L}} \forall y_1\mathsf{F} \to \forall y_m \cdots \forall y_1\mathsf{F}$ by induction on m. The case $m = 1$ is a tautology. Assuming the result for m, by $\forall$-Monotonicity $\vdash_{\mathrm{L}} \forall y_{m+1}\forall y_1\mathsf{F} \to \forall y_{m+1}\forall y_m \cdots \forall y_1\mathsf{F}$, and $\vdash_{\mathrm{L}} \forall y_1\mathsf{F} \to \forall y_{m+1}\forall y_1\mathsf{F}$ by (1), so $\vdash_{\mathrm{L}} \forall y_1\mathsf{F} \to \forall y_{m+1}\forall y_m \cdots \forall y_1\mathsf{F}$ by PC, giving the result for $m + 1$. Hence it holds for all m.

It now suffices to show that $\vdash_{\mathrm{L}} \forall x_n \cdots \forall x_1\mathsf{F} \to \forall y_1\mathsf{F}$. This goes by induction on n. The case $n = 1$ was shown in part (2). Assuming the result for

n, then $\forall$-Monotonicity gives $\vdash_{\mathrm{L}} \forall x_{n+1}\forall x_n \cdots \forall x_1 \mathsf{F} \to \forall x_{n+1}\forall y_1 \mathsf{F}$. Hence $\vdash_{\mathrm{L}} \forall x_{n+1}\forall x_n \cdots \forall x_1 \mathsf{F} \to \forall y_1 \mathsf{F}$, since $\vdash_{\mathrm{L}} \forall x_{n+1}\forall y_1 \mathsf{F} \to \forall y_1 \mathsf{F}$ by (1). ⊣

Here now is the principal use we make of Commuting Quantifiers.

LEMMA 2.5.5. *Let* L *be a quantified modal logic that includes the scheme* CQ. *Let* φ *be a* UI-*instance of the form*

$$\forall z\psi \to \psi(\mathsf{c}/z),$$

and let $\mathsf{c}_1, \dots, \mathsf{c}_k$ *be a list of distinct constants that includes* c. *If* $y_1, \dots, y_k$ *are new variables not occurring in* $\forall z\varphi$, *then*

$$\vdash_{\mathrm{L}} \forall y_k \cdots \forall y_1 \varphi(y_1/\mathsf{c}_1)\cdots(y_k/\mathsf{c}_k). \tag{2.5.2}$$

PROOF. First we assume that c is c_1. The general case is then dealt with by applying CQ.

Now Lemma 1.9.1 showed that $\vdash_{\mathrm{L}} \forall y_1\varphi(y_1/\mathsf{c})$, i.e. $\vdash_{\mathrm{L}} \forall y_1\varphi(y_1/\mathsf{c}_1)$. By the rule Sub of Lemma 1.2.3 we then get

$$\vdash_{\mathrm{L}} \forall y_1\varphi(y_1/\mathsf{c}_1)(y_2/\mathsf{c}_2).$$

Hence $\vdash_{\mathrm{L}} \forall y_2\forall y_1\varphi(y_1/\mathsf{c}_1)(y_2/\mathsf{c}_2)$ by rule UG. Repeated applications of Sub and then UG in this way leads to the desired result (2.5.2).

Now suppose instead that c is c_j with $j > 1$. Then apply the above argument to the sequence

$$\mathsf{c}_j, \mathsf{c}_1, \dots, \mathsf{c}_{j-1}, \mathsf{c}_{j+1}, \dots, \mathsf{c}_k,$$

to get

$$\begin{aligned}&\vdash_{\mathrm{L}} \forall y_k \cdots \forall y_{j+1}\forall y_{j-1}\cdots\forall y_1\forall y_j \\ &\qquad \varphi(y_j/\mathsf{c}_j)(y_1/\mathsf{c}_1)\cdots(y_{j-1}/\mathsf{c}_{j-1})(y_{j+1}/\mathsf{c}_{j+1})\cdots(y_k/\mathsf{c}_k).\end{aligned} \tag{2.5.3}$$

Now we make two observations. First, the substitution operators (y_i/c_i) can be applied to φ is any order with the same outcome, as the c_i's are all distinct and no y_h is equal to any c_i or occurs in φ. Second, *by repeated use of* CQ, we can interchange any two members of a string of $\forall$-quantifiers, in the sense that

$$\vdash_{\mathrm{L}} \cdots\forall x\cdots\forall x'\cdots\varphi \leftrightarrow \cdots\forall x'\cdots\forall x\cdots\varphi.$$

Using these two observations we then derive (2.5.2) from (2.5.3). ⊣

Now we come to a proof of the existence of the C-complete sets we need. Recall that $\mathrm{UI}(\tau)$ is the set of all UI-instances $\forall x\varphi \to \varphi(\tau/x)$ in which τ is the instantiating term. We define

$$\mathrm{UI}(C) = \bigcup_{\tau\in C} \mathrm{UI}(\tau).$$

THEOREM 2.5.6. *Let* L *be a quantified modal logic that includes* CQ. *Let* C *be an infinite set of constants, none of which occur in the set* Σ *of formulas. Then* $\mathrm{UI}(C) \cup \Sigma$ *is* C-*complete in* L.

PROOF. Suppose $\mathrm{UI}(C) \cup \Sigma \vdash_{\mathrm{L}} \varphi(\mathsf{c}/x)$ for all $\mathsf{c} \in C$. We have to show $\Sigma \vdash_{\mathrm{L}} \forall x\varphi$. We use basic proof-rules for $\forall$, including those of parts (1) and (2) of Lemma 1.2.1.

Since C is infinite, we can choose a $\mathsf{c} \in C$ that does not occur in φ. Then there exist $\varphi_0, \ldots, \varphi_{p-1} \in \mathrm{UI}(C)$ and $\psi_0, \ldots, \psi_{q-1} \in \Sigma$ with

$$\vdash_{\mathrm{L}} \bigwedge_{i<p} \varphi_i \to \Big(\bigwedge_{j<q} \psi_j \to \varphi(\mathsf{c}/x) \Big).$$

Let this L-theorem be called χ, and let $\mathsf{c}_1, \ldots, \mathsf{c}_n$ be all the constants from C, other than c, that occur in χ. Take distinct new variables $y_1, \ldots, y_n, y$ not occurring in χ and consider the substitution

$$\sigma = (y/\mathsf{c})(y_1/\mathsf{c}_1) \cdots (y_n/\mathsf{c}_n).$$

Since $\vdash_{\mathrm{L}} \chi$, by the rule Sub (Lemma 1.2.3) we get $\vdash_{\mathrm{L}} \chi\sigma$. Now σ commutes with the propositional connectives, and has $\psi_j\sigma = \psi_j$ for all $j < q$ because $\psi_j \in \Sigma$ so none of $\mathsf{c}, \mathsf{c}_1, \ldots, \mathsf{c}_n$ occur in ψ_j by hypothesis. Hence from $\vdash_{\mathrm{L}} \chi\sigma$ we get

$$\vdash_{\mathrm{L}} \bigwedge_{i<p} (\varphi_i\sigma) \to \Big(\bigwedge_{j<q} \psi_j \to \varphi(\mathsf{c}/x)\sigma \Big).$$

Writing $\forall\vec{y}$ for the quantifier string $\forall y_n \cdots \forall y_1 \forall y$, we then derive

$$\vdash_{\mathrm{L}} \forall\vec{y}\Big(\bigwedge_{i<p} \varphi_i\sigma \Big) \to \forall\vec{y}\Big(\bigwedge_{j<q} \psi_j \to \varphi(\mathsf{c}/x)\sigma \Big).$$

Now for each $i < p$, φ_i belongs to $\mathrm{UI}(C)$, and so $\vdash_{\mathrm{L}} \forall\vec{y}(\varphi_i\sigma)$ by Lemma 2.5.5, which requires CQ, applied to the list of constants $\mathsf{c}, \mathsf{c}_1, \ldots, \mathsf{c}_n$. It follows that $\vdash_{\mathrm{L}} \forall\vec{y}\big(\bigwedge_{i<p} \varphi_i\sigma \big)$, so this L-theorem can be detached from the one above to yield $\vdash_{\mathrm{L}} \forall\vec{y}\Big(\bigwedge_{j<q} \psi_j \to \varphi(\mathsf{c}/x)\sigma \Big)$, i.e.

$$\vdash_{\mathrm{L}} \forall y_n \cdots \forall y_1 \forall y \Big(\bigwedge_{j<q} \psi_j \to \varphi(\mathsf{c}/x)\sigma \Big). \tag{2.5.4}$$

Our strategy next is to show that from $\mathrm{UI}(C)$ we can deductively eliminate $\forall y_n, \ldots, \forall y_1$ in turn from the L-theorem (2.5.4) and reduce the substitution $(\mathsf{c}/x)\sigma$ to (y/x)—see (2.5.7) below.

To show this, note that since $\mathsf{c}_n \in C$, for any formula ψ we have

$$\mathrm{UI}(C) \vdash_{\mathrm{L}} \forall y_n \psi \to \psi(\mathsf{c}_n/y_n). \tag{2.5.5}$$

Taking ψ to be $\forall y_{n-1} \cdots \forall y_1 \forall y \Big(\bigwedge_{j<q} \psi_j \to \varphi(\mathsf{c}/x)\sigma \Big)$ here, from (2.5.4) and (2.5.5) and Modus Ponens we get

$$\mathrm{UI}(C) \vdash_{\mathrm{L}} \forall y_{n-1} \cdots \forall y_1 \forall y \Big(\bigwedge_{j<q} \psi_j \to \varphi(\mathsf{c}/x)\sigma \Big)(\mathsf{c}_n/y_n).$$

Hence

$$\mathrm{UI}(C) \vdash_{\mathrm{L}} \forall y_{n-1} \cdots \forall y_1 \forall y \Big(\bigwedge_{j<q} \psi_j \to \varphi(\mathsf{c}/x)\sigma(\mathsf{c}_n/y_n) \Big)$$

as y_n does not occur in any ψ_j.

Repeating this argument for $\mathsf{c}_2, \dots, \mathsf{c}_n$ in turn, we reach

$$\mathrm{UI}(C) \vdash_{\mathrm{L}} \forall y \Big(\bigwedge_{j<q} \psi_j \to \varphi(\mathsf{c}/x)\sigma(\mathsf{c}_n/y_n) \cdots (\mathsf{c}_1/y_1) \Big). \tag{2.5.6}$$

But the substitution $(\mathsf{c}/x)\sigma(\mathsf{c}_n/y_n) \cdots (\mathsf{c}_1/y_1)$ is

$$(\mathsf{c}/x)(y/\mathsf{c})(y_1/\mathsf{c}_1) \cdots (y_n/\mathsf{c}_n)(\mathsf{c}_n/y_n) \cdots (\mathsf{c}_1/y_1),$$

which has the same effect on φ as (y/x), because none of $\mathsf{c}, y, y_1, \dots, y_n$ occur in φ. Therefore (2.5.6) states that

$$\mathrm{UI}(C) \vdash_{\mathrm{L}} \forall y \Big(\bigwedge_{j<q} \psi_j \to \varphi(y/x) \Big). \tag{2.5.7}$$

Since y is not in any of the ψ_j's, this leads to

$$\mathrm{UI}(C) \vdash_{\mathrm{L}} \bigwedge_{j<q} \psi_j \to \forall y \varphi(y/x).$$

Hence $\mathrm{UI}(C) \cup \Sigma \vdash_{\mathrm{L}} \forall y \varphi(y/x)$, as the ψ_j's belong to Σ. But $\vdash_{\mathrm{L}} \forall y \varphi(y/x) \to \forall x \varphi$ by Lemma 1.2.1(4), as y does not occur in φ, so this finally leads to $\mathrm{UI}(C) \cup \Sigma \vdash_{\mathrm{L}} \forall x \varphi$ as required. ⊣

A careful analysis of this proof provides a sufficient condition for the consistency of $\mathrm{UI}(C) \cup \Sigma$:

COROLLARY 2.5.7. *Let* L *be a quantified modal logic that includes* CQ. *Let* C *be a set of constants, none of which occur in* Σ. *If* $\Sigma \nvdash_{\mathrm{L}} \forall x \mathsf{F}$ *for some variable* x, *then* $\mathrm{UI}(C) \cup \Sigma$ *is* L*-consistent.*

PROOF. Note that $\Sigma \nvdash_{\mathrm{L}} \forall x \mathsf{F}$ holds for some variable x iff it holds for all, as $\vdash_{\mathrm{L}} \forall x \mathsf{F} \leftrightarrow \forall y \mathsf{F}$ by Lemma 2.5.4(2).

We work contrapositively and suppose $\mathrm{UI}(C) \cup \Sigma$ is not L-consistent. Then $\mathrm{UI}(C) \cup \Sigma \vdash_{\mathrm{L}} \mathsf{F}$. Take $\varphi = \mathsf{F}$ in the proof of the Theorem. We do not need to choose the constant c (so C need not be infinite) or the variable y. The substitution σ becomes $(y_1/\mathsf{c}_1) \cdots (y_n/\mathsf{c}_n)$ and the formulas $\varphi(\mathsf{c}/x)$ and $\varphi(\mathsf{c}/x)\sigma$ are just F. So at line (2.5.4) we get

$$\vdash_{\mathrm{L}} \forall y_n \cdots \forall y_1 \Big(\bigwedge_{j<q} \psi_j \to \mathsf{F} \Big),$$

with each ψ_j in Σ and no y_i occurring in any ψ_j. Hence

$$\vdash_{\mathrm{L}} \bigwedge_{j<q} \psi_j \to \forall y_n \cdots \forall y_1 \mathsf{F},$$

showing that $\Sigma \vdash_{\mathrm{L}} \forall y_n \cdots \forall y_1 \mathsf{F}$. But then $\Sigma \vdash_{\mathrm{L}} \forall x \mathsf{F}$ for any x by Lemma 2.5.4(3). ⊣

2.6. Completeness with CBF and BF

Let L be a logic including CQ, CBF and BF, in a *countable* signature $\mathcal{L}$ (we show how to relax this size limitation at the end of the section). We now construct a characteristic Kripkean constant-domains model for L, by restricting the worlds of the model $\mathcal{M}^{\circ}_{\mathrm{L}}$ of Section 2.3, which itself has the same worlds as the model $\mathcal{M}^{\flat}_{\mathrm{L}}$ of Section 2.2 (but a different accessibility relation).

Recall that $\mathcal{M}^{\flat}_{\mathrm{L}}$ is based on the set of pairs

$$W^{\flat}_{\mathrm{L}} = \{(\Gamma, C) : \Gamma \in W_{\mathrm{L}} \text{ and } C \subseteq D_{\mathrm{L}}(\Gamma)\},$$

where W_{L} is the set of all L-maximal sets of $\mathcal{L}$-formulas, and

$$D_{\mathrm{L}}(\Gamma) = \{\tau \in U_{\mathcal{L}} : \mathrm{UI}(\tau) \subseteq \Gamma\}.$$

Also, a typical member w of $W^{\flat}_{\mathrm{L}}$ is denoted $w = (\Gamma_w, C_w)$. Now we define

$$W^{\natural}_{\mathrm{L}} = \{w \in W^{\flat}_{\mathrm{L}} : \Gamma_w \text{ is } C_w\text{-complete in L}\}.$$

A new structure

$$\mathcal{S}^{\natural}_{\mathrm{L}} = (W^{\natural}_{\mathrm{L}}, R^{\natural}_{\mathrm{L}}, Prop^{\natural}_{\mathrm{L}}, U_{\mathcal{L}}, D^{\natural}_{\mathrm{L}}),$$

is specified on $W^{\natural}_{\mathrm{L}}$ as follows:

- $wR^{\natural}_{\mathrm{L}}u$ iff $\Gamma_w R_{\mathrm{L}} \Gamma_u$ and $C_w = C_u$. (hence $R^{\natural}_{\mathrm{L}}$ is the restriction of the relation R°_{L} to $W^{\natural}_{\mathrm{L}}$).
- $|\varphi|^{\natural}_{\mathrm{L}} = |\varphi|^{\flat}_{\mathrm{L}} \cap W^{\natural}_{\mathrm{L}} = \{w \in W^{\natural}_{\mathrm{L}} : \varphi \in \Gamma_w\}$.
- $Prop^{\natural}_{\mathrm{L}} = \{\, |\varphi|^{\natural}_{\mathrm{L}} : \varphi \text{ is an } \mathcal{L}\text{-sentence}\}$.
- $D^{\natural}_{\mathrm{L}}(w) = C_w$.

A premodel $\mathcal{M}^{\natural}_{\mathrm{L}}$ for $\mathcal{L}$ on $\mathcal{S}^{\natural}_{\mathrm{L}}$ is given by:

- $|\mathsf{c}|^{\mathcal{M}^{\natural}_{\mathrm{L}}} = \mathsf{c}$.
- $|F|^{\mathcal{M}^{\natural}_{\mathrm{L}}}(\tau_1, \ldots, \tau_n) = F\tau_1 \cdots \tau_n$.
- $|P|^{\mathcal{M}^{\natural}_{\mathrm{L}}}(\tau_1, \ldots, \tau_n) = |P\tau_1 \cdots \tau_n|^{\natural}_{\mathrm{L}}$.

In essence, $\mathcal{S}^{\natural}_{\mathrm{L}}$ restricts the constant-domains structure $\mathcal{S}^{\circ}_{\mathrm{L}}$ of Section 2.3 to the set of worlds w whose Γ_w-component is C_w-complete, while $\mathcal{M}^{\natural}_{\mathrm{L}}$ restricts $\mathcal{M}^{\circ}_{\mathrm{L}}$. This allows $\mathcal{M}^{\natural}_{\mathrm{L}}$ to provide the Kripkean interpretation of $\forall$.

The closure properties of the L-maximal sets Γ_w ensure that

$$|\mathsf{F}|^{\natural}_{\mathrm{L}} = \emptyset, \quad |\varphi|^{\natural}_{\mathrm{L}} \cap |\psi|^{\natural}_{\mathrm{L}} = |\varphi \wedge \psi|^{\natural}_{\mathrm{L}}, \quad \text{and} \quad W^{\natural}_{\mathrm{L}} - |\varphi|^{\natural}_{\mathrm{L}} = |\neg\varphi|^{\natural}_{\mathrm{L}},$$

so $Prop^\natural_{\mathrm{L}}$ is a Boolean set algebra. To show that $Prop^\natural_{\mathrm{L}}$ is closed under $[R^\natural_{\mathrm{L}}]$, we use both the Barcan Formula and its Converse to prove

LEMMA 2.6.1. $[R^\natural_{\mathrm{L}}]|\varphi|^\natural_{\mathrm{L}} = |\Box\varphi|^\natural_{\mathrm{L}} \in Prop^\natural_{\mathrm{L}}$.

PROOF. Let $w \in |\Box\varphi|^\natural_{\mathrm{L}}$. If $wR^\natural_{\mathrm{L}}u$, then $\Gamma_w R_{\mathrm{L}} \Gamma_u$, so $\varphi \in \Gamma_u$ as $\Box\varphi \in \Gamma_w$, hence $u \in |\varphi|^\natural_{\mathrm{L}}$. This shows that $w \in [R^\natural_{\mathrm{L}}]|\varphi|^\natural_{\mathrm{L}}$.

For the converse, we need the fact that L includes BF and CBF. Suppose $w \in W^\natural_{\mathrm{L}}$ has $w \notin |\Box\varphi|^\natural_{\mathrm{L}}$. Then $\Box\varphi \notin \Gamma_w$, so by (1.9.1), $\Box^-\Gamma_w \nvdash_{\mathrm{L}} \varphi$ where $\Box^-\Gamma_w = \{\psi : \Box\psi \in \Gamma_w\}$. Hence $\Box^-\Gamma_w \cup \{\neg\varphi\}$ is L-consistent.

Now since Γ_w is L-maximal, $\Box^-\Gamma_w$ is identical to $\Box^{-\mathrm{L}}\Gamma_w = \{\psi : \Gamma_w \vdash_{\mathrm{L}} \Box\psi\}$, which is C_w-complete by the BF-Lemma 2.5.3, because Γ_w is C_w-complete by definition of $W^\natural_{\mathrm{L}}$.

Since $\Box^-\Gamma_w$ is C_w-complete, so too is $\Box^-\Gamma_w \cup \{\neg\varphi\}$ by Lemma 2.5.1. Therefore by Theorem 2.5.2, $\Box^-\Gamma_w \cup \{\neg\varphi\}$ has an L-maximal extension Δ that is C_w-complete. Then $\Gamma_w R_{\mathrm{L}} \Delta$ as $\Box^-\Gamma_w \subseteq \Delta$. But L includes CBF, so $\Gamma_w R_{\mathrm{L}} \Delta$ implies $D_{\mathrm{L}}(\Gamma_w) \subseteq D_{\mathrm{L}}(\Delta)$, as shown in Corollary 2.1.4.

Now let $u = (\Delta, C_w)$. We have $C_w \subseteq D_{\mathrm{L}}(\Gamma_w) \subseteq D_{\mathrm{L}}(\Delta)$, with Δ being C_w-complete in L, so $u \in W^\natural_{\mathrm{L}}$. Also, $\Gamma_w R_{\mathrm{L}} \Gamma_u = \Delta$ and $C_w = C_u$, which means that $wR^\natural_{\mathrm{L}}u$. But $\neg\varphi \in \Gamma_u$, so $\varphi \notin \Gamma_u$ by L-consistency, hence $u \notin |\varphi|^\natural_{\mathrm{L}}$. Since $wR^\natural_{\mathrm{L}}u$, this shows $w \notin [R^\natural_{\mathrm{L}}]|\varphi|^\natural_{\mathrm{L}}$, completing the proof. ⊣

Next we show how the structure of $\mathcal{S}^\natural_{\mathrm{L}}$ reflects the Kripkean semantics for the quantifier $\forall$.

LEMMA 2.6.2. *If $\forall x\varphi$ is a sentence, then in $\mathcal{S}^\natural_{\mathrm{L}}$,*

$$|\forall x\varphi|^\natural_{\mathrm{L}} = \bigcap_{\tau \in U_{\mathcal{L}}} E\tau \Rightarrow |\varphi(\tau/x)|^\natural_{\mathrm{L}} = \bigsqcap_{\tau \in U_{\mathcal{L}}} E\tau \Rightarrow |\varphi(\tau/x)|^\natural_{\mathrm{L}}.$$

PROOF. In $\mathcal{S}^\natural_{\mathrm{L}}$, the the existence sets are given by

$$E\tau = \{w \in W^\natural_{\mathrm{L}} : \tau \in D^\natural_{\mathrm{L}}(w)\} = \{w \in W^\natural_{\mathrm{L}} : \tau \in C_w\}.$$

Now let $w \in |\forall x\varphi|^\natural_{\mathrm{L}}$. Then $\forall x\varphi \in \Gamma_w$. For any $\tau \in U_{\mathcal{L}}$, if $w \in E\tau$ then $\tau \in C_w \subseteq D_{\mathrm{L}}(\Gamma_w)$, so $\forall x\varphi \to \varphi(\tau/x) \in \Gamma_w$, hence as $\forall x\varphi \in \Gamma_w$ we get $\varphi(\tau/x) \in \Gamma_w$, giving $w \in |\varphi(\tau/x)|^\natural_{\mathrm{L}}$; thus $w \in E\tau \Rightarrow |\varphi(\tau/x)|^\natural_{\mathrm{L}}$.

Conversely, let $w \in E\tau \Rightarrow |\varphi(\tau/x)|^\natural_{\mathrm{L}}$ for all $\tau \in U_{\mathcal{L}}$. Then whenever $\tau \in C_w$, we have $w \in E\tau$, hence $w \in |\varphi(\tau/x)|^\natural_{\mathrm{L}}$, so $\varphi(\tau/x) \in \Gamma_w$. *But Γ_w is C_w complete*, so then $\Gamma_w \vdash_{\mathrm{L}} \forall x\varphi$, hence $\forall x\varphi \in \Gamma_w$ as Γ_w is L-maximal, and thus $w \in |\forall x\varphi|^\natural_{\mathrm{L}}$.

This proves that

$$|\forall x\varphi|^\natural_{\mathrm{L}} = \bigcap_{\tau \in U_{\mathcal{L}}} E\tau \Rightarrow |\varphi(\tau/x)|^\natural_{\mathrm{L}}.$$

Now $|\forall x\varphi|^{\natural}_{\mathrm{L}} \in Prop^{\natural}_{\mathrm{L}}$ by definition, so $\bigcap_{\tau\in U_{\mathcal{L}}} E\tau \Rightarrow |\varphi(\tau/x)|^{\natural}_{\mathrm{L}}$ is admissible in $\mathcal{S}^{\natural}_{\mathrm{L}}$. But in general, if $\bigcap \mathcal{Z}$ is admissible, then it is equal to $\prod \mathcal{Z}$ (Lemma 1.5.1). Hence

$$\bigcap_{\tau\in U_{\mathcal{L}}} E\tau \Rightarrow |\varphi(\tau/x)|^{\natural}_{\mathrm{L}} = \prod_{\tau\in U_{\mathcal{L}}} E\tau \Rightarrow |\varphi(\tau/x)|^{\natural}_{\mathrm{L}}. \qquad \dashv$$

COROLLARY 2.6.3. *For any formula of the form* $\forall x\varphi$, *and any* $f \in U_{\mathcal{L}}{}^{\omega}$,

$$|(\forall x\varphi)^f|^{\natural}_{\mathrm{L}} = \bigcap_{\tau\in U_{\mathcal{L}}} E\tau \Rightarrow |\varphi^{f[\tau/x]}|^{\natural}_{\mathrm{L}} = \prod_{\tau\in U_{\mathcal{L}}} E\tau \Rightarrow |\varphi^{f[\tau/x]}|^{\natural}_{\mathrm{L}}.$$

PROOF. Replace φ by $\varphi^{f\setminus x}$ in the Lemma just proved, and use the equations $(\forall x\varphi)^f = \forall x(\varphi^{f\setminus x})$ and $\varphi^{f\setminus x}(\tau/x) = \varphi^{f[\tau/x]}$ from Lemma 1.9.3 (see the proof of part (5) of that Lemma). $\dashv$

We are now in a position to prove a Truth Lemma for the premodel $\mathcal{M}^{\natural}_{\mathrm{L}}$:

THEOREM 2.6.4 (Truth Is Membership). *Let* φ *be any formula. Then for all* $f \in U_{\mathcal{L}}{}^{\omega}$,

$$|\varphi|^{\mathcal{M}^{\natural}_{\mathrm{L}}} f = |\varphi^f|^{\natural}_{\mathrm{L}},$$

and hence for all $w \in W^{\natural}_{\mathrm{L}}$,

$$\mathcal{M}^{\natural}_{\mathrm{L}}, w, f \models \varphi \quad \textit{iff} \quad \varphi^f \in w.$$

PROOF. This follows the same pattern as the proof of Theorem 1.9.4 for $\mathcal{M}_{\mathrm{L}}$ (see also Theorem 2.2.3 for $\mathcal{M}^{\flat}_{\mathrm{L}}$). The new parts are the inductive cases for $\Box$ and $\forall$. The case of $\Box$ is taken care of by Lemma 2.6.1. For the case of $\forall$, assuming that the result holds for φ, we have

$$\begin{aligned} |\forall x\varphi|^{\mathcal{M}^{\natural}_{\mathrm{L}}} f &= \prod_{\tau\in U_{\mathcal{L}}} E\tau \Rightarrow |\varphi|^{\mathcal{M}^{\natural}_{\mathrm{L}}} f[\tau/x] && \text{semantics of } \forall \\ &= \prod_{\tau\in U_{\mathcal{L}}} E\tau \Rightarrow |\varphi^{f[\tau/x]}|^{\natural}_{\mathrm{L}} && \text{induction hypothesis on } \varphi \\ &= |(\forall x\varphi)^f|_{\mathrm{L}} && \text{Corollary 2.6.3,} \end{aligned}$$

so it holds for $\forall x\varphi$. $\dashv$

This analysis of $\mathcal{M}^{\natural}_{\mathrm{L}}$ has been obtained without any appeal to CQ. That axiom scheme comes into play now in showing that L is complete for validity in $\mathcal{M}^{\natural}_{\mathrm{L}}$. The completeness proof requires our CQ-dependent results from the previous section about consistency and C-completeness of sets of the form $\mathrm{UI}(C) \cup \Sigma$.

THEOREM 2.6.5. $\mathcal{M}^{\natural}_{\mathrm{L}}$ *is a Kripkean model that characterises* L:

$$\vdash_{\mathrm{L}} \varphi \quad \textit{iff} \quad \mathcal{M}^{\natural}_{\mathrm{L}} \models \varphi.$$

PROOF. First, $\mathcal{M}^\natural_{\mathrm{L}}$ is a model: each truth set $|\varphi|^{\mathcal{M}^\natural_{\mathrm{L}}} f$ is admissible, because it is equal to $|\varphi^f|^\natural_{\mathrm{L}} \in Prop^\natural_{\mathrm{L}}$.

Secondly, $\mathcal{M}^\natural_{\mathrm{L}}$ is Kripkean: applying Theorem 2.6.4 to Corollary 2.6.3 gives

$$|\forall x\varphi|^{\mathcal{M}^\natural_{\mathrm{L}}} f = \bigcap_{\tau \in U_{\mathcal{L}}} E\tau \Rightarrow |\varphi|^{\mathcal{M}^\natural_{\mathrm{L}}} f[\tau/x],$$

which is the definition (1.6.5) of "Kripkean" for $\mathcal{M}^\natural_{\mathrm{L}}$.

Next, L is sound for validity in $\mathcal{M}^\natural_{\mathrm{L}}$: combining Theorem 2.6.4 with the corresponding Theorem 1.9.4 for $\mathcal{M}_{\mathrm{L}}$, we have

$$\mathcal{M}^\natural_{\mathrm{L}}, w, f \models \varphi \quad \text{iff} \quad \varphi^f \in \Gamma_w \quad \text{iff} \quad \mathcal{M}_{\mathrm{L}}, \Gamma_w, f \models \varphi.$$

Thus if $\vdash_{\mathrm{L}} \varphi$, then φ is valid in $\mathcal{M}_{\mathrm{L}}$ (Theorem 1.9.6), so $\mathcal{M}_{\mathrm{L}}, \Gamma_w, f \models \varphi$ and hence $\mathcal{M}^\natural_{\mathrm{L}}, w, f \models \varphi$ for every $w \in W^\natural_{\mathrm{L}}$ and $f \in U_{\mathcal{L}}{}^\omega$. Therefore φ is valid in $\mathcal{M}^\natural_{\mathrm{L}}$.

Lastly, L is complete for validity in $\mathcal{M}^\natural_{\mathrm{L}}$ (this is where we use our results that depend on CQ). Suppose $\nvdash_{\mathrm{L}} \varphi$. Then we have to show that φ is falsifiable in $\mathcal{M}^\natural_{\mathrm{L}}$. As in the proof of Theorem 1.9.6, there is some $f \in U_{\mathcal{L}}{}^\omega$ such that $\nvdash_{\mathrm{L}} \varphi^f$. If we can find some $w \in W^\natural_{\mathrm{L}}$ with $\varphi^f \notin \Gamma_w$, then we will have $\mathcal{M}^\natural_{\mathrm{L}}, w, f \not\models \varphi$, completing our task.

Let $\Sigma = \{\neg\varphi^f\}$, which is L-consistent as $\nvdash_{\mathrm{L}} \varphi^f$. We distinguish two cases. In the first case, $\Sigma \vdash_{\mathrm{L}} \forall x\mathsf{F}$ for all $x \in \mathsf{InVar}$. Since $\vdash_{\mathrm{L}} \forall x\mathsf{F} \to \forall x\psi$ by $\forall$-Monotonicity, this gives $\Sigma \vdash_{\mathrm{L}} \forall x\psi$ for all formulas ψ. Let Γ be any L-maximal extension of Σ. Then $\Gamma \vdash_{\mathrm{L}} \forall x\psi$ for every x and ψ. This implies that Γ is C-complete for every C, including $C = \emptyset$. Put $w = (\Gamma, \emptyset)$. Then $w \in W^\natural_{\mathrm{L}}$ because $\emptyset \subseteq D_{\mathrm{L}}(\Gamma)$ and Γ is $\emptyset$-complete. But $\neg\varphi^f \in \Gamma = \Gamma_w$, so $\varphi^f \notin \Gamma_w$ by L-consistency. Thus we have found a w as required.

[Actually in this case $D_{\mathrm{L}}(\Gamma) = \emptyset$, since for every τ the sentence $\forall x\mathsf{F} \to \mathsf{F}$ belongs to $\mathrm{UI}(\tau)$, but not to Γ or else $\Gamma \vdash_{\mathrm{L}} \mathsf{F}$, hence $\tau \notin D_{\mathrm{L}}(\Gamma)$.]

The other case is that $\Sigma \nvdash_{\mathrm{L}} \forall x\mathsf{F}$ for some x (and hence for all x by Lemma 2.5.4(2)). Given our general assumption that $\mathcal{L}$ has infinitely many constants, as Σ contains one formula there is an infinite set C of constants that do not occur in Σ. Then $\mathrm{UI}(C) \cup \Sigma$ is C-complete by Theorem 2.5.6 and L-consistent by Corollary 2.5.7. Hence by Theorem 2.5.2 there is a C-complete L-maximal set Γ extending $\mathrm{UI}(C) \cup \Sigma$. Put $w = (\Gamma, C)$. Then $C \subseteq D_{\mathrm{L}}(\Gamma_w)$ as $\mathrm{UI}(C) \subseteq \Gamma$, so $w \in W^\natural_{\mathrm{L}}$. As above we get $\varphi^f \notin \Gamma_w$, so again we have a w as required, and that finishes the proof. ⊣

THEOREM 2.6.6. *The logic* QK + CQ + CBF + BF *is characterised by the set of all Kripkean models on its* $\mathcal{S}^\natural$*-structure, as well as being characterised by the class of all Kripkean models on constant-domains model structures.*

PROOF. Let L = QK + CQ + CBF + BF.

If $\vdash_{\mathrm{L}} \varphi$, then for any constant-domains structure $\mathcal{S}$, the logic $\mathrm{L}^K_{\mathcal{S}}$ defined by the Kripkean models on $\mathcal{S}$ (Theorem 1.7.8(2)) includes the schemes CQ, CBF, and BF, so includes L, hence $\varphi \in \mathrm{L}^K_{\mathcal{S}}$.

But if φ is valid in all Kripkean models on constant-domains model structures, then φ is valid in all Kripkean models on the constant-domains structure $\mathcal{S}^\natural_{\mathrm{L}}$.

Finally, if φ is valid in all Kripkean models on $\mathcal{S}^\natural_{\mathrm{L}}$, then φ is valid in $\mathcal{M}^\natural_{\mathrm{L}}$, and so $\vdash_{\mathrm{L}} \varphi$. ⊣

COROLLARY 2.6.7. *The logic* QK + CQ + CBF + BF *is characterised by the class of all full model structures with constant domains.*

PROOF. Again let L = QK + CQ + CBF + BF.

Soundness: Let $\vdash_{\mathrm{L}} \varphi$. If $\mathcal{S}$ has constant domains and is full, then every model on $\mathcal{S}$ is Kripkean and so validates φ by the Theorem.

Completeness: Suppose $\nvdash_{\mathrm{L}} \varphi$. Then φ is not valid in the Kripkean model $\mathcal{M}^\natural_{\mathrm{L}}$ on $\mathcal{S}^\natural_{\mathrm{L}}$. Hence by Corollary 1.6.8, φ is not valid in the full structure $(\mathcal{S}^\natural_{\mathrm{L}})^+$, which has the same domain function as $\mathcal{S}^\natural_{\mathrm{L}}$, hence has constant domains. ⊣

We now want to give a Kripkean-models characterisation of the logic QS + CQ+CBF+BF for an arbitrary S, so we examine the general frame underlying $\mathcal{S}^\natural_{\mathrm{L}}$ in this case.

LEMMA 2.6.8. *If* L *is any quantified modal logic that extends the logic* QS + CQ + CBF + BF, *then* S *is validated by the general frame* $\mathcal{G}^\natural_{\mathrm{L}}$ *underlying* $\mathcal{S}^\natural_{\mathrm{L}}$.

PROOF. Here $\mathcal{G}^\natural_{\mathrm{L}} = (W^\natural_{\mathrm{L}}, R^\natural_{\mathrm{L}}, Prop^\natural_{\mathrm{L}})$. The proof that $\mathcal{G}^\natural_{\mathrm{L}}$ validates S is formally just like the proof in Lemma 2.2.9 that $\mathcal{G}^\flat_{\mathrm{L}}$ validates S, but for clarity we repeat the details with ♮ replacing ♭.

If $\mathcal{M}$ is any propositional model on $\mathcal{G}^\natural_{\mathrm{L}}$, then $\mathcal{M}$ assigns to each $p \in \mathsf{PropVar}$ a set $|p|^{\mathcal{M}} \in Prop^\natural_{\mathrm{L}}$. Hence $|p|^{\mathcal{M}} = |\varphi_p|^\natural_{\mathrm{L}}$ for some $\mathcal{L}$-sentence φ_p. Let $A \mapsto A^*$ be the mapping of propositional formulas to $\mathcal{L}$-formulas induced by the uniform substitution of φ_p for p. Then an inductive proof shows that $|A|^{\mathcal{M}} = |A^*|^\natural_{\mathrm{L}}$ for all propositional A. The induction step for $\Box$ is taken care of by Lemma 2.6.1.

Now if $\vdash_{\mathrm{S}} A$, then $\vdash_{\mathrm{QS}} A^*$, so $\vdash_{\mathrm{L}} A^*$ and hence for every $w \in W^\natural_{\mathrm{L}}$, $A^* \in \Gamma_w$ and so $w \in |A^*|^\natural_{\mathrm{L}}$. Therefore $|A|^{\mathcal{M}} = |A^*|^\natural_{\mathrm{L}} = W^\natural_{\mathrm{L}}$, i.e. $\mathcal{M} \models A$. This proves that $\mathcal{G}^\natural_{\mathrm{L}}$ validates S.

For the algebraically minded, an alternative argument is to observe that $|\varphi|_{\mathrm{L}} \mapsto |\varphi|^\natural_{\mathrm{L}}$ is a well-defined isomorphism between $Prop_{\mathrm{L}}$ and $Prop^\natural_{\mathrm{L}}$ when these are viewed as modal algebras. Since $Prop_{\mathrm{L}}$ validates S, so too does $Prop^\natural_{\mathrm{L}}$. ⊣

THEOREM 2.6.9. *The logic* QS + CQ + CBF + BF *is characterised by the set of all Kripkean models on its* $\mathcal{S}^\natural$*-structure, as well as being characterised by* *the class of all Kripkean models on constant-domain model structures whose underlying general frame validates* S.

PROOF. Let $L = QS + CQ + CBF + BF$.

If $\vdash_L \varphi$, then if a structure $\mathcal{S}$ has a general frame validating S, then QS is valid in every model on $\mathcal{S}$ by Theorem 1.7.11; thus if $\mathcal{S}$ also has constant domains, then every *Kripkean* model on $\mathcal{S}$ validates CQ, CBF, and BF as well as QS, so validates L; hence validates φ.

But if φ is valid in all Kripkean models on constant-domains model structures whose general frame validates S, then φ is valid in all Kripkean models on $\mathcal{S}_L^\natural$, by Lemma 2.6.8.

Finally, if φ is valid in all Kripkean models on $\mathcal{S}_L^\natural$, then φ is valid in $\mathcal{M}_L^\natural$, and so $\vdash_L \varphi$. $\dashv$

These characterisations depended on the background signature being countable, since that assumption is required by Theorem 2.5.2, which provides the C-complete maximal sets needed in the construction and analysis of $\mathcal{M}_L^\natural$. But results like this can be lifted to uncountable signatures, by a method that we now sketch.

Let $\mathcal{L}$ be an uncountable signature, including infinitely many constants, and let L be the logic $QS + CQ + CBF + BF$ in the set of $\mathcal{L}$-formulas. Now take an $\mathcal{L}$-formula φ that is valid in all Kripkean $\mathcal{L}$-models on $\mathcal{S}_L^\natural$. We want to show that $\vdash_L \varphi$.

Let $\mathcal{L}_\varphi$ be a countable subsignature of $\mathcal{L}$, still with infinitely many constants, that includes all the (finitely many) members of $\mathcal{L}$ that appear in φ. Also, for each positive integer n, choose some particular n-ary function symbol F_n to be in $\mathcal{L}_\varphi$ if $\mathcal{L}$ has n-ary function symbols, and let c_0 be some fixed constant of $\mathcal{L}_\varphi$. Define L_φ to be the set of all $\mathcal{L}_\varphi$-formulas that are L-theorems. Then L_φ is just the logic $QS + CQ + CBF + BF$ in the set of $\mathcal{L}_\varphi$-formulas.

Now suppose $\mathcal{M}$ is any Kripkean $\mathcal{L}_\varphi$-model on $\mathcal{S}_L^\natural$. Define an $\mathcal{L}$-premodel $\mathcal{M}'$ by declaring it to be identical to $\mathcal{M}$ except on symbols ζ that are not in $\mathcal{L}_\varphi$. For such ζ, put $|\zeta|^{\mathcal{M}'} = |c_0|^{\mathcal{M}}$ if ζ is a constant; put $|\zeta|^{\mathcal{M}'} = |F_n|^{\mathcal{M}}$ if ζ is an n-ary function symbol; and if ζ is an n-ary predicate symbol, let $|\zeta|^{\mathcal{M}'}$ be the n-ary function on $U_\mathcal{L}$ with constant value $W_L^\natural$. For each $\mathcal{L}$-formula ψ, let ψ' be the $\mathcal{L}_\varphi$-formula resulting from replacing any constant of ψ not in $\mathcal{L}_\varphi$ by c_0; any n-ary function symbol of ψ not in $\mathcal{L}_\varphi$ by F_n; and any atomic formula of ψ whose predicate symbol is not in $\mathcal{L}_\varphi$ by $\top$. Then in general, $|\psi|^{\mathcal{M}'} = |\psi'|^{\mathcal{M}}$. From this it can be shown that $\mathcal{M}'$ is a Kripkean $\mathcal{L}$-model, because $\mathcal{M}$ is a Kripkean $\mathcal{L}_\varphi$-model. But φ is valid in all Kripkean $\mathcal{L}$-models on $\mathcal{S}_L^\natural$, and hence is valid in $\mathcal{M}'$. Since $\varphi' = \varphi$, and hence $|\varphi|^{\mathcal{M}'} = |\varphi|^{\mathcal{M}}$, this implies that φ is valid in $\mathcal{M}$. Altogether we have now that φ is valid in every Kripkean $\mathcal{L}_\varphi$-model on $\mathcal{S}_L^\natural$. Since $\mathcal{L}_\varphi$ is countable, the proof of the above

Theorem applies to show that φ is an L_φ-theorem. Hence it is an L-theorem as required.

On the other hand, if we know that φ is an L-theorem, then it is immediate that is valid in all Kripkean $\mathcal{L}$-models on $\mathcal{S}_L^\natural$ by the soundness part of the last Theorem. That soundness argument does not depend on the cardinality of the signature $\mathcal{L}$.

In this way we see that L is characterised by the set of all Kripkean $\mathcal{L}$-models on $\mathcal{S}_L^\natural$. The argument can also be carried through with $\mathcal{S}_L^\natural$ replaced by any class $\mathscr{C}$ of model structures such that QS + CQ + CBF + BF is characterised by the set of all Kripkean models on members of $\mathscr{C}$ for any countable signature. The conclusion of the argument will be that the same characterisation by $\mathscr{C}$ holds for arbitrary signatures.

This method applies to other characterisations of logics that we derive below under the restriction to countable signatures, to lift those result to any larger signature.

2.7. Completeness with UI and BF

A general characterisation by Kripkean models can also be given for logics that include Universal Instantiation as well as the Barcan Formula. Recalling that CBF is derivable from UI, let L be any logic including CQ, UI and BF, again in a countable signature $\mathcal{L}$.

We construct a characteristic model for L that is Kripkean and has one universal domain. This is much simpler than the $\mathcal{M}_L^\natural$ construction, because in the presence of UI the canonical structure $\mathcal{S}_L$ already has one universal domain, and indeed $D_L(w) = U_\mathcal{L}$ for all $w \in W_L$ (Lemma 2.4.2). So we just need to restrict the set of worlds of $\mathcal{M}_L$, similarly to the way we restricted $W_L^\flat$ to $W_L^\natural$, to make a Kripkean model in this case.

Define a new structure

$$\mathcal{S}_L^K = (W_L^K, R_L^K, Prop_L^K, U_\mathcal{L}, D_L^K)$$

as follows:

- $W_L^K = \{w \in W_L : w \text{ is } U_\mathcal{L}\text{-complete}\}$.
- $wR_L^K u$ iff $\{\varphi : \Box\varphi \in w\} \subseteq u$
  (hence R_L^K is the restriction of the relation R_L to W_L^K).
- $|\varphi|_L^K = |\varphi|_L \cap W_L^K = \{w \in W_L^K : \varphi \in w\}$.
- $Prop_L^K = \{\, |\varphi|_L^K : \varphi \text{ is an } \mathcal{L}\text{-sentence}\}$.
- $D_L^K(w) = U_\mathcal{L}$.

A premodel $\mathcal{M}_L^K$ for $\mathcal{L}$ on $\mathcal{S}_L^K$ is given by:

- $|\mathsf{c}|^{\mathcal{M}_L^K} = \mathsf{c}$.
- $|F|^{\mathcal{M}_L^K}(\tau_1, \ldots, \tau_n) = F\tau_1 \cdots \tau_n$.

- $|P|^{\mathcal{M}_{\mathrm{L}}^K}(\tau_1,\ldots,\tau_n) = |P\tau_1\cdots\tau_n|_{\mathrm{L}}^K$.

The analysis of $\mathcal{M}_{\mathrm{L}}^K$ is a simpler version of the analysis of $\mathcal{M}_{\mathrm{L}}^{\natural}$. The closure properties of L-maximal sets ensure that $Prop_{\mathrm{L}}^K$ is a Boolean set algebra in the usual way. Closure of $Prop_{\mathrm{L}}^K$ under the operation $[R_{\mathrm{L}}^K]$ follows from

LEMMA 2.7.1. $[R_{\mathrm{L}}^K]|\varphi|_{\mathrm{L}}^K = |\Box\varphi|_{\mathrm{L}}^K \in Prop_{\mathrm{L}}^K$.

PROOF. The inclusion of $|\Box\varphi|_{\mathrm{L}}^K$ in $[R_{\mathrm{L}}^K]|\varphi|_{\mathrm{L}}^K$ follows directly from the definition of R_{L}^K. For the converse inclusion, if $w \in W_{\mathrm{L}}^K$ has $w \notin |\Box\varphi|_{\mathrm{L}}^K$, we proceed as in the proof of Lemma 2.6.1. We have $\Box\varphi \notin w$, so $\{\psi : \Box\psi \in w\} \cup \{\neg\varphi\}$ is L-consistent.

Now $\{\psi : \Box\psi \in w\}$ is $U_{\mathcal{L}}$-complete by the BF-Lemma 2.5.3, because w is $U_{\mathcal{L}}$-complete by definition of W_{L}^K. Hence $\{\psi : \Box\psi \in w\} \cup \{\neg\varphi\}$ is $U_{\mathcal{L}}$-complete by Lemma 2.5.1 and so has an L-maximal $U_{\mathcal{L}}$-complete extension u by Theorem 2.5.2. Then $wR_{\mathrm{L}}u$ as $\{\psi : \Box\psi \in w\} \subseteq u$, and $u \in W_{\mathrm{L}}^K$, so $wR_{\mathrm{L}}^K u$. But $\neg\varphi \in u$, so $u \notin |\varphi|_{\mathrm{L}}^K$, showing that $w \notin [R_{\mathrm{L}}^K]|\varphi|_{\mathrm{L}}^K$. ⊣

Since $\mathcal{S}_{\mathrm{L}}^K$ has one universal domain, we have $E\tau = W_{\mathrm{L}}^K$ for all $\tau \in U_{\mathcal{L}}$, and so the existence sets become redundant in describing the interpretation of $\forall$. We have

LEMMA 2.7.2. *If $\forall x\varphi$ is a sentence, then in $\mathcal{S}_{\mathrm{L}}^K$,*

$$|\forall x\varphi|_{\mathrm{L}}^K = \bigcap_{\tau\in U_{\mathcal{L}}} |\varphi(\tau/x)|_{\mathrm{L}}^K = \prod_{\tau\in U_{\mathcal{L}}} |\varphi(\tau/x)|_{\mathrm{L}}^K.$$

Hence for any formula of the form $\forall x\varphi$, and any $f \in {U_{\mathcal{L}}}^{\omega}$,

$$|(\forall x\varphi)^f|_{\mathrm{L}}^K = \bigcap_{\tau\in U_{\mathcal{L}}} |\varphi^{f[\tau/x]}|_{\mathrm{L}}^K = \prod_{\tau\in U_{\mathcal{L}}} |\varphi^{f[\tau/x]}|_{\mathrm{L}}^K.$$

PROOF. (cf. Lemma 2.6.2 and its Corollary.) Each $w \in W_{\mathrm{L}}^K$ is $U_{\mathcal{L}}$-complete, hence

$$\text{if } \varphi(\tau/x) \in w \text{ for all } \tau \in U_{\mathcal{L}}, \text{ then } \forall x\varphi \in w.$$

The converse implication holds as w includes all instances of UI, so altogether this implies

$$|\forall x\varphi|_{\mathrm{L}}^K = \bigcap_{\tau\in U_{\mathcal{L}}} |\varphi(\tau/x)|_{\mathrm{L}}^K.$$

Now if $\forall x\varphi$ is a sentence, then $|\forall x\varphi|_{\mathrm{L}}^K \in Prop_{\mathrm{L}}^K$, so $\bigcap_{\tau\in U_{\mathcal{L}}} |\varphi(\tau/x)|_{\mathrm{L}}^K$ is admissible, and therefore is equal to $\prod_{\tau\in U_{\mathcal{L}}} |\varphi(\tau/x)|_{\mathrm{L}}^K$.

This proves the first statement of the Lemma. The second follows from the first by using the equations $(\forall x\varphi)^f = \forall x(\varphi^{f\setminus x})$ and $\varphi^{f\setminus x}(\tau/x) = \varphi^{f[\tau/x]}$ from Lemma 1.9.3. ⊣

The last two Lemmas lead in a now familiar way to a Truth Lemma for the premodel $\mathcal{M}_{\mathrm{L}}^K$:

THEOREM 2.7.3 (Truth Is Membership). *Let φ be any formula. Then for all $f \in U_{\mathcal{L}}{}^{\omega}$,*

$$|\varphi|^{\mathcal{M}_{\mathrm{L}}^{K}} f = |\varphi^{f}|_{\mathrm{L}}^{K},$$

and hence for all $w \in W_{\mathrm{L}}^{K}$,

$$\mathcal{M}_{\mathrm{L}}^{K}, w, f \models \varphi \quad \textit{iff} \quad \varphi^{f} \in w. \qquad \dashv$$

As usual, the truth lemma for $\mathcal{M}_{\mathrm{L}}^{K}$ leads to a proof that it characterises L. But in the presence of UI, the completeness proof is much simpler than the case of $\mathcal{M}_{\mathrm{L}}^{\natural}$, and does not require the CQ-dependent results of Section 2.5, as we now show.

THEOREM 2.7.4. *$\mathcal{M}_{\mathrm{L}}^{K}$ is a Kripkean model that characterises* L:

$$\vdash_{\mathrm{L}} \varphi \quad \textit{iff} \quad \mathcal{M}_{\mathrm{L}}^{K} \models \varphi.$$

PROOF. (cf. the proof of Theorem 2.6.5.)

$\mathcal{M}_{\mathrm{L}}^{K}$ is a model: each truth set $|\varphi|^{\mathcal{M}_{\mathrm{L}}^{K}} f$ is equal to $|\varphi^{f}|_{\mathrm{L}}^{K} \in \mathit{Prop}_{\mathrm{L}}^{K}$, so is admissible.

$\mathcal{M}_{\mathrm{L}}^{K}$ is Kripkean: applying Theorem 2.7.3 to the second part of Lemma 2.7.2 gives

$$|\forall x \varphi|^{\mathcal{M}_{\mathrm{L}}^{K}} f = \bigcap_{\tau \in U_{\mathcal{L}}} |\varphi|^{\mathcal{M}_{\mathrm{L}}^{K}} f[\tau/x],$$

which is the same as $\bigcap_{\tau \in U_{\mathcal{L}}} E\tau \Rightarrow |\varphi|^{\mathcal{M}_{\mathrm{L}}^{K}} f[\tau/x]$ in this one-universal-domain model.

L is sound for validity in $\mathcal{M}_{\mathrm{L}}^{K}$: we have

$$\mathcal{M}_{\mathrm{L}}^{K}, w, f \models \varphi \quad \text{iff} \quad \varphi^{f} \in w \quad \text{iff} \quad \mathcal{M}_{\mathrm{L}}, w, f \models \varphi,$$

hence if $\vdash_{\mathrm{L}} \varphi$, then φ is valid in $\mathcal{M}_{\mathrm{L}}$ so we get $\mathcal{M}_{\mathrm{L}}^{K}, w, f \models \varphi$ for every $w \in W_{\mathrm{L}}^{K}$ and $f \in U_{\mathcal{L}}{}^{\omega}$, hence φ is valid in $\mathcal{M}_{\mathrm{L}}^{K}$.

L is complete for validity in $\mathcal{M}_{\mathrm{L}}^{K}$: if $\nvdash_{\mathrm{L}} \varphi$, then there is some $f \in U_{\mathcal{L}}{}^{\omega}$ with $\nvdash_{\mathrm{L}} \varphi^{f}$, and hence $\{\neg\varphi^{f}\}$ is L-consistent. But $\{\neg\varphi^{f}\}$ is also $U_{\mathcal{L}}$-complete in L. For if $\{\neg\varphi^{f}\} \vdash_{\mathrm{L}} \psi(\tau/x)$ for all $\tau \in U_{\mathcal{L}}$, then choosing some constant c not in φ^{f} or ψ we have $\vdash_{\mathrm{L}} \neg\varphi^{f} \to \psi(\mathsf{c}/x)$, hence $\vdash_{\mathrm{L}} \neg\varphi^{f} \to \forall x\psi$ by the rule $\forall$GC of Lemma 1.2.3, so $\{\neg\varphi^{f}\} \vdash_{\mathrm{L}} \forall x\psi$. Hence by Theorem 2.5.2 there is a $U_{\mathcal{L}}$-complete L-maximal set w extending $\{\neg\varphi^{f}\}$. Then $w \in W_{\mathrm{L}}^{K}$ with $\varphi^{f} \notin w$, so that $\mathcal{M}_{\mathrm{L}}^{K}, w, f \not\models \varphi$, showing that φ is not valid in $\mathcal{M}_{\mathrm{L}}^{K}$. $\dashv$

THEOREM 2.7.5. *The logic* QK + CQ + UI + BF *is characterised by the set of all Kripkean models on its $\mathcal{S}^{K}$-structure, as well as being characterised by the class of all Kripkean models on one-universal-domain model structures.*

PROOF. Parallel to Theorem 2.6.6, using the Theorem just proved in place of Theorem 2.6.5. $\dashv$

COROLLARY 2.7.6. *The logic* QK + CQ + UI + BF *is characterised by the class of all full model structures with one universal domain.*

PROOF. Parallel to Corollary 2.6.7. ⊣

To give a Kripkean-models characterisation of the logic QS + CQ + UI + BF for an arbitrary S, we need the following result about the general frame underlying $\mathcal{S}^K_{\mathrm{L}}$.

LEMMA 2.7.7. *If* L *is any quantified modal logic that extends the logic* QS + CQ + UI + BF, *then* S *is validated by the general frame underlying* $\mathcal{S}^K_{\mathrm{L}}$.

PROOF. Parallel to Lemma 2.6.8. The algebraic version is that the map $|\varphi|_{\mathrm{L}} \mapsto |\varphi|^K_{\mathrm{L}}$ from $Prop_{\mathrm{L}}$ to $Prop^K_{\mathrm{L}}$ is a well-defined isomorphism between modal algebras. Since $Prop_{\mathrm{L}}$ validates S, so too does $Prop^K_{\mathrm{L}}$. ⊣

Now applying this Lemma, in a similar manner to Theorem 2.6.9, we can show

THEOREM 2.7.8. *The logic* QS + CQ + UI + BF *is characterised by the set of all Kripkean models on its* $\mathcal{S}^K$*-structure, as well as being characterised by the class of all Kripkean models on one-universal-domain model structures whose underlying general frame validates* S. ⊣

2.8. S-frame Incompleteness Revisited

In the last two sections we have seen how certain quantified logics that include QS and the Barcan Formula are characterised by Kripkean models whose underlying *general* frame validates S. But what about the underlying Kripke frames?

We saw earlier that if S is canonical, then certain logics L had a characteristic model based on a Kripke frame validating S. In particular, for canonical S:

- If L extends QS, then S is validated by the Kripke frame $\mathcal{F}_{\mathrm{L}} = (W_{\mathrm{L}}, R_{\mathrm{L}})$ underlying the model structure $\mathcal{S}_{\mathrm{L}}$ (Theorem 1.10.6).
- If L extends QS + CBF, then S is validated by the Kripke frame $\mathcal{F}^{\circ}_{\mathrm{L}} = (W^{\flat}_{\mathrm{L}}, R^{\circ}_{\mathrm{L}})$ underlying the constant-domains model structure $\mathcal{S}^{\circ}_{\mathrm{L}}$ (Corollary 2.3.9).

This raises the question:

> if S is canonical and L extends QS + CQ + CBF + BF, does the Kripke frame $\mathcal{F}^{\natural}_{\mathrm{L}} = (W^{\natural}_{\mathrm{L}}, R^{\natural}_{\mathrm{L}})$ underlying $\mathcal{S}^{\natural}_{\mathrm{L}}$ validate S?

If so, then in particular QS + CQ + CBF + BF would be characterised by the class of all constant-domain Kripkean models whose underlying Kripke frame validates S.

Similarly we could ask:

> if S is canonical and L extends QS + CQ + UI + BF, does the Kripke frame $\mathcal{F}^K_{\mathrm{L}} = (W^K_{\mathrm{L}}, R^K_{\mathrm{L}})$ underlying $\mathcal{S}^K_{\mathrm{L}}$ validate S?

If so, then in particular QS + CQ + UI + BF would be characterised by the class of all one-universal-domain Kripkean models whose underlying Kripke frame validates S.

Related questions arise from the characterisation of certain extensions of QK + BF by *full* structures, in Corollary 2.6.7 and Corollary 2.7.6. For canonical S we can ask:

> is the logic QS + CQ + CBF + BF characterised by the class of all full constant-domains structures whose Kripke frame validates S ?

and similarly

> is the logic QS + CQ + UI + BF characterised by the class of all full one-universal-domain structures whose Kripke frame validates S ?

In fact the answer to all four of these questions is negative in general. We can show that by taking S to be the canonical propositional logic S4M discussed in Subsection 1.11.2. There we exhibited a formula (1.11.4) that is

- valid in every Kripkean model that is based on an S4M-frame (Lemma 1.11.2), but
- is not a theorem of QS4M + CQ + UI + BF; hence not a theorem of QS4M + CQ + CBF + BF.

So, putting L = QS4M + CQ + CBF + BF, if $\mathcal{F}_{\mathrm{L}}^{\natural}$ did validate S4M, then $\mathcal{M}_{\mathrm{L}}^{\natural}$ would be a Kripkean model on an S4M-frame, so would validate the non-L-theorem (1.11.4). But that would contradict the fact that $\mathcal{M}_{\mathrm{L}}^{\natural}$ characterises L. Thus $\mathcal{F}_{\mathrm{L}}^{\natural}$ cannot validate S4M after all, showing that the answer to the first question is negative.

Replacing CBF by UI here shows that the answer to the second question is negative: $\mathcal{F}_{\mathrm{L}}^{K}$ does not validate S4M when L = QS4M + CQ + UI + BF.

Altogether, the logic QS4M + CQ + CBF + BF is

- characterised by the constant-domains Kripkean models whose underlying *general* frame validates S4M, but
- not characterised by the constant-domains Kripkean models whose underlying *Kripke* frame validates S4M.

Likewise, the logic QS4M + CQ + UI + BF is

- characterised by the one-universal-domain Kripkean models whose underlying *general* frame validates S4M, but
- not characterised by the one-universal-domain Kripkean models whose underlying *Kripke* frame validates S4M.

The other two questions about full structures on S-frames have negative answers in the following strong form.

THEOREM 2.8.1. *Let* L *be any quantified modal logic that includes* QS4M *and is included in* QS4M + CQ + UI + BF. *Then* L *is not characterised by any class of full structures.*

PROOF. Let $\mathcal{S}$ be a full structure validating L. Then $\mathcal{S}$ validates QS4M. By Theorem 1.7.13, the Kripke frame of $\mathcal{S}$ validates S4M.[26] Since $\mathcal{S}$ is full, every model on $\mathcal{S}$ is Kripkean, so validates (1.11.4) by Lemma 1.11.2.

This shows that every full structure validating L must also validate (1.11.4), which is not an L-theorem. Hence L cannot be characterised by any class of full structures. ⊣

EXAMPLE 2.8.2. At the end of Section 1.6 we claimed that there are logics that are characterised by the set of all Kripkean models on some structure, but are not characterised by any class of full structures.

An example is L = QS4M + CQ + CBF + BF. This is characterised by the set of all Kripkean models on the structure $\mathcal{S}^{\natural}_{\mathrm{L}}$ (Theorem 2.6.9), but is not characterised by any class of full structures at all, as shown by the Theorem just proved.

Another example is L = QS4M + CQ + UI + BF, with the structure $\mathcal{S}^{K}_{\mathrm{L}}$ (Theorem 2.7.5). ⊣

We can put a finger on just why the Kripke frames of $\mathcal{S}^{K}_{\mathrm{L}}$ and $\mathcal{S}^{\natural}_{\mathrm{L}}$ can fail to be S-frames, while the frames of $\mathcal{S}_{\mathrm{L}}$ and $\mathcal{S}^{\circ}_{\mathrm{L}}$ must validate S, when S is canonical and L is a suitable extension of QS.

The frame $\mathcal{F}_{\mathrm{L}}$ of $\mathcal{S}_{\mathrm{L}}$ contains *every* L-maximal set of $\mathcal{L}$-formulas. This is essential to the proof (in Theorem 1.10.5) that $\mathcal{F}_{\mathrm{L}}$ is isomorphic to an inner subframe of the canonical propositional S-frame $\mathcal{F}_{\mathrm{S}}$, and hence that S-validity is preserved in passing from $\mathcal{F}_{\mathrm{S}}$ to $\mathcal{F}_{\mathrm{L}}$.

However, the frame $\mathcal{F}^{K}_{\mathrm{L}}$ of $\mathcal{S}^{K}_{\mathrm{L}}$ only contains *some* of the L-maximal sets, the ones that are $U_{\mathcal{L}}$-complete. $\mathcal{F}^{K}_{\mathrm{L}}$ is a subframe of $\mathcal{F}_{\mathrm{L}}$, since $W^{K}_{\mathrm{L}} \subseteq W_{\mathrm{L}}$ and R^{K}_{L} is the restriction of R_{L} to W^{K}_{L}. But $\mathcal{F}^{K}_{\mathrm{L}}$ is not an *inner* subframe of $\mathcal{F}_{\mathrm{L}}$: if it were, then validity of S would be preserved in passing from $\mathcal{F}_{\mathrm{L}}$ to $\mathcal{F}^{K}_{\mathrm{L}}$ when L = QS + CQ + UI + BF—which we have just seen is false for S=S4M.

Likewise, $\mathcal{F}^{\natural}_{\mathrm{L}}$ is only a subframe of $\mathcal{F}^{\circ}_{\mathrm{L}}$, not an inner subframe in general, and validity of S is not always preserved in passing from $\mathcal{F}^{\circ}_{\mathrm{L}}$ to $\mathcal{F}^{\natural}_{\mathrm{L}}$.

We can draw from these observations some positive answers to our four questions for certain canonical logics S. Define S to be *preserved by subframes* if the class of all Kripke frames that validate S is closed under subframes. For example, if the class of S-frames is defined by universal first-order conditions on frames—like reflexiveness, transitivity, symmetry or linearity—then S is preserved by subframes. The logic S4M is not preserved by subframes: one of the defining conditions for S4M-frames is (1.11.3), which involves an existential assertion that is preserved by inner subframes, but not by all subframes.

---

[26]Theorem 1.7.13 requires the signature to have infinitely many monadic predicates in general. In fact in the present case, only one monadic predicate is required to show that the Kripke frame of $\mathcal{S}$ validates S4M, because the axioms defining S4M have a single propositional variable.

THEOREM 2.8.3. *Let* S *be a canonical propositional modal logic that is preserved by subframes. Then*

(1) *The quantified modal logic* QS + CQ + CBF + BF *is characterised by the class of all Kripkean models on constant-domains model structures whose underlying Kripke frame validates* S.
(2) *This logic is also characterised by the class of all full constant-domains structures whose Kripke frame validates* S.

PROOF. With L = QS + CQ + CBF + BF, the proofs of (1) and (2) proceeds similarly to those of Theorem 2.6.9 and Corollary 2.6.7, but using the fact that the frame $\mathcal{F}_{\mathrm{L}}^{\natural}$ underlying $\mathcal{S}_{\mathrm{L}}^{\natural}$ validates S because it is a subframe of $\mathcal{F}_{\mathrm{L}}^{\circ}$, which validates S by Corollary 2.3.9. ⊣

A number of the more well known canonical propositional modal logics are covered by this result: their frames are defined by simple universal properties that are preserved by subframes. Thus

- QS4 + CQ + CBF + BF is characterised by the class of all constant-domains Kripkean models whose accessibility relation is reflexive and transitive.
- QB+CQ+CBF+BF is characterised by the class of all constant-domains Kripkean models with a symmetric accessibility relation.
- QS5 + CQ + CBF + BF is characterised by the class of all constant-domains Kripkean models that are based on an equivalence relation.

Replacing CBF by UI in this analysis, and using the fact that $\mathcal{F}_{\mathrm{L}}^{K}$ is a subframe of $\mathcal{F}_{\mathrm{L}}$, we obtain the follow results, applying to the same propositional logics S.

THEOREM 2.8.4. *Let* S *be a canonical propositional modal logic that is preserved by subframes. Then*

(1) *The quantified modal logic* QS + CQ + UI + BF *is characterised by the class of all Kripkean models on one-universal-domain model structures whose underlying Kripke frame validates* S.
(2) *This logic is also characterised by the class of all full one-universal-domain structures whose Kripke frame validates* S. ⊣

To conclude this discussion of S-frame completeness and incompleteness, observe that for certain logics L that include QS + CQ + CBF + BF, what we have is actually a *trade-off* between the two objectives of

(i) a characteristic L-model based on an S-frame; and
(ii) a characteristic L-model that is Kripkean.

If S is canonical, then (i) is provided by $\mathcal{M}_{\mathrm{L}}$ (Theorem 1.9.6, Theorem 1.10.6(1)), while (ii) is provided by $\mathcal{M}_{\mathrm{L}}^{\natural}$ (Theorem 2.6.5). For some L, the model $\mathcal{M}_{\mathrm{L}}^{\natural}$ fulfils both objectives, e.g. when S is S4, B, or S5.

But in some other cases, there is no single model that provides both (i) and (ii), since each is incompatible with the other. This is illustrated by our ubiquitous example of L = QS4M + CQ + CBF + BF. We saw that this logic is not characterised by Kripkean models on S4M-frames, since all such models validate the non-L-theorem (1.11.4). So in this case

(i) $\mathcal{M}_{\mathrm{L}}$ is a characteristic L-model based on an S4M-frame, hence cannot be Kripkean; while

(ii) $\mathcal{M}^{\natural}_{\mathrm{L}}$ is a characteristic L-model that is Kripkean, hence cannot be based on an S4M-frame.

Chapter 3

THE EXISTENCE PREDICATE

Recall that in any model structure, an existence set

$$Ea = \{w \in W : a \in Dw\},$$

represents the proposition "a exists". If every existence set is admissible, we will call the structure *E-admissible*. We already know that this condition has consequences for validity: in particular it ensures that the Commuting Quantifiers axiom CQ is valid (Theorem 1.7.16). But the formal languages we have been using lack the means to refer directly to existence sets in a model, and so lack the means to force Ea to be admissible.

To provide such a means, we introduce a monadic predicate constant $\mathcal{E}$, allowing the formation of atomic formulas $\mathcal{E}\tau$ that can be read "τ exists". In any premodel, $\mathcal{E}$ is interpreted by the function $|\mathcal{E}|^{\mathcal{M}} : U \to \wp W$ having $|\mathcal{E}|^{\mathcal{M}}a = Ea$. Thus $|\mathcal{E}|^{\mathcal{M}}$ is determined by the underlying model structure, independently of the particular premodel $\mathcal{M}$. The semantics of formula $\mathcal{E}\tau$ then has

$$|\mathcal{E}\tau|^{\mathcal{M}}f = |\mathcal{E}|^{\mathcal{M}}(|\tau|^{\mathcal{M}}f) = E(|\tau|^{\mathcal{M}}f),$$

so that

$$\mathcal{M}, w, f \models \mathcal{E}\tau \quad \text{iff} \quad |\tau|^{\mathcal{M}}f \in Dw,$$

correctly formalising the "τ exists" reading.

Now for each individual $a \in U$, picking an f and an x with $fx = a$ gives $|\mathcal{E}x|^{\mathcal{M}}f = Ea$. Thus every existence set is a truth set, so in a *model*, i.e. in an $\mathcal{M}$ in which every truth set is admissible, we get E-admissibility: $Ea \in Prop$ for all $a \in U$. That shows that a model structure $\mathcal{S}$ must be E-admissible in order for there to be any models on $\mathcal{S}$ for languages with a monadic constant $\mathcal{E}$ interpreted in this way.

In this chapter we assume that our languages contain the existence predicate $\mathcal{E}$, which is taken as not belonging to the ambient signature $\mathcal{L}$, but is a constant symbol, like the propositional connectives and $\forall$, with a fixed interpretation. We study axiomatisation questions for some resulting logics, and will see that the ability to formally express existence assertions simplifies the definition of a "logic", and the construction of canonical models. It allows us to axiomatise the logic characterised by Kripkean models without any restrictions on their domain function D, and also to further clarify the relationships between the Barcan Formulas and the properties of expanding and contracting domains.

3.1. Axiomatising Existence

In the presence of $\mathcal{E}$, the only quantifier postulates that need to be assumed are the *Existing Instantiation* scheme

EI: $\forall x\varphi \to (\mathcal{E}\tau \to \varphi(\tau/x))$, where τ is free for x in φ;

and the modified $\forall$-Introduction rule

$\mathcal{E}\forall$-Intro: $\dfrac{\varphi \to (\mathcal{E}x \to \psi)}{\varphi \to \forall x\psi}$, if x is not free in φ.

The EI scheme is sometimes conveniently taken in the PC-equivalent form $\mathcal{E}\tau \to (\forall x\varphi \to \varphi(\tau/x))$, and the premiss of $\mathcal{E}\forall$-Intro may similarly be presented as $\mathcal{E}\tau \to (\varphi \to \psi)$. For example, from the tautology

$$\mathcal{E}x \to (\top \to \mathcal{E}x)$$

we immediately derive

$$\top \to \forall x\mathcal{E}x$$

by $\mathcal{E}\forall$-Intro. Hence $\forall x\mathcal{E}x$ is derivable. This is as it should be, since $\forall x\mathcal{E}x$ is valid. Given the actualist intepretation of $\forall$, the sentence $\forall x\mathcal{E}x$ does not make the assertion "every x exists", but only the tautological "every existing x exists". Indeed an existence assertion $\mathcal{E}x$ is redundant within the scope of the actualist quantifiers $\forall x$ and $\exists x$, in the following sense.

THEOREM 3.1.1. *If a set of formulas* L *includes the scheme* EI *and is closed under* PC *and the rule* $\mathcal{E}\forall$-Intro, *then it includes the schemes*

(1) $\forall x(\mathcal{E}x \to \varphi) \leftrightarrow \forall x\varphi$,
(2) $\exists x(\mathcal{E}x \wedge \varphi) \leftrightarrow \exists x\varphi$.

PROOF. For (1), replacing φ by $\mathcal{E}x \to \varphi$ in EI and taking $\tau = x$ gives

$$\forall x(\mathcal{E}x \to \varphi) \to (\mathcal{E}x \to (\mathcal{E}x \to \varphi)),$$

from which, by PC, we derive $\forall x(\mathcal{E}x \to \varphi) \to (\mathcal{E}x \to \varphi)$. This immediately yields $\forall x(\mathcal{E}x \to \varphi) \to \forall x\varphi$ by $\mathcal{E}\forall$-Intro.

For the converse of this last formula, we have $\forall x\varphi \to (\mathcal{E}x \to \varphi)$ as an instance of EI, from which $\forall x\varphi \to (\mathcal{E}x \to (\mathcal{E}x \to \varphi))$, follows by PC, and this in turn yields $\forall x\varphi \to \forall x(\mathcal{E}x \to \varphi)$ by $\mathcal{E}\forall$-Intro. (1) is derived from these by PC.

(2) is the dual of (1) and equivalent to it as a scheme: replace φ by $\neg\varphi$ in (1) and apply PC-principles and Replacement of Provable Equivalents (Lemma 1.2.1(3)) to obtain (2). ⊣

For a given signature $\mathcal{L}$, a *quantified modal logic* is now defined to be any set L of $\mathcal{L}$-formulas that includes all Boolean tautologies and instances of the schemes K and EI and is closed under the rules MP, N, $\mathcal{E}\forall$-Intro, TI and GC. Such an L will often be called an *$\mathcal{E}$-logic* to emphasise the current definition.

Our first task is to verify that a logic in this new sense is still a logic in the sense defined for $\mathcal{E}$-free languages in Section 1.2:

THEOREM 3.1.2. *If a set* L *includes the scheme* EI *and is closed under* PC *and the rule* $\mathcal{E}\forall$-Intro, *then it includes the schemes* AI, UD *and* VQ, *and is closed under* UG.

PROOF. By "closed under PC" we mean that L is closed under tautological consequence: if $\varphi_1, \dots, \varphi_n \in$ L and $\varphi_1 \wedge \cdots \wedge \varphi_n \to \psi$ is a tautology, then $\psi \in$ L.

The following proof sequence shows L includes Actual Instantiation, given that y is free for x in φ:

1. $\mathcal{E}y \to (\forall x\varphi \to \varphi(y/x))$	EI
2. $\top \to (\mathcal{E}y \to (\forall x\varphi \to \varphi(y/x)))$	1, PC
3. $\top \to \forall y(\forall x\varphi \to \varphi(y/x))$	2, $\mathcal{E}\forall$-Intro
4. $\forall y(\forall x\varphi \to \varphi(y/x))$	3, PC.

For Vacuous Quantification, given x not free in φ we simply derive

1. $\varphi \to (\mathcal{E}x \to \varphi)$	tautology
2. $\varphi \to \forall x\varphi$	1, $\mathcal{E}\forall$-Intro.

The closure of L under the Universal Generalisation rule is similarly direct:

1. φ	given
2. $\top \to (\mathcal{E}x \to \varphi)$	1, PC
3. $\top \to \forall x\varphi$	2, $\mathcal{E}\forall$-Intro
4. $\forall x\varphi$	3, PC.

Universal Distribution is less direct. First we show that L is closed under the $\forall$-Monotonicity rule

$$\frac{\varphi \to \psi}{\forall x\varphi \to \forall x\psi}.$$

1. $\varphi \to \psi$	given
2. $\forall x\varphi \to (\mathcal{E}x \to \varphi)$	EI
3. $\forall x\varphi \to (\mathcal{E}x \to \psi)$	1,2, PC
4. $\forall x\varphi \to \forall x\psi$	3, $\mathcal{E}\forall$-Intro.

Next we show that L includes the $\forall\wedge$-*Distribution* scheme

$$\forall x\varphi \wedge \forall x\psi \to \forall x(\varphi \wedge \psi).$$

1. $\mathcal{E}x \to (\forall x\varphi \to \varphi)$ EI
2. $\mathcal{E}x \to (\forall x\psi \to \psi)$ EI
3. $\mathcal{E}x \to (\forall x\varphi \wedge \forall x\psi \to \varphi \wedge \psi)$ 1,2, PC
4. $\forall x\varphi \wedge \forall x\psi \to \forall x(\varphi \wedge \psi)$ 3, $\mathcal{E}\forall$-Intro.

Now we can derive UD:

1. $(\varphi \to \psi) \wedge \varphi \to \psi$ tautology
2. $\forall x((\varphi \to \psi) \wedge \varphi) \to \forall x\psi$ 1, $\forall$-Monotonicity
3. $\forall x(\varphi \to \psi) \wedge \forall x\varphi \to \forall x((\varphi \to \psi) \wedge \varphi)$ $\forall\wedge$-Distribution
4. $\forall x(\varphi \to \psi) \to (\forall x\varphi \to \forall x\psi)$ 2, 3, PC. $\dashv$

The existence predicate allows us to obtain a proof-theoretic analogue of the fact that CQ is valid when existence sets are admissible. This depends on the original $\forall$-postulates, as well as the new ones involving $\mathcal{E}$:

THEOREM 3.1.3. *Every $\mathcal{E}$-logic includes the scheme of Commuting Quantifiers.*

PROOF. For distinct variables x, y we derive

1. $\mathcal{E}y \to (\forall y\varphi \to \varphi)$ EI
2. $\forall x\mathcal{E}y \to (\forall x\forall y\varphi \to \forall x\varphi)$ 1, UG, UD, PC
3. $\mathcal{E}y \to \forall x\mathcal{E}y$ VQ, $y \neq x$.
4. $\mathcal{E}y \to (\forall x\forall y\varphi \to \forall x\varphi)$ 2, 3, PC
5. $\forall x\forall y\varphi \to \forall y\forall x\varphi$ 4, $\mathcal{E}\forall$-Intro, y not free in $\forall x\forall y\varphi$. $\dashv$

Another useful fact about $\mathcal{E}$-logics is that they are closed under the rule

$$\frac{\varphi \to (\mathcal{E}\mathsf{c} \to \psi(\mathsf{c}/x))}{\varphi \to \forall x\psi}, \tag{3.1.1}$$

where x is not free in φ, and the constant c does not occur in φ or ψ. This follows by GC and $\mathcal{E}\forall$-Intro.

The soundness of the new $\forall$-postulates will now be demonstrated. Note first that the semantics of atomic formulas $\mathcal{E}\tau$ treats $\mathcal{E}$ just like any other monadic predicate symbol, and so the results from Section 1.6 about the truth relation continue to hold, and in particular the Substitution Lemma 1.6.2.

THEOREM 3.1.4. *In any premodel $\mathcal{M}$, the Existing Instantiation scheme is valid. If $\mathcal{M}$ is a model, then the rule $\mathcal{E}\forall$-*Intro *preserves validity in $\mathcal{M}$.*

PROOF. For EI, suppose τ is free for x in φ. Then given any $f \in U^\omega$, the semantics of $\forall$ entails that for each $a \in U$,

$$|\forall x\varphi|^{\mathcal{M}} f \subseteq Ea \Rightarrow |\varphi|^{\mathcal{M}} f[a/x].$$

Now put $a = |\tau|^{\mathcal{M}} f$. Then $Ea = |\mathcal{E}\tau|^{\mathcal{M}} f$ by the semantics of $\mathcal{E}$, and $|\varphi|^{\mathcal{M}} f[a/x] = |\varphi|^{\mathcal{M}} f[|\tau|^{\mathcal{M}} f/x] = |\varphi(\tau/x)|^{\mathcal{M}} f$ by the Substitution Lemma 1.6.2, so we get

$$|\forall x\varphi|^{\mathcal{M}} f \subseteq |\mathcal{E}\tau|^{\mathcal{M}} f \Rightarrow |\varphi(\tau/x)|^{\mathcal{M}} f = |\mathcal{E}\tau \to \varphi(\tau/x)|^{\mathcal{M}} f.$$

Hence $\mathcal{M} \models \forall x\varphi \to (\mathcal{E}\tau \to \varphi(\tau/x))$.

For the rule $\mathcal{E}\forall$-Intro, suppose $\mathcal{M} \models \varphi \to (\mathcal{E}x \to \psi)$, with x not free in φ. Given any $f \in U^\omega$ and $a \in U$, the assignments f and $f[a/x]$ agree on all free variables of φ, so by Lemma 1.6.1,

$$|\varphi|^{\mathcal{M}} f = |\varphi|^{\mathcal{M}} f[a/x] \subseteq |\mathcal{E}x|^{\mathcal{M}} f[a/x] \Rightarrow |\psi|^{\mathcal{M}} f[a/x],$$

with the set inclusion holding as $\varphi \to (\mathcal{E}x \to \psi)$ is valid in $\mathcal{M}$. But $|\mathcal{E}x|^{\mathcal{M}} f[a/x] = Ea$, so we get

$$|\varphi|^{\mathcal{M}} f \subseteq Ea \Rightarrow |\psi|^{\mathcal{M}} f[a/x]$$

for all $a \in U$. Now if $\mathcal{M}$ is a model, then $|\varphi|^{\mathcal{M}} f \in Prop$, so then

$$|\varphi|^{\mathcal{M}} f \subseteq \bigsqcap_{a \in U} \left(Ea \Rightarrow |\psi|^{\mathcal{M}} f[a/x]\right) = |\forall x\psi|^{\mathcal{M}} f.$$

Hence $\mathcal{M} \models \varphi \to \forall x\psi$ as required. ⊣

Combining with our earlier soundness results, we conclude

THEOREM 3.1.5. *Let $\mathcal{S}$ be any model structure. Then for languages including the existence predicate $\mathcal{E}$:*

(1) *The set $\mathrm{L}_{\mathcal{S}} = \{\varphi : \mathcal{S} \models \varphi\}$ of formulas valid in all models on $\mathcal{S}$ is an $\mathcal{E}$-logic.*

(2) *The set $\mathrm{L}^K_{\mathcal{S}}$ of formulas valid in all Kripkean models on $\mathcal{S}$ is an $\mathcal{E}$-logic.* ⊣

To discuss completeness theorems, let L be an $\mathcal{E}$-logic for a signature $\mathcal{L}$. We can still use the canonical structure

$$\mathcal{S}_{\mathrm{L}} = (W_{\mathrm{L}}, R_{\mathrm{L}}, Prop_{\mathrm{L}}, U_{\mathcal{L}}, D_{\mathrm{L}})$$

based on the set W_{L} of all L-maximal set of $\mathcal{L}$-formulas, and the canonical model $\mathcal{M}_{\mathrm{L}} = (\mathcal{S}_{\mathrm{L}}, |-|^{\mathcal{M}_{\mathrm{L}}})$, as developed in Section 1.9. The presence of $\mathcal{E}$ allows some simplification in the description of the domain function D_{L} and existence sets:

LEMMA 3.1.6. (1) *For all $w \in W_{\mathrm{L}}$, $D_{\mathrm{L}}(w) = \{\tau \in U_{\mathcal{L}} : \mathcal{E}\tau \in w\}$.*

(2) *For all $\tau \in U_{\mathcal{L}}$, $E\tau = |\mathcal{E}\tau|_{\mathrm{L}}$.*

PROOF. (1): recall that $\tau \in D_{\mathrm{L}}(w)$ iff $\mathrm{UI}(\tau) \subseteq w$, where $\mathrm{UI}(\tau)$ is the set of all UI-instances $\forall x\varphi \to \varphi(\tau/x)$. Thus if $\tau \in D_{\mathrm{L}}(w)$, then $(\forall x\mathcal{E}x \to \mathcal{E}\tau) \in w$, so $\mathcal{E}\tau \in w$ because $\vdash_{\mathrm{L}} \forall x\mathcal{E}x$ as noted earlier. Conversely, since every formula $\mathcal{E}\tau \to (\forall x\varphi \to \varphi(\tau/x))$ is in w by EI, if $\mathcal{E}\tau \in w$ we get $\mathrm{UI}(\tau) \subseteq w$, hence $\tau \in D_{\mathrm{L}}(w)$.

(2): recall that $|\mathcal{E}\tau|_{\mathrm{L}} = \{w \in W_{\mathrm{L}} : \mathcal{E}\tau \in w\}$. But by definition, $w \in E\tau$ iff $\tau \in D_{\mathrm{L}}(w)$, which is equivalent to $\mathcal{E}\tau \in w$ by (1). ⊣

Part (2) of this Lemma shows that $E\tau \in \mathit{Prop}_{\mathrm{L}}$ whenever τ belongs to the universe of $\mathcal{S}_{\mathrm{L}}$. So $\mathcal{S}_{\mathrm{L}}$ is E-admissible, i.e. all its existence sets are admissible, as is necessary for there to be models on $\mathcal{S}_{\mathrm{L}}$ for the language with $\mathcal{E}$.

As a further application of part (2), if τ is any $\mathcal{L}$-term (open or closed), and $f \in U_{\mathcal{L}}{}^{\omega}$, then

$$|\mathcal{E}\tau|^{\mathcal{M}_{\mathrm{L}}} f = E(|\tau|^{\mathcal{M}_{\mathrm{L}}} f) = E(\tau^f) = |\mathcal{E}(\tau^f)|_{\mathrm{L}} = |(\mathcal{E}\tau)^f|_{\mathrm{L}}.$$

This is the case $\varphi = \mathcal{E}\tau$ of the Truth Lemma $|\varphi|^{\mathcal{M}_{\mathrm{L}}} f = |\varphi^f|_{\mathrm{L}}$ (Theorem 1.9.4). The proof that this holds for all formulas φ, and consequently that $\mathcal{M}_{\mathrm{L}}$ is a model that characterises L in the $\mathcal{E}$-language, all then proceed as before.[27]

If S is any set of propositional modal formulas, we denote by QES the *smallest* $\mathcal{E}$-logic that contains every $\mathcal{L}$-formula that is a substitution-instance of a member of S. Proceeding by the arguments of Section 1.10, we have:

- QES is characterised by validity in its canonical structure $\mathcal{S}_{\mathrm{QES}}$, and by validity in all E-admissible model structures whose underlying general frame validates S (see Theorem 1.10.2).
- If S is a canonical propositional logic, then QES is characterised by the class of all E-admissible model structures whose underlying Kripke frame validates S (see Theorem 1.10.6).

It is worth recording that inclusion of the Universal Instantiation scheme UI in an $\mathcal{E}$-logic L makes the existence predicate redundant. For, from UI we get $\vdash_{\mathrm{L}} \forall x\mathcal{E}x \to \mathcal{E}\tau$ for any term τ, and hence $\vdash_{\mathrm{L}} \mathcal{E}\tau$ since $\vdash_{\mathrm{L}} \forall x\mathcal{E}x$ in general. It follows that in $\mathcal{S}_{\mathrm{L}}$, $|\mathcal{E}\tau|_{\mathrm{L}} = W_{\mathrm{L}}$. This connects with Lemma 2.4.2, where we saw that if $\vdash_{\mathrm{L}}$ UI, then in $\mathcal{S}_{\mathrm{L}}$ there is one universal domain $U_{\mathcal{L}}$, with $E\tau = W_{\mathrm{L}}$ for all $\tau \in U_{\mathcal{L}}$.

3.2. Completeness for Kripkean $\mathcal{E}$-Models

The $\mathcal{E}$-free logic of Kripkean models with constant domains was axiomatised in Section 2.6, using the Barcan Formula and its converse. For Kripkean

[27] In an $\mathcal{E}$-logic, the proof of the key result $|\forall x\varphi|_{\mathrm{L}} = \prod_{\tau \in U_{\mathcal{L}}} E\tau \Rightarrow |\varphi(\tau/x)|_{\mathrm{L}}$ (Lemma 1.9.2) can be given more simply, using EI and the rule (3.1.1): an instructive exercise for the reader, who can also consult Lemma 5.6.14 for assistance if needed.

models on arbitrary structures, we do not have an axiomatisation of the $\mathcal{E}$-free logic, but we can give one for the language including $\mathcal{E}$. The expressive power of the existence predicate allows the formulation of inference rules that solve the problem of constructing suitably "$\forall$-complete" maximal sets.

For this purpose we introduce the notion of a *template*, which can be thought of, roughly, as an expression of the form

$$\varphi_0 \to \Box(\varphi_1 \to \Box(\varphi_2 \to \cdots \to \Box(\varphi_{n-1} \to \#)\cdots), \tag{3.2.1}$$

where the φ_i's are formulas, and the symbol # is a place holder for a formula. Replacing # by a formula turns the whole expression into a genuine formula. The placeholder # by itself is a template (when $n = 0$).

Introducing the *strict implication* connective $\prec$ by defining $\varphi \prec \psi$ to be $\Box(\varphi \to \psi)$, we can rewrite (3.2.1) as

$$\varphi_0 \to (\varphi_1 \prec (\varphi_2 \prec \cdots \prec (\varphi_{n-1} \prec \#)\cdots). \tag{3.2.2}$$

Formally, the set Tem of templates is defined to be the smallest set of symbol-sequences such that

(1) $\# \in \mathsf{Tem}$;
(2) if $\rho \in \mathsf{Tem}$, then $\varphi \to \rho \in \mathsf{Tem}$ for all formulas φ; and
(3) if $\rho \in \mathsf{Tem}$, then $\Box\rho \in \mathsf{Tem}$.

Each template ρ has a single occurrence of the symbol #. We write $\rho(\psi)$ for the formula obtained from ρ by replacing # by the formula ψ. This substitution operation satisfies (indeed can be defined inductively by)

$$\#(\psi) = \psi$$
$$(\varphi \to \rho)(\psi) = \varphi \to \rho(\psi)$$
$$(\Box\rho)(\psi) = \Box\rho(\psi),$$

and we use these equations repeatedly without further comment.

By a free occurrence of a variable x in a template ρ we mean a free occurrence of x in some formula φ that occurs in ρ. We can now formulate the *Template $\forall$-Introduction* rule

$$\text{T}\forall\text{-Intro:} \quad \frac{\rho(\mathcal{E}x \to \psi)}{\rho(\forall x\psi)}, \quad \text{if } \rho \in \mathsf{Tem} \text{ and } x \text{ is not free in } \rho.$$

The $\mathcal{E}\forall$-Intro rule is itself just the instance of T$\forall$-Intro in which ρ is the template $\varphi \to \#$.

Inference rules of this type were introduced by R. H. Thomason [1970], who used the rule

$$\frac{\rho(\Box\psi)}{\rho(\Box\forall x\psi)}, \quad \text{where } x \text{ is not free in } \rho, \tag{3.2.3}$$

with ρ in the strict-implication form (3.2.2). This was applied to a completeness theorem for a certain quantified extension of S4 with an identity predicate and a (definable) existence predicate, relative to Kripkean models. The same rule with ρ in the form (3.2.2) was used in [Hughes and Cresswell 1996] to construct a Kripkean canonical model for extensions of QEK. A version of T$\forall$-Intro with ρ in the (3.2.2) form was used similarly in [Garson 2001]. The more general class of expressions Tem was introduced in [Goldblatt 1982] (under the name "admissible forms") for the purpose of axiomatising certain dynamic logics.

It is worth clarifying the relationship between Tem and (3.2.1). The point is that the expressions in (3.2.1) provide a set of "normal forms" for templates. Every template is equivalent to one of the form (3.2.1), which helps to explain the naturalness of this form. To see this, note that from (2) and (3) we get

(4) if $\rho \in$ Tem, then $\varphi \to \Box\rho \in$ Tem for all formulas φ.

But if $\rho \in$ Tem, then by (4), $\top \to \Box\rho$ is a template, and this is PC-equivalent to $\Box\rho$. So every template is equivalent to one generated from # using only (2) and (4).

Now iteration of (2) will produce templates of the form

$$\varphi_0 \to (\varphi_1 \to \cdots \to (\varphi_{n-1} \to \rho)\cdots).$$

This is PC-equivalent to

$$(\varphi_0 \wedge \varphi_1 \wedge \cdots \wedge \varphi_{n-1}) \to \rho.$$

Thus iterated use of (2) can be replaced by a single use. Similarly, a template $\varphi_0 \to (\varphi_1 \to \Box\rho)$, produced by applying (4) and then (2) to ρ, is equivalent to $\varphi_0 \wedge \varphi_1 \to \Box\rho$, produced by a single application of (4). So once (4) is used, (2) is not subsequently required. (2) is needed to form $\varphi \to$ # from #, but a template $\varphi \to \Box$#, formed directly from # by (4), is equivalent to $\varphi \to \Box(\top \to \#)$, generated by (2) then (4). Hence every template is equivalent to one generated from # by a single application of (2) followed by a finite number of applications of (4), as in (3.2.1).

The following technical fact holds for any quantified modal logic L:

LEMMA 3.2.1. *Let* $\vdash_L \varphi_1 \to \varphi_2$. *Then* $\vdash_L \rho(\varphi_1) \to \rho(\varphi_2)$ *for all templates* ρ.

PROOF. We prove $\vdash_L \rho(\varphi_1) \to \rho(\varphi_2)$ by induction on the formation of $\rho \in$ Tem. If ρ is #, then $\rho(\varphi_i) = \varphi_i$ and the result is immediate.

Now assume inductively the result for ρ. Then

$$\vdash_L (\varphi \to \rho(\varphi_1)) \to (\varphi \to \rho(\varphi_2))$$

by PC, and $\varphi \to \rho(\varphi_i) = (\varphi \to \rho)(\varphi_i)$, so the result holds for $\varphi \to \rho$. Also $\vdash_L \Box\rho(\varphi_1) \to \Box\rho(\varphi_2)$ by modal principles (rule N, axiom K, PC), and $\Box\rho(\varphi_i) = (\Box\rho)(\varphi_i)$, so the result holds for $\Box\rho$. ⊣

THEOREM 3.2.2. *If* L *is closed under* T$\forall$-Intro, *then it is closed under the rule* (3.2.3).

Proof. Suppose $\rho \in \text{Tem}$ with x not free in ρ. Let ρ' be got from ρ by replacing # by $\Box$#. Then ρ' is a template, since $\Box\# \in \text{Tem}$ and ρ' is got from $\Box$# by the same sequence of applications of the template formation rules (1)–(3) that generates ρ from #. Also x is not free in ρ'.

Note that in general $\rho'(\psi) = \rho(\Box\psi)$. Thus if $\vdash_{\mathrm{L}} \rho(\Box\psi)$, then $\vdash_{\mathrm{L}} \rho'(\psi)$, hence as $\psi \to (\mathcal{E}x \to \psi)$ is a tautology, we get $\vdash_{\mathrm{L}} \rho'(\mathcal{E}x \to \psi)$ by the above Lemma. So if L is closed under T$\forall$-Intro, this gives $\vdash_{\mathrm{L}} \rho'(\forall x\psi)$, i.e. $\vdash_{\mathrm{L}} \rho(\Box\forall x\psi)$. Hence L is closed under rule (3.2.3). ⊣

Now we consider the soundness of Template $\forall$-Introduction.

Lemma 3.2.3. *Let ρ be any template in which the variable x is not free. Then in a Kripkean model $\mathcal{M}$, if $\mathcal{M}, w, f \not\models \rho(\forall x\psi)$, then for some $a \in U$, $\mathcal{M}, w, f[a/x] \not\models \rho(\mathcal{E}x \to \psi)$.*

Proof. By induction on the formation of ρ. For the case $\rho = \#$, if $\mathcal{M}$, $w, f \not\models \forall x\psi$, then since $\mathcal{M}$ is Kripkean, there is some $a \in Dw$ with $\mathcal{M}$, $w, f[a/x] \not\models \psi$ (1.6.6). Then $\mathcal{M}, w, f[a/x] \models \mathcal{E}x$, hence $\mathcal{M}, w, f[a/x] \not\models \mathcal{E}x \to \psi$, so the result holds in this case.

Now assume the result for a template ρ. Suppose x is not free in $\varphi \to \rho$. If $\mathcal{M}, w, f \not\models \varphi \to \rho(\forall x\psi)$, then $\mathcal{M}, w, f \models \varphi$ and $\mathcal{M}, w, f \not\models \rho(\forall x\psi)$. But x is not free in ρ, so by induction hypothesis on ρ there is some $a \in U$ with $\mathcal{M}, u, f[a/x] \not\models \rho(\mathcal{E}x \to \psi)$. Also x is not free in φ, so f and $f[a/x]$ agree on all free variables of φ, hence $\mathcal{M}, w, f[a/x] \models \varphi$ by Lemma 1.6.1. Thus $\mathcal{M}, u, f[a/x] \not\models \varphi \to \rho(\mathcal{E}x \to \psi)$, so the result holds for the template $\varphi \to \rho$.

Next suppose x is not free in $\Box\rho$, hence not free in ρ. If $\mathcal{M}, w, f \not\models \Box\rho(\forall x\psi)$, then $\mathcal{M}, u, f \not\models \rho(\forall x\psi)$ for some u with wRu. By induction hypothesis on ρ, there is some $a \in U$ with $\mathcal{M}, u, f[a/x] \not\models \rho(\mathcal{E}x \to \psi)$. Then $\mathcal{M}, w, f[a/x] \not\models \Box\rho(\mathcal{E}x \to \psi)$, so the result holds for the template $\Box\rho$. ⊣

Corollary 3.2.4 (Soundness for T$\forall$-Intro). *The rule* T$\forall$-Intro *is sound for validity in all Kripkean models, i.e. if x is not free in ρ and $\mathcal{M}$ is a Kripkean model with $\mathcal{M} \models \rho(\mathcal{E}x \to \psi)$, then $\mathcal{M} \models \rho(\forall x\psi)$.* ⊣

Now we take up the question of axiomatising $\mathcal{E}$-logics that are characterised by Kripkean models. Fix an $\mathcal{E}$-logic L that is closed under the rule T$\forall$-Intro. We are going to construct a canonical Kripkean model that characterises L. This modifies the construction of the previous section, restricting the set of worlds of $\mathcal{M}_{\mathrm{L}}$ to those that have the right closure property to reflect the Kripkean semantics of $\forall$.

As usual for canonical model constructions, we assume that the background signature $\mathcal{L}$ has infinitely many constants. As usual for Kripkean canonical models, we assume that $\mathcal{L}$ is countable.

A set Σ of $\mathcal{L}$-formulas is called *$\mathcal{E}\forall$-complete in* L if, for any formula φ, any individual variable x, and any template ρ, if $\Sigma \vdash_{\mathrm{L}} \rho(\mathcal{E}\tau \to \varphi(\tau/x))$ for all $\tau \in U_{\mathcal{L}}$, then $\Sigma \vdash_{\mathrm{L}} \rho(\forall x\varphi)$.

LEMMA 3.2.5. *Let* Σ *be* $\mathcal{E}\forall$*-complete. Then*

(1) $\Sigma \cup \Gamma$ *is* $\mathcal{E}\forall$*-complete for every finite set* Γ *of formulas.*

(2) $\Box^{-\mathrm{L}}\Sigma = \{\varphi : \Sigma \vdash_{\mathrm{L}} \Box\varphi\}$ *is* $\mathcal{E}\forall$*-complete.*

PROOF. (1): Let $\Sigma \cup \Gamma \vdash_{\mathrm{L}} \rho(\mathcal{E}\tau \to \varphi(\tau/x))$ for all $\tau \in U_{\mathcal{L}}$. If ψ is the conjunction of the members of Γ, then $\Sigma \vdash_{\mathrm{L}} \psi \to \rho(\mathcal{E}\tau \to \varphi(\tau/x))$ for all $\tau \in U_{\mathcal{L}}$. Applying the $\mathcal{E}\forall$-completeness of Σ to the template $\psi \to \rho$ then gives $\Sigma \vdash_{\mathrm{L}} \psi \to \rho(\forall x\psi)$. Hence $\Sigma \cup \Gamma \vdash_{\mathrm{L}} \rho(\forall x\psi)$.

(2): Let $\Box^{-\mathrm{L}}\Sigma \vdash_{\mathrm{L}} \rho(\mathcal{E}\tau \to \varphi(\tau/x))$ for all $\tau \in U_{\mathcal{L}}$. Then by (2.5.1), $\Sigma \vdash_{\mathrm{L}} \Box\rho(\mathcal{E}\tau \to \varphi(\tau/x))$ for all $\tau \in U_{\mathcal{L}}$. Applying the $\mathcal{E}\forall$-completeness of Σ to the template $\Box\rho$ then gives $\Sigma \vdash_{\mathrm{L}} \Box\rho(\forall x\psi)$, hence $\Box^{-\mathrm{L}}\Sigma \vdash_{\mathrm{L}} \rho(\forall x\psi)$. ⊣

The version of "Lindenbaum's Lemma" that applies to $\mathcal{E}$-logics is the following result on the existence of $\mathcal{E}\forall$-complete sets.

THEOREM 3.2.6. *Every* L*-consistent* $\mathcal{E}\forall$*-complete set of formulas has an* L*-maximal* $\mathcal{E}\forall$*-complete extension.*

PROOF. This is an adaptation of Theorem 2.5.2, using part (1) of the Lemma just proved. We sketch the details.

Let Σ_0 be L-consistent and $\mathcal{E}\forall$-complete. Since the signature is countable, there is an enumeration $\{\psi_n : n \in \omega\}$ of the set of all formulas of the form $\rho(\forall x\varphi)$. Suppose inductively that Σ_n has been defined to be L-consistent, with $\Sigma_n - \Sigma_0$ finite. If $\Sigma_n \vdash_{\mathrm{L}} \psi_n$, put $\Sigma_{n+1} = \Sigma_n \cup \{\psi_n\}$. Otherwise, where $\psi_n = \rho(\forall x\varphi)$ for some ρ, x and φ, put

$$\Sigma_{n+1} = \Sigma_n \cup \{\neg\rho(\mathcal{E}\tau \to \varphi(\tau/x))\}$$

for some term $\tau \in U_{\mathcal{L}}$ with $\Sigma_n \nvdash_{\mathrm{L}} \rho(\mathcal{E}\tau \to \varphi(\tau/x))$. Such a τ exists because Σ_n is $\mathcal{E}\forall$-complete by the above Lemma.

Then $\Sigma = \bigcup_{n\in\omega} \Sigma_n$ is L-consistent, and any L-maximal extension of Σ is $\mathcal{E}\forall$-complete. ⊣

To define a Kripkean canonical model for L, we first cut down the structure $\mathcal{S}_{\mathrm{L}}$ of the previous section to its $\mathcal{E}\forall$-complete members, putting

$$\mathcal{S}^K_{\mathrm{L}} = (W^K_{\mathrm{L}}, R^K_{\mathrm{L}}, Prop^K_{\mathrm{L}}, U_{\mathcal{L}}, D^K_{\mathrm{L}}),$$

where:

- $W^K_{\mathrm{L}} = \{w \in W_{\mathrm{L}} : w \text{ is } \mathcal{E}\forall\text{-complete}\}$.
- $wR^K_{\mathrm{L}}u$ iff $\{\varphi : \Box\varphi \in w\} \subseteq u$.
  (hence R^K_{L} is the restriction of the relation R_{L} to W^K_{L}).
- $|\varphi|^K_{\mathrm{L}} = |\varphi|_{\mathrm{L}} \cap W^K_{\mathrm{L}} = \{w \in W^K_{\mathrm{L}} : \varphi \in w\}$.
- $Prop^K_{\mathrm{L}} = \{\, |\varphi|^K_{\mathrm{L}} : \varphi \text{ is an } \mathcal{L}\text{-sentence}\}$.
- $D^K_{\mathrm{L}}(w) = \{\tau \in U_{\mathcal{L}} : \mathcal{E}\tau \in w\}$.

A premodel $\mathcal{M}^K_{\mathrm{L}}$ for $\mathcal{L}$ on $\mathcal{S}^K_{\mathrm{L}}$ is defined, in the usual way:

- $|\mathsf{c}|^{\mathcal{M}^K_{\mathrm{L}}} = \mathsf{c}$.
- $|F|^{\mathcal{M}^K_{\mathrm{L}}}(\tau_1, \ldots, \tau_n) = F\tau_1 \cdots \tau_n$.

- $|P|^{\mathcal{M}_L^K}(\tau_1, \ldots, \tau_n) = |P\tau_1 \cdots \tau_n|_L^K$.

Then $|\tau|^{\mathcal{M}_L^K} f = \tau^f$ for all terms τ and $f \in U_{\mathcal{L}}{}^{\omega}$.

In the structure $\mathcal{S}_L^K$ we get

$$E\tau = |\mathcal{E}\tau|_L^K \in Prop_L^K$$

for all individuals $\tau \in U_{\mathcal{L}}$, so $\mathcal{S}_L^K$ is E-admissible. $Prop_L^K$ is a Boolean set algebra in the usual way, and its closure under the operation $[R_L^K]$ follows from

LEMMA 3.2.7. $[R_L^K]|\varphi|_L^K = |\Box\varphi|_L^K \in Prop_L^K$.

PROOF. This is very like Lemma 2.7.1 (cf. also Lemma 2.6.1), but using Lemma 3.2.5 on $\mathcal{E}\forall$-completeness in place of Lemma 2.5.1 and the BF-Lemma 2.5.3 on C-completeness. We give the main details to check the points of difference.

If $w \in W_L^K$ has $w \notin |\Box\varphi|_L^K$, then $\Box\varphi \notin w$, so $\varphi \notin \Box^- w = \{\psi : \Box\psi \in w\}$, hence $\Box^- w \cup \{\neg\varphi\}$ is L-consistent.

Now since w is L-maximal, $\Box^- w$ is identical to $\Box^{-L} w = \{\psi : w \vdash_L \Box\psi\}$, which is $\mathcal{E}\forall$-complete by Lemma 3.2.5(2), because w is $\mathcal{E}\forall$-complete by definition of W_L^K.

Since $\Box^- w$ is $\mathcal{E}\forall$-complete, so too is $\Box^- w \cup \{\neg\varphi\}$ by Lemma 3.2.5(1). Therefore by Theorem 3.2.6, $\Box^- w \cup \{\neg\varphi\}$ has an L-maximal extension u that is $\mathcal{E}\forall$-complete. Then $wR_L u$ as $\Box^- w \subseteq u$, and $u \in W_L^K$, so $wR_L^K u$. But $\neg\varphi \in u$, so $u \notin |\varphi|_L^K$, showing that $w \notin [R_L^K]|\varphi|_L^K$.

The converse holds by definition of R_L^K. ⊣

Here is how the structure of $\mathcal{S}_L^K$ reflects the Kripkean semantics for the quantifier $\forall$.

LEMMA 3.2.8. *If $\forall x\varphi$ is a sentence, then in $\mathcal{S}_L^K$,*

$$|\forall x\varphi|_L^K = \bigcap_{\tau \in U_{\mathcal{L}}} E\tau \Rightarrow |\varphi(\tau/x)|_L^K = \bigsqcap_{\tau \in U_{\mathcal{L}}} E\tau \Rightarrow |\varphi(\tau/x)|_L^K.$$

Hence for any formula of the form $\forall x\varphi$, and any $f \in U_{\mathcal{L}}{}^{\omega}$,

$$|(\forall x\varphi)^f|_L^K = \bigcap_{\tau \in U_{\mathcal{L}}} E\tau \Rightarrow |\varphi^{f[\tau/x]}|_L^K = \bigsqcap_{\tau \in U_{\mathcal{L}}} E\tau \Rightarrow |\varphi^{f[\tau/x]}|_L^K.$$

PROOF. Let $w \in W_L^K$. By the Existing Instantiation axiom,

$$\text{if } \forall x\varphi \in w, \text{ then } \mathcal{E}\tau \to \varphi(\tau/x) \in w \text{ for all } \tau \in U_{\mathcal{L}}.$$

But the converse of this implication holds by $\mathcal{E}\forall$-completeness of w, using the template $\rho = \#$. This shows that

$$|\forall x\varphi|_L^K = \bigcap_{\tau \in U_{\mathcal{L}}} |\mathcal{E}\tau \to \varphi(\tau/x)|_L^K.$$

Now $|\mathcal{E}\tau \to \varphi(\tau/x)|_L^K = |\mathcal{E}\tau|_L^K \Rightarrow |\varphi(\tau/x)|_L^K = E\tau \Rightarrow |\varphi(\tau/x)|_L^K$, so we get

$$|\forall x\varphi|_L^K = \bigcap_{\tau \in U_{\mathcal{L}}} E\tau \Rightarrow |\varphi(\tau/x)|_L^K$$

as required. Then $|\forall x\varphi|_{\mathrm{L}}^{K} = \prod_{\tau\in U_{\mathcal{L}}} E\tau \Rightarrow |\varphi(\tau/x)|_{\mathrm{L}}^{K}$ follows because $|\forall x\varphi|_{\mathrm{L}}^{K}$ is admissible.

The second part of the Lemma follows from the first as usual by the equations $(\forall x\varphi)^{f} = \forall x(\varphi^{f\setminus x})$ and $\varphi^{f\setminus x}(\tau/x) = \varphi^{f[\tau/x]}$. ⊣

The last two results lead to the usual Truth Lemma for the premodel $\mathcal{M}_{\mathrm{L}}^{K}$:

THEOREM 3.2.9 (Truth Is Membership). *Let φ be any formula. Then for all $f \in U_{\mathcal{L}}{}^{\omega}$,*

$$|\varphi|^{\mathcal{M}_{\mathrm{L}}^{K}} f = |\varphi^{f}|_{\mathrm{L}}^{K},$$

and hence for all $w \in W_{\mathrm{L}}^{K}$,

$$\mathcal{M}_{\mathrm{L}}^{K}, w, f \models \varphi \quad \textit{iff} \quad \varphi^{f} \in w. \qquad \dashv$$

From this we can establish that $\mathcal{M}_{\mathrm{L}}^{K}$ is a characteristic Kripkean model for L. This proof is like that of Theorem 2.7.4 (cf. also Theorem 2.6.5), but the completeness part involves an extra proof rule for logics with $\mathcal{E}$, namely

$$\frac{\rho(\mathcal{E}\mathsf{c} \to \psi(\mathsf{c}/x))}{\rho(\forall x\psi)}, \text{ if } x \text{ is not free in } \rho, \text{ and } \mathsf{c} \text{ is not in } \rho \text{ or } \psi. \qquad (3.2.4)$$

This rule is derivable in L by composing the T∀-Intro rule with the rule

$$\frac{\rho(\mathcal{E}\mathsf{c} \to \psi(\mathsf{c}/x))}{\rho(\mathcal{E}x \to \psi)} \quad \text{if } x \text{ is not free in } \rho, \text{ and } \mathsf{c} \text{ is not in } \rho \text{ or } \psi,$$

which is itself just a special case of the rule GC.

THEOREM 3.2.10. *$\mathcal{M}_{\mathrm{L}}^{K}$ is a Kripkean model that characterises* L:

$$\vdash_{\mathrm{L}} \varphi \quad \textit{iff} \quad \mathcal{M}_{\mathrm{L}}^{K} \models \varphi.$$

PROOF. $\mathcal{M}_{\mathrm{L}}^{K}$ is a model: each truth set $|\varphi|^{\mathcal{M}_{\mathrm{L}}^{K}} f$ is equal to $|\varphi^{f}|_{\mathrm{L}}^{K} \in \mathit{Prop}_{\mathrm{L}}^{K}$, so is admissible.

$\mathcal{M}_{\mathrm{L}}^{K}$ is Kripkean: applying Theorem 3.2.9 to the second part of Lemma 3.2.8 directly gives the Kripkean condition

$$|\forall x\varphi|^{\mathcal{M}_{\mathrm{L}}^{K}} f = \bigcap_{\tau\in U_{\mathcal{L}}} E\tau \Rightarrow |\varphi|^{\mathcal{M}_{\mathrm{L}}^{K}} f[\tau/x].$$

L is sound for validity in $\mathcal{M}_{\mathrm{L}}^{K}$: we have

$$\mathcal{M}_{\mathrm{L}}^{K}, w, f \models \varphi \quad \text{iff} \quad \varphi^{f} \in w \quad \text{iff} \quad \mathcal{M}_{\mathrm{L}}, w, f \models \varphi,$$

hence if $\vdash_{\mathrm{L}} \varphi$, then φ is valid in $\mathcal{M}_{\mathrm{L}}$ so we get $\mathcal{M}_{\mathrm{L}}^{K}, w, f \models \varphi$ for every $w \in W_{\mathrm{L}}^{K}$ and $f \in U_{\mathcal{L}}{}^{\omega}$. Therefore φ is valid in $\mathcal{M}_{\mathrm{L}}^{K}$.

L is complete for validity in $\mathcal{M}_{\mathrm{L}}^{K}$: if $\nvdash_{\mathrm{L}} \varphi$, then there is some $f \in U_{\mathcal{L}}{}^{\omega}$ with $\nvdash_{\mathrm{L}} \varphi^{f}$, and hence $\{\neg\varphi^{f}\}$ is L-consistent. But $\{\neg\varphi^{f}\}$ is also $\mathcal{E}\forall$-complete in L. For if $\{\neg\varphi^{f}\} \vdash_{\mathrm{L}} \rho(\mathcal{E}\tau \to \psi(\tau/x))$ for all $\tau \in U_{\mathcal{L}}$, then choosing some constant c not in φ^{f} or ρ or ψ we have $\vdash_{\mathrm{L}} \neg\varphi^{f} \to \rho(\mathcal{E}\mathsf{c} \to \psi(\mathsf{c}/x))$, hence $\vdash_{\mathrm{L}} \neg\varphi^{f} \to \rho(\forall x\psi)$ by the rule (3.2.4) for the case of the template $\neg\varphi^{f} \to \rho$, so $\{\neg\varphi^{f}\} \vdash_{\mathrm{L}} \forall x\psi$. Hence by Theorem 3.2.6 there is a $\mathcal{E}\forall$-complete L-maximal set w extending $\{\neg\varphi^f\}$. Then $w \in W_L^K$ with $\varphi^f \notin w$, so that $\mathcal{M}_L^K, w, f \not\models \varphi$, showing that φ is not valid in $\mathcal{M}_L^K$. ⊣

We will use the name L + T∀-Intro for the smallest $\mathcal{E}$-logic that includes a given logic L and is closed under the rule T∀-Intro.

THEOREM 3.2.11. *The logic* QEK + T∀-*Intro is characterised by each of the following*:

(1) *The set of all Kripkean models on its* $\mathcal{S}^K$*-structure.*
(2) *The class of all Kripkean models.*
(3) *The class of all full model structures.*

PROOF. For any structure $\mathcal{S}$, by Theorem 3.1.5(2) and Corollary 3.2.4, the set $L_{\mathcal{S}}^K$ of formulas valid in all Kripkean models on $\mathcal{S}$ is an $\mathcal{E}$-logic that is closed under the rule T∀-Intro. Hence $L_{\mathcal{S}}^K$ includes QEK+T∀-Intro.

This accounts for the Soundness parts of (1)–(3), with (3) using the fact that every model on a full structure is Kripkean.

For the completeness parts, let L = QEK + T∀-Intro. For (1): if φ is valid in all Kripkean models on $\mathcal{S}_L^K$, then from Theorem 3.2.10, $\mathcal{M}_L^K \models \varphi$ and hence $\vdash_L \varphi$.

For (2), if φ is valid in all Kripkean models, then $\mathcal{M}_L^K \models \varphi$, hence $\vdash_L \varphi$ as in (1).

For (3), if $\nvdash_L \varphi$, then φ is not valid in the Kripkean model $\mathcal{M}_L^K$ on $\mathcal{S}_L^K$, hence by the method of Corollary 1.6.8, φ is not valid in the full structure $(\mathcal{S}_L^K)^+$. ⊣

To characterise the logic QES + T∀-Intro for an arbitrary S, we need the following result about the general frame underlying $\mathcal{S}_L^K$.

LEMMA 3.2.12. *If* L *is any* $\mathcal{E}$*-logic that extends the logic* QES + T∀-*Intro, then* S *is validated by the general frame underlying* $\mathcal{S}_L^K$.

PROOF. As for the proof of Lemma 2.7.7. ⊣

Using this result, in a similar manner to Theorem 2.6.9, we can show

THEOREM 3.2.13. *The logic* QES + T∀-*Intro is characterised by the set of all Kripkean models on its* $\mathcal{S}^K$*-structure, as well as being characterised by the class of all Kripkean models whose underlying general frame validates* S. ⊣

Previous experience tells us that we should not expect that the logic QES + T∀-Intro is characterised by Kripkean models on S-frames, or that the $\mathcal{S}^K$-structure of this logic is based on an S-frame, even when S is canonical.

EXAMPLE 3.2.14. This negative expectation is confirmed by yet again using the example of S=S4M. The sentence (1.11.4) is valid in every Kripkean model based on an S4M-frame (Lemma 1.11.2), but is not a theorem of QES4M + T∀-Intro. The latter is shown by the counter-model $\mathcal{M}$ to (1.11.4) given in Subsection 1.11.2. It has the following properties:

- $\mathcal{M}$ is based on a full structure, so is Kripkean and E-admissible.

- $\mathcal{M}$ is based on an S4-frame.
- $\mathcal{M}$ validates the McKinsey axiom, even though it is not based on an S4M-frame. The proof of that required each truth set $|\varphi|^{\mathcal{M}}f$ to be either a finite or a cofinite subset of W. The one new case to consider for this in our present language is when φ is of the form $\mathcal{E}\tau$. But $\mathcal{M}$ has one universal domain, and so $|\mathcal{E}\tau|^{\mathcal{M}}f = W \in Prop$ for every f.

Thus $\mathcal{M}$ validates QES4M + T$\forall$-Intro, showing that this logic does not have (1.11.4) as a theorem. It follows also that the $\mathcal{S}^K$-structure of the logic is not based on an S4M-frame, or else its canonical $\mathcal{M}^K$-model would be a Kripkean model on an S4M-frame, hence would validate (1.11.4), contradicting the fact that this model characterises QES4M + T$\forall$-Intro. ⊣

On the positive side, we can note again that the Kripke frame underlying $\mathcal{S}_{\mathrm{L}}^K$ is a subframe of the Kripke frame $\mathcal{F}_{\mathrm{L}}$ underlying $\mathcal{S}_{\mathrm{L}}$, albeit not an inner subframe. Moreover, $\mathcal{F}_{\mathrm{L}}$ validates S when S is canonical and L includes QES, by the same proof as for QS in Theorem 1.10.6. So we can derive, similarly to Theorem 2.8.3, a suitable characterisation of QES + T$\forall$-Intro for canonical S when S is preserved by subframes.

THEOREM 3.2.15. *Let* S *be a canonical propositional modal logic that is preserved by subframes. Then*

(1) *The logic* QES + T$\forall$*-Intro is characterised by the class of all Kripkean models whose underlying Kripke frame validates* S.
(2) *This logic is also characterised by the class of all full structures whose Kripke frame validates* S. ⊣

3.3. Necessity of (Non)Existence

The existence predicate can be used to formulate simple axioms that express the structural properties of expanding and contracting domains, namely

NE: $\mathcal{E}x \to \Box\mathcal{E}x$ *Necessity of Existence*

NNE: $\neg\mathcal{E}x \to \Box\neg\mathcal{E}x$ *Necessity of Non-Existence.*

The scheme NE is valid in any structure that has expanding domains. In fact we have

THEOREM 3.3.1. *For any structure* $\mathcal{S}$, *the following are equivalent.*

(1) $\mathcal{S}$ *has expanding domains.*
(2) NE *is valid in all premodels on* $\mathcal{S}$.
(3) NE *is valid in some premodel on* $\mathcal{S}$.

PROOF. (1) implies (2): Expanding domains is equivalent to the condition $Ea \subseteq [R]Ea$ (2.1.1), which ensures that $|\mathcal{E}x|^{\mathcal{M}}f \subseteq |\Box\mathcal{E}x|^{\mathcal{M}}f$ for any premodel $\mathcal{M}$ on $\mathcal{S}$.

(2) implies (3): Immediate from the fact that $\mathcal{S}$ does have premodels, since the signature can be interpreted arbitrarily.

(3) implies (1): Suppose there is an $\mathcal{M}$ validating NE. Given any $a \in U$, take an f with $fx = a$. Then

$$Ea = |\mathcal{E}x|^{\mathcal{M}} f \subseteq |\Box \mathcal{E}x|^{\mathcal{M}} f = [R]Ea.$$

Hence $\mathcal{S}$ has expanding domains. ⊣

If an $\mathcal{E}$-logic L includes NE, then its canonical structure $\mathcal{S}_{\mathrm{L}}$ has expanding domains. For, using NE we can infer $\vdash_{\mathrm{L}} \mathcal{E}\tau \to \Box\mathcal{E}\tau$ for every $\tau \in U_{\mathcal{L}}$ by Term Instantiation. Thus if $\tau \in D_{\mathrm{L}}(w)$ in $\mathcal{S}_{\mathrm{L}}$, i.e. $\mathcal{E}\tau \in w$, then $\Box\mathcal{E}\tau \in w$, hence every u with $wR_{\mathrm{L}}u$ has $\mathcal{E}\tau \in u$ and so $\tau \in D_{\mathrm{L}}(u)$.

These observations allow the completeness theorems of this chapter to be extended to systems with NE as follows.

- QES + NE is characterised by validity in the expanding-domains structure $\mathcal{S}_{\mathrm{QES+NE}}$, and by validity in all E-admissible model structures with expanding domains whose underlying general frame validates S.
- If S is a canonical propositional logic, then QES + NE is characterised by the class of all E-admissible model structures with expanding domains whose underlying Kripke frame validates S.
- The logic QEK + T∀-Intro + NE is characterised by each of the following:
  - The set of all Kripkean models on its $\mathcal{S}^K$-structure, which has expanding domains.
  - The class of all Kripkean models with expanding domains.
  - The class of all full model structures with expanding domains.
- For any S, QES + T∀-Intro + NE is characterised by the set of all Kripkean models on its expanding-domains $\mathcal{S}^K$-structure, as well as being characterised by the class of all Kripkean models on expanding-domains structures whose underlying general frame validates S.
- If S is a canonical propositional modal logic that is preserved by subframes, then:
  - The logic QES + T∀-Intro + NE is characterised by the class of all Kripkean models on expanding-domains structures whose underlying Kripke frame validates S.
  - This logic is also characterised by the class of all full structures with expanding domains whose Kripke frame validates S. ⊣

These characterisations would be unchanged if NE was replaced by the Converse Barcan Formula, since, as we already saw, CBF is valid in all expanding-domain structures (Theorem 2.1.1), and the inclusion of CBF in L ensures that $\mathcal{S}_{\mathrm{L}}$ has expanding domains (Corollary 2.1.4). This equivalence of NE and CBF can be readily demonstrated proof-theoretically:

THEOREM 3.3.2. *For any $\mathcal{E}$-logic* L, *the following are equivalent.*

(1) $\vdash_{\mathrm{L}}$ CBF.

(2) $\vdash_L \forall x\Box\mathcal{E}x$.
(3) $\vdash_L$ NE.

PROOF. (1) implies (2): If (1), then $\vdash_L \Box\forall x\mathcal{E}x \to \forall x\Box\mathcal{E}x$ as an instance of CBF. But $\vdash_L \forall x\mathcal{E}x$ for any L, as noted at the beginning of Section 3.1, so $\vdash_L \Box\forall x\mathcal{E}x$ by Necessitation. Hence (2) follows in this case by Modus Ponens.

(2) implies (3): We have $\vdash_L \forall x\Box\mathcal{E}x \to (\mathcal{E}x \to \Box\mathcal{E}x)$ as an instance of Existing Instantiation. So if (2) holds we get the required $\vdash_L \mathcal{E}x \to \Box\mathcal{E}x$ by Modus Ponens.

(3) implies (1): If $\vdash_L$ NE, then $\vdash_L \Box\forall x\varphi \to \forall x\Box\varphi$ by the proof sequence

1. $\mathcal{E}x \to (\forall x\varphi \to \varphi)$	EI
2. $\Box\mathcal{E}x \to (\Box\forall x\varphi \to \Box\varphi)$	1, N, K, PC
3. $\mathcal{E}x \to (\Box\forall x\varphi \to \Box\varphi)$	NE, 2, PC
4. $\Box\forall x\varphi \to \forall x\Box\varphi$	3, $\mathcal{E}\forall$-Intro. ⊣

The scheme NNE has the equivalent contrapositive form

$\Diamond\mathcal{E}x \to \mathcal{E}x$ (*possible existence implies existence*).

Similarly to Theorem 3.3.1, we can show

THEOREM 3.3.3. *For any structure $\mathcal{S}$, the following are equivalent.*

(1) *$\mathcal{S}$ has contracting domains.*
(2) NNE *is valid in all premodels on $\mathcal{S}$.*
(3) NNE *is valid in some premodel on $\mathcal{S}$.* ⊣

If a logic L includes NNE, then its canonical structure $\mathcal{S}_L$ has contracting domains. For if wR_Lu and $\mathcal{E}\tau \in u$, then $\Diamond\mathcal{E}\tau \in w$ and hence $\mathcal{E}\tau \in w$ by NNE. From this it can be shown that all of the above characterisations of logics with NE remain true if NE is replaced by NNE and "expanding domains" is replaced by "contracting domains". The characterisations also all remain true if NE is replaced by NE + NNE and "expanding domains" is replaced by "constant domains".

Now for the $\mathcal{E}$-free language we saw from the $\mathcal{S}^\flat$ construction in the previous chapter that every logic of the form QS is characterised by structures with *contracting* domains (Theorem 2.2.10). Also, QS + CBF is characterised by structures with *constant* domains, by the $\mathcal{S}^\circ$ construction (Theorem 2.3.6). These results fail to lift to the $\mathcal{E}$-logics QES. The point is that in the absence of $\mathcal{E}$ we cannot capture the contracting-domains condition proof-theoretically, and we were able to impose this condition without changing the set of valid formulas, hence without changing the logics. But in the presence of $\mathcal{E}$, if QES was characterised by contracting-domain structures, then since such structures validate NNE we would have QES $\vdash$ NNE. However that can fail for some S. Likewise, if QES + CBF was characterised by constant-domain structures, then we would have QES + CBF $\vdash$ NNE, which also can fail for some S.

EXAMPLE 3.3.4. QES4 + CBF $\nvdash$ NNE.

Let $\mathcal{M}$ be the two-world model in Example 2.3.4. This is based on an S4-frame, and has $Prop = \wp W$, so is E-admissible and is a model for the language with $\mathcal{E}$. The structure has expanding domains, so validates CBF. But it does not have have contracting domains, so $\mathcal{M}$ falsifies NNE.

Altogether this is a model that validates QES4+CBF but not NNE, showing QES4 + CBF $\nvdash$ NNE. $\dashv$

We can strengthen the result of this Example to

$$\text{QES} + \text{CBF} \nvdash \text{NNE}$$

where S is the propositional modal logic characterised by the Kripke frame of the model $\mathcal{M}$. This logic S is an extension of both S4.3 and S4M (cf. Example 2.3.4).

But we cannot strengthen the result as far as S=S5, because the logic QES5 + CBF does derive NNE, or equivalently QES5 + NE $\vdash$ NNE. In fact NE and NNE are deductively equivalent over any $\mathcal{E}$-logic that has the schemes $\Diamond\Box\mathcal{E}x \to \mathcal{E}x$ and $\mathcal{E}x \to \Box\Diamond\mathcal{E}x$, instances of the Brouwerian axiom and its dual. For, from $\mathcal{E}x \to \Box\mathcal{E}x$ we derive $\Diamond\mathcal{E}x \to \Diamond\Box\mathcal{E}x$ by general modal principles, and hence $\Diamond\mathcal{E}x \to \mathcal{E}x$ by $\Diamond\Box\mathcal{E}x \to \mathcal{E}x$. Conversely, from $\Diamond\mathcal{E}x \to \mathcal{E}x$ we get $\Box\Diamond\mathcal{E}x \to \Box\mathcal{E}x$, and hence $\mathcal{E}x \to \Box\mathcal{E}x$ by $\mathcal{E}x \to \Box\Diamond\mathcal{E}x$.

3.4. Independence of BF From NNE

Using the $\mathcal{S}^{\natural}$ construction we showed that every logic of the form

$$\text{QS} + \text{CQ} + \text{CBF} + \text{BF}$$

is characterised by *Kripkean* models with constant domains (see Theorem 2.6.9). Now CQ is derivable in $\mathcal{E}$-logics, so in asking whether this characterisation lifts to $\mathcal{E}$-logics we might ask

> Is QES + CBF + BF characterised by Kripkean models with constant domains?

If this were so, then since constant-domain models validate NNE, we would have NNE as a theorem of QES + CBF + BF. But that need not be not true:

EXAMPLE 3.4.1. QES4 + CBF + BF $\nvdash$ NNE.

Let $\mathcal{S}$ be the two-world structure underlying the model in Example 2.3.4. Recall that this has $W = \{w, u\}$, $Prop = \wp W$, $U = \{a, b\}$, $Dw = \{a\}$, $Du = \{a, b\}$, and $R = \{(w, w), (w, u), (u, u)\}$. $\mathcal{S}$ is based on an S4-frame, is full and hence E-admissible, and has expanding domains. Thus $\mathcal{S}$ validates

QES4 + CBF. As it does not have contracting domains, every model on $\mathcal{S}$ falsifies NNE (any f with $fx = b$ will falsify $\neg\mathcal{E}x \to \Box\neg\mathcal{E}x$ at w).

We use a different model on $\mathcal{S}$ to the one in Example 2.3.4, because we now want to validate BF. Define $\mathcal{M}$ on $\mathcal{S}$ by declaring $|P|^{\mathcal{M}} : U^n \to Prop$ to be the function with constant value W, for each n-ary $P \in \mathcal{L}$. This makes every atomic formula $P\tau_1 \cdots \tau_n$ valid in $\mathcal{M}$.

In this model, the individuals a and b are semantically indistinguishable at world u: it makes no difference to the truth of a formula at the "dead-end" world u which of a and b are assigned as values of its free variables. This may be intuitively evident, but we formalise it by the assertion that for each formula φ,

$$\mathcal{M}, u, f[a/x] \models \varphi \quad \text{iff} \quad \mathcal{M}, u, f[b/x] \models \varphi \tag{3.4.1}$$

for every f and x. This can be proved by induction on the formation of φ. The case $\varphi = P\tau_1 \cdots \tau_n$ holds because such a φ is valid in $\mathcal{M}$. The case $\varphi = \mathcal{E}\tau$ holds because every member of U exists at u, hence $\mathcal{E}\tau$ is unfalsifiable at u. The inductive cases for the connectives $\wedge$ and $\neg$ are straightforward. The inductive case of $\Box$ holds because in general $\mathcal{M}, u, g \models \Box\varphi$ iff $\mathcal{M}, u, g \models \varphi$, since u is R-related only to itself.

For the case of $\forall$, assume the result for φ and consider $\forall y\varphi$. First, if $y = x$, then the assertions that $u, f[a/x] \models \forall x\varphi$ and $u, f[b/x] \models \forall x\varphi$ are both equivalent to $u, f \models \forall x\varphi$, since x is not free in $\forall x\varphi$ (Lemma 1.6.1). If $y \neq x$, and $u, f[a/x] \models \forall y\varphi$, then $u, f[a/x][a/y] \models \varphi$ and $u, f[a/x][b/y] \models \varphi$. But $f[a/x][a/y] = f[a/y][a/x]$, so $u, f[a/y][a/x] \models \varphi$, and thus $u, f[a/y][b/x] \models \varphi$ by induction hypothesis on φ; hence $u, f[b/x][a/y] \models \varphi$. Similarly, from $u, f[a/x][b/y] \models \varphi$ we derive $u, f[b/x][b/y] \models \varphi$. This shows that $u, f[b/x] \models \forall y\varphi$. Interchanging a and b in this argument we get the converse that $u, f[b/x] \models \forall y\varphi$ implies $u, f[a/x] \models \forall y\varphi$, completing the proof of (3.4.1) for all formulas.

We can now prove that $\mathcal{M}$ validates the Barcan Formula. Since R is reflexive, $\mathcal{M}$ validates $\Box\varphi \to \varphi$, and hence $\forall x\Box\varphi \to \forall x\varphi$. So if $u, f \models \forall x\Box\varphi$, then $u, f \models \forall x\varphi$, hence $u, f \models \Box\forall x\varphi$ as u is R-related only to itself. Thus $\forall x\Box\varphi \to \Box\forall x\varphi$ cannot be falsified at u. But it cannot be falsified at w either, for if $w, f \models \forall x\Box\varphi$, then $w, f \models \forall x\varphi$ and also $w, f[a/x] \models \Box\varphi$. From the latter we get $u, f[a/x] \models \varphi$ as wRu, so also $u, f[b/x] \models \varphi$ by (3.4.1). Thus $u, f \models \forall x\varphi$. Since $w, f \models \forall x\varphi$ we now have $w, f \models \Box\forall x\varphi$ as required.

Altogether, $\mathcal{M}$ validates QES4 + CBF + BF but falsifies NNE, showing that NNE is not a theorem of this logic.

Again we can replace S4 here by the extension of S4.3 + M that is characterised by the Kripke frame of $\mathcal{S}$ (Example 2.3.4). ⊣

Since CBF is equivalent to NE, we can write this result as

$$\text{QES4} + \text{NE} + \text{BF} \nvdash \text{NNE}.$$

The result shows that BF is not strong enough to ensure the derivation of the contracting-domains axiom NNE over QES4 + NE. In fact for some logics S, including S4, we also have

$$\text{QES} + \text{NE} + \text{NNE} \nvdash \text{BF},$$

so BF and NNE are *mutually independent* over QES + NE.

To show this, we construct a proof-theoretic translation from the $\mathcal{E}$-logic QES + NE + NNE into QS + UI, the smallest $\mathcal{E}$-free logic including QS and the Universal Instantiation axiom $\forall x\varphi \to \varphi(\tau/x)$. To motivate this, recall from Section 2.4 that QS + UI is characterised by structures with one universal domain. Such structures have $Ea = W$ for every individual a, so they validate any existence formula $\mathcal{E}\tau$. In effect they make $\mathcal{E}\tau$ equivalent to the constant formula $\top$.

Now for each formula φ in the language with $\mathcal{E}$, let $\varphi^\top$ be the $\mathcal{E}$-free formula obtained from φ by replacing each of its atomic formulas of the form $\mathcal{E}\tau$ by $\top$.

LEMMA 3.4.2. *For any set* S *of propositional modal formulas,*

$$\text{QES} + \text{NE} + \text{NNE} \vdash \varphi \quad \textit{implies} \quad \text{QS} + \text{UI} \vdash \varphi^\top.$$

PROOF. Let $\text{L} = \{\varphi : \text{QS} + \text{UI} \vdash \varphi^\top\}$. It suffices to show that L is an $\mathcal{E}$-logic including QES and containing NE and NNE, for then L includes QES + NE + NNE, which gives the result.

We use the fact that the translation $\varphi \mapsto \varphi^\top$ commutes with the connectives and quantifiers: $(\varphi \to \psi)^\top = \varphi^\top \to \psi^\top$, $(\Box\varphi)^\top = \Box(\varphi^\top)$, $(\forall x\varphi)^\top = \forall x(\varphi^\top)$ etc. Also important is that it commutes with variable-substitution: $(\varphi(\tau/x))^\top = \varphi^\top(\tau/x)$.

Now if φ is an $\mathcal{L}$-instance of a PC-tautology, got by substituting certain $\mathcal{L}$-formulas ψ_i for propositional variables p_i, then $\varphi^\top$ is an instance of the same tautology, got by substituting $\psi_i^\top$ for p_i. Hence $\text{QS} + \text{UI} \vdash \varphi^\top$ and so $\varphi \in \text{L}$.

Similarly, if φ is an $\mathcal{L}$-instance of a member of S, then $\varphi^\top$ is an instance of the same member of S, so $\text{QS} \vdash \varphi^\top$ and hence $\varphi \in \text{L}$.

If φ is an instance of the modal axiom K, then $\varphi^\top$ is also an instance of K, so again $\text{QS} \vdash \varphi^\top$ and hence $\varphi \in \text{L}$.

If φ is the instance $\forall x\varphi \to (\mathcal{E}\tau \to \varphi(\tau/x))$ of axiom EI, then $\varphi^\top$ is $\forall x\varphi^\top \to (\top \to \varphi^\top(\tau/x))$, which is derivable in QS + UI from the instance $\forall x\varphi^\top \to \varphi^\top(\tau/x)$ of UI.

If φ is the NE-instance $\mathcal{E}x \to \Box\mathcal{E}x$, then $\varphi^\top$ is $\top \to \Box\top$, tautologically derivable in QS from the theorem $\Box\top$.

If φ is the NNE-instance $\neg\mathcal{E}x \to \Box\neg\mathcal{E}x$, then $\varphi^\top$ is the tautological $\neg\top \to \Box\neg\top$.

This completes the proof that all the axiom schemes of QES + NE + NNE are included in L. It remains to show L is closed under the inference rules of this logic.

Modus Ponens and *Necessitation*: If $\varphi, \varphi \to \psi \in$ L, then φ^{T} and $(\varphi \to \psi)^{\mathsf{T}} = \varphi^{\mathsf{T}} \to \psi^{\mathsf{T}}$ are QS + UI-theorems, hence so is ψ^{T} by MP, therefore $\psi \in$ L. Also $(\Box\varphi)^{\mathsf{T}} = \Box(\varphi^{\mathsf{T}})$ is an QS + UI-theorem by rule N, so $\Box\varphi \in$ L.

$\mathcal{E}\forall$-Introduction: Suppose $\varphi \to (\mathcal{E}x \to \psi)$ belongs to L, with x is not free in φ. Then QS + UI $\vdash \varphi^{\mathsf{T}} \to (\top \to \psi^{\mathsf{T}})$, hence QS + UI $\vdash \varphi^{\mathsf{T}} \to \psi^{\mathsf{T}}$. Therefore QS + UI $\vdash \varphi^{\mathsf{T}} \to \forall x \psi^{\mathsf{T}}$ by the rule $\forall$-Intro (Lemma 1.2.1(1)), as x is not free in φ^{T}. But $\varphi^{\mathsf{T}} \to \forall x \psi^{\mathsf{T}} = (\varphi \to \forall x \psi)^{\mathsf{T}}$, so $\varphi \to \forall x \psi$ belongs to L as required.

Term Instantiation: Suppose $\varphi \in$ L and τ is free for x in φ. Then in QS+UI, from φ^{T} we infer $\varphi^{\mathsf{T}}(\tau/x)$, i.e. $(\varphi(\tau/x))^{\mathsf{T}}$, using TI, as τ is still free for x in φ^{T}. Hence $\varphi(\tau/x) \in$ L.

Generalisation on Constants: Suppose $\varphi(\mathsf{c}/x) \in$ L and c is not in φ. Then in QS + UI, from $(\varphi(\mathsf{c}/x))^{\mathsf{T}}$, i.e. $\varphi^{\mathsf{T}}(\mathsf{c}/x)$, we infer φ^{T} by GC, as c is not in φ^{T}. Hence $\varphi \in$ L. ⊣

THEOREM 3.4.3. (1) QES + NE + NNE $\vdash$ BF *implies* QS + UI $\vdash$ BF.
(2) QES4 + NE + NNE $\nvdash$ BF.

PROOF. (1): By "QS+UI $\vdash$ BF" we mean that QS+UI includes the scheme BF *in the $\mathcal{E}$-free language*. Now if QS + UI $\nvdash$ BF, then QS + UI $\nvdash \varphi$ for some $\mathcal{E}$-free instance φ of BF. But $\varphi^{\mathsf{T}} = \varphi$, so QES + NE + NNE $\nvdash \varphi$ by the Lemma.

(2): from (1), as QS4 + UI $\nvdash$ BF was shown in Example 2.4.4. ⊣

For another example in which QES + NE + NNE $\nvdash$ BF we could take S to be the extension of S4.3 + M characterised by the Kripke frame of the model in Example 2.3.4, since in that case we have QS + UI $\nvdash$ BF, as explained in Remark 2.4.5.

3.5. What is the Role of the Barcan Formula?

By the "role" of this eponymous axiom we mean its metalogical relationships: the semantic conditions that characterise its validity, and the consequences of adopting it as an axiom. In this section we give some proof-theoretic results that further clarify the role of BF, at least for logics with an existence predicate.

It appears that to give a generic argument for the validity of BF in a model we need to assume that the model (i) gives $\forall$ the Kripkean interpretation and (ii) has contracting domains (Lemma 2.2.7). But these conditions, while sufficient, are not necessary for validity of BF. In fact it is possible for an $\mathcal{E}$-logic L that includes BF to have a canonical model $\mathcal{M}_{\mathrm{L}}$ that is neither Kripkean nor has contracting domains, yet validates BF since it validates all L-theorems:

EXAMPLE 3.5.1. Let L = QES4M + CBF + BF, so L includes BF by definition.[28] Now in Example 3.4.1 we showed that QES+CBF+BF $\nvdash$ NNE, where S is the logic characterised by the two-world Kripke frame of that example. S includes S4M, so QES + CBF + BF includes L, and hence L $\nvdash$ NNE.

It follows that $\mathcal{M}_{\mathrm{L}}$ cannot have contracting domains, or else it would validate NNE, contrary to the fact that it falsifies all non-theorems of L.

To show that $\mathcal{M}_{\mathrm{L}}$ is non-Kripkean, note first that $\mathcal{M}_{\mathrm{L}}$ is based on a Kripke frame validating the canonical logic S4M (see Theorem 1.10.6). So if $\mathcal{M}_{\mathrm{L}}$ were Kripkean, it would validate the sentence (1.11.4) by Lemma 1.11.2, and hence this sentence would be an L-theorem. However (1.11.4) is *not* an L-theorem, because the the counter-model $\mathcal{M}$ to (1.11.4) given in Subsection 1.11.2 validates L. We already explained in Example 3.2.14 that this model validates QES4M. The model has one universal domain and so also validates CBF and BF, as noted in Subsection 1.11.2. ⊣

The role played by the Barcan Formula in our completeness theorems was performed in the BF-Lemma 2.5.3, where it was used to obtain certain C-complete sets that are needed to construct a canonical model that is Kripkean. But the deductive principle that relates most closely to the Kripkean interpretation is the inference rule T∀-Intro. The logic of all Kripkean models is axiomatisable as QEK + T∀-Intro (Theorem 3.2.11), while in general QES + T∀-Intro is the logic of all Kripkean models on S-structures, i.e. models structures whose general frame validates S (Theorem 3.2.13). Similarly, the deductive principle that relates most closely to the contracting-domains condition is the axiom NNE: the logic of all contracting-domains S-structures is QES + NNE. These results are additive: the logic of all Kripkean models on contracting-domains S-structures is QES + T∀-Intro + NNE.

All this suggests that to understand the function of BF better we should compare it to the combination of T∀-Intro and NNE. Our first observation is that this combination does have the deductive strength to derive BF. Over logics of the form QES, that can be seen to follow from the semantic characterisation just mentioned. Kripkean models on contracting-domains S-structures all validate BF, so

$$\text{QES} + \text{T}\forall\text{-Intro} + \text{NNE} \vdash \text{BF}.$$

In fact this relationship holds over any $\mathcal{E}$-logic, and to prove it we do not need the full strength of T∀-Intro, but only the rule

$$\Box\forall\text{-Intro:}\quad \frac{\varphi \to \Box(\mathcal{E}x \to \psi)}{\varphi \to \Box\forall x\psi},\quad \text{if } x \text{ is not free in } \varphi.$$

This is the instance of T∀-Intro given by the template $\rho = \varphi \to \Box\#$.

[28] CBF (or equivalently NE) is not needed in this example, but the stronger we make the L the more effective the example is.

THEOREM 3.5.2. *For any* $\mathcal{E}$*-logic* L,

$$\mathrm{L} + \Box\forall\text{-Intro} + \mathrm{NNE} \vdash \mathrm{BF}.$$

Hence

$$\mathrm{L} + \mathrm{T}\forall\text{-Intro} + \mathrm{NNE} \vdash \mathrm{BF}.$$

PROOF. By the derivation

1. $\forall x\Box\varphi \to (\mathcal{E}x \to \Box\varphi)$	EI
2. $\Box\varphi \to \Box(\mathcal{E}x \to \varphi)$	tautology, N, K
3. $\forall x\Box\varphi \to (\mathcal{E}x \to \Box(\mathcal{E}x \to \varphi))$	1, 2, PC
4. $\Box\neg\mathcal{E}x \to \Box(\mathcal{E}x \to \varphi)$	tautology, N, K
5. $\neg\mathcal{E}x \to \Box(\mathcal{E}x \to \varphi)$	NNE, 4, PC
6. $\forall x\Box\varphi \to \Box(\mathcal{E}x \to \varphi)$	3, 5, PC
7. $\forall x\Box\varphi \to \Box\forall x\varphi$	6, $\Box\forall$-Intro. ⊣

The proof just given shows that the result does not even require the full generality of $\Box\forall$-Intro, but only its one instance

$$\frac{\forall x\Box\varphi \to \Box(\mathcal{E}x \to \varphi)}{\forall x\Box\varphi \to \Box\forall x\varphi}.$$

Now in the opposite direction, starting with BF, we already know that BF does not have the deductive strength to derive NNE over logics in general. For instance, we saw in Example 3.4.1 that

$$\mathrm{QES4} + \mathrm{CBF} + \mathrm{BF} \nvdash \mathrm{NNE}.$$

On the other hand, BF does have the strength to derive T$\forall$-Intro. To prove that we need a preliminary fact:

LEMMA 3.5.3. *If an* $\mathcal{E}$*-logic* L *includes the Barcan Formula, then for any template* ρ *in which* x *is not free,*

$$\vdash_{\mathrm{L}} \forall x\rho(\mathcal{E}x \to \psi) \to \rho(\forall x\psi).$$

PROOF. By induction on the formation of ρ.

For the base case $\rho = \#$, from the UD axiom

$$\forall x(\mathcal{E}x \to \psi) \to (\forall x\mathcal{E}x \to \forall x\psi)$$

and the L-theorem $\forall x\mathcal{E}x$ we obtain

$$\vdash_{\mathrm{L}} \forall x(\mathcal{E}x \to \psi) \to \forall x\psi$$

by PC, which is the required result in this case.

Now assume the result for a template ρ. Suppose x is not free in $\varphi \to \rho$. Then x is not free in φ, so $\vdash_{\mathrm{L}} \varphi \to \forall x\varphi$ by VQ. Together with the UD axiom

$$\forall x(\varphi \to \rho(\mathcal{E}x \to \psi)) \to (\forall x\varphi \to \forall x\rho(\mathcal{E}x \to \psi)),$$

this yields

$$\vdash_{\mathrm{L}} \forall x(\varphi \to \rho(\mathcal{E}x \to \psi)) \to (\varphi \to \forall x\rho(\mathcal{E}x \to \psi)),$$

by PC. From this, using the induction hypothesis on ρ and PC, we get

$$\vdash_{\mathrm{L}} \forall x(\varphi \to \rho(\mathcal{E}x \to \psi)) \to (\varphi \to \rho(\forall x\psi)),$$

showing that the result holds for $\varphi \to \rho$.

Finally we show the result holds for $\Box\rho$, which is where we need the Barcan Formula. By modal principles applied to the induction hypothesis on ρ,

$$\vdash_{\mathrm{L}} \Box\forall x\rho(\mathcal{E}x \to \psi) \to \Box\rho(\forall x\psi).$$

But by BF

$$\vdash_{\mathrm{L}} \forall x\Box\rho(\mathcal{E}x \to \psi) \to \Box\forall x\rho(\mathcal{E}x \to \psi),$$

hence by PC

$$\vdash_{\mathrm{L}} \Box\forall x\rho(\mathcal{E}x \to \psi) \to \Box\rho(\forall x\psi),$$

giving the result for $\Box\rho$. ⊣

We now extend the use of the symbol $\vdash$ by writing

$$\mathrm{L} \vdash \mathrm{T}\forall\text{-Intro}$$

to mean that the logic L is closed under the rule T∀-Intro.

THEOREM 3.5.4. *For any $\mathcal{E}$-logic* L,

$$\mathrm{L} + \mathrm{BF} \vdash \mathrm{T}\forall\text{-Intro}.$$

PROOF. Suppose that $\mathrm{L} + \mathrm{BF} \vdash \rho(\mathcal{E}x \to \psi)$ with x not free in ρ. Then by UG, $\mathrm{L} + \mathrm{BF} \vdash \forall x\rho(\mathcal{E}x \to \psi)$. Hence $\mathrm{L} + \mathrm{BF} \vdash \rho(\forall x\psi)$ by the Lemma just proved and MP, since of course $\mathrm{L} + \mathrm{BF}$ includes BF. ⊣

Combining this result with Theorem 3.5.2, we infer, for any $\mathcal{E}$-logic L, that

$$\mathrm{L} + \mathrm{BF} + \mathrm{NNE} = \mathrm{L} + \mathrm{T}\forall\text{-Intro} + \mathrm{NNE}.$$

In particular,

$$\mathrm{QES} + \mathrm{BF} + \mathrm{NNE} = \mathrm{QES} + \mathrm{T}\forall\text{-Intro} + \mathrm{NNE},$$

so the logic of all Kripkean models on contracting-domains S-structures is more simply axiomatised as QES + BF + NNE.

Moreover, for logics L that already include NNE we have

$$\mathrm{L} + \mathrm{BF} = \mathrm{L} + \mathrm{T}\forall\text{-Intro}.$$

To sum up: we have seen that for logics that include the contracting-domains axiom NNE, the Barcan Formula may remain independent, but it is always equivalent to the Kripkean-models rule T∀-Intro. In the context of contracting-domains models, the Barcan formula is characterised by the standard Kripkean interpretation of the quantifier ∀, and is sufficient to ensure that this interpretation holds in canonical models.

But our quest to understand BF is not over. We have not axiomatised the $\mathcal{E}$-free logic of all Kripkean-models with contracting domains. Is this QK + CQ + BF? In the absence of a solution to that, we can ask if there is *some* natural semantic characterisation of QK+CQ+BF, and of QS+CQ+BF more generally? And what of QS + BF and QES + BF? Answers to these questions will be given at the end of the next chapter.

Chapter 4

PROPOSITIONAL FUNCTIONS AND PREDICATE SUBSTITUTION

Substitution for *predicate letters* is a syntactically intricate inference procedure that is derivable in standard systems of non-modal quantificational logic. It allows a new theorem to be deduced from a given theorem φ by replacing an atomic formula $P\tau_1 \cdots \tau_n$ within φ by another formula ψ, of arbitrary complexity, involving the terms $\tau_1, \ldots, \tau_n$. The intricacy comes in stating the restrictions that must be placed on free variables of φ and ψ for the substitution to be allowed.

Church [1956, p. 289] gives some historical notes on this rule, pointing out that it was inadequately stated in early works of Hilbert and Ackerman, Carnap, and Quine; and first correctly stated, but not in full generality, in Hilbert and Bernays's *Grundlagen der Mathematik* in 1934. Church himself calls the predicate letter P a *functional variable*, viewing it as a variable whose values are *propositional functions* of individuals. In non-modal logic, this means that P is interpreted as an n-ary function $U^n \to \{\mathsf{truth}, \mathsf{falsehood}\}$ from individuals to truth-values.

Intuitively, a logic that is closed under substitution for predicate letters is one whose theorems represent "universal laws", expressing properties that hold of all predicates, i.e. hold no matter what interpretation is given to the predicate letters, hence hold no matter what formulas are (correctly) substituted for them.

Now in the modal setting we have identified propositions with sets of worlds, rather than with the binary truth-values, and have interpreted formulas as propositional functions of the type $|\varphi|^{\mathcal{M}} : U^\omega \to \mathit{Prop}$. But just as it is questionable whether an arbitrarily chosen set of worlds need be admissible as a proposition, so too we might question whether an arbitrary function $U^\omega \to \mathit{Prop}$ should count as the value of a formula. Thus Russell, who viewed a propositional function as being given by an expression containing a variable, such as "x is human", felt that "a correlation of entities to propositions which is wholely arbitrary is unsatisfactory" [Russell 1975, p. 92]. In relation to the assertion 'fx is true for all values of x', his objection was that

> "before we can know what 'fx' means, we have to know 'fa' and 'fb' and 'fc' and so on, throughout the whole universe. General propositions thus lose their *raison d'être* since what they assert can only be set forth by enumeration of all the separate cases."

In this chapter we extend the notion of a model structure by adding a set *PropFun* of *admissible propositional functions* of the type $U^\omega \to \mathit{Prop}$. This produces the notion of a *functional model structure*. *PropFun* will be required to have appropriate closure properties, corresponding to the connectives and quantifiers, that ensure that in any model $\mathcal{M}$, the functions $|\varphi|^{\mathcal{M}}$ are all admissible.

In quantified modal logic, systems of the form QS or QES are closed under substitution for predicate letters. We will show in this chapter that this rule is sound for validity in any functional model structure, but not always sound for validity in a single model. The situation is parallel to that in propositional modal logic, where the rule of substitution for propositional variables is sound for validity in a frame, but not always for validity in a model.

We will construct for each logic L a canonical functional model structure, and show that this structure characterises L if, and only if, L is closed under the rule of substitution for predicate letters.

4.1. Functional Model Structures

In a model structure $(W, R, \mathit{Prop}, U, D)$, we will now use the symbols α, β, α' etc. for functions of the type $U^\omega \to \wp W$. We call these *propositional functions*. Corresponding to the connectives $\wedge$, $\neg$, $\Box$ there are operations $\alpha \cap \beta$, $-\alpha$, $[R]\alpha$ on such functions, defined by "point-wise" lifting of the corresponding operations on *Prop*. For $f \in U^\omega$ these have:

$$(\alpha \cap \beta)f = \alpha f \cap \beta f.$$
$$(-\alpha)f = -(\alpha f).$$
$$([R]\alpha)f = [R](\alpha f).$$

We can also lift the inclusion relation $\subseteq$ between subsets of W to a partial ordering $\alpha \subseteq \beta$ on propositional functions, by putting

$$\alpha \subseteq \beta \quad \text{iff} \quad \alpha \cap \beta = \alpha,$$

or equivalently,

$$\alpha \subseteq \beta \quad \text{iff} \quad \text{for all } f \in U^\omega,\ \alpha f \subseteq \beta f. \qquad (4.1.1)$$

Under these definitions, the set of propositional functions becomes a Boolean algebra with modal operator $[R]$. The Boolean join $\alpha \cup \beta$ is the pointwise lifting of set union, i.e. $(\alpha \cup \beta)f = \alpha f \cup \beta f$, and the Boolean implication

$\alpha \Rightarrow \beta = -\alpha \cup \beta$ is the pointwise lifting of the Boolean set implication, i.e. $(\alpha \Rightarrow \beta)f = \alpha f \Rightarrow \beta f$.

Corresponding to the modality $\Diamond$ is the operation $\alpha \mapsto \langle R\rangle\alpha = -[R] - \alpha$. This has $(\langle R\rangle\alpha)f = \langle R\rangle(\alpha f)$.

Corresponding to the primitive sentence F is the function $\alpha_\emptyset : U^\omega \to \wp W$, with constant value $\alpha_\emptyset f = \emptyset$. This is the least propositional function under the partial ordering $\subseteq$. The $\subseteq$-greatest function is α_W, defined by $\alpha_W f = W$ for all $f \in U^\omega$. It is equal to $-\alpha_\emptyset$.

Corresponding to $\forall$ we define for each individual variable x an operation $\forall_x$ that assigns to each $\alpha : U^\omega \to \wp W$ the function $\forall_x\alpha : U^\omega \to \wp W$ defined by

$$(\forall_x\alpha)f = \bigsqcap_{a\in U} \big(Ea \Rightarrow \alpha f[a/x]\big).$$

We can also define $\exists_x\alpha$ to be $-\forall_x - \alpha$, and show that

$$(\exists_x\alpha)f = \bigsqcup_{a\in U} \big(Ea \cap \alpha f[a/x]\big).$$

The updating operation $f \mapsto f[a/x]$ on U^ω is lifted to propositional functions, defining the updating $\alpha[a/x] : U^\omega \to \wp W$ of $\alpha : U^\omega \to \wp W$ by putting

$$\alpha[a/x]f = \alpha(f[a/x])$$

for all $f \in U^\omega$.

A *functional model structure* is a system

$$\mathfrak{S} = (W, R, Prop, U, D, PropFun)$$

such that:

- $(W, R, Prop, U, D)$ is a model structure.
- *PropFun* is a set of functions from U^ω to *Prop*, called the *admissible propositional functions* of $\mathfrak{S}$.
- *PropFun* contains $\alpha_\emptyset$ and is closed under the operations $\cap$, $-$, $[R]$ and $\forall_x$ for all $x \in \mathsf{InVar}$, defined point-wise on U^ω as above.
- $\mathfrak{S}$ is *updatable*, meaning that admissibility of its propositional functions is preserved by updatings, i.e.

  $\alpha \in PropFun$ implies $\alpha[a/x] \in PropFun$ for all $x \in \mathsf{InVar}$ and $a \in U$.

We will assume that our language contains the existence predicate $\mathcal{E}$. For a given signature $\mathcal{L}$, a *premodel* $\mathcal{M} = (\mathfrak{S}, |{-}|^{\mathcal{M}})$ for $\mathcal{L}$ based on $\mathfrak{S}$ is given by an interpretation function $|{-}|^{\mathcal{M}}$ specified exactly as before (at the beginning of Section 1.6). Then $\mathcal{M}$ assigns a function $|\varphi|^{\mathcal{M}} : U^\omega \to \wp W$ to each $\mathcal{L}$-formula φ, exactly as before, including formulas containing $\mathcal{E}$. It is evident from the inductive definition of $|\varphi|^{\mathcal{M}}$ and the definitions of the new operations on propositional functions that:

$$
\begin{aligned}
|\mathsf{F}|^{\mathcal{M}} &= \alpha_{\emptyset}. \\
|\varphi \wedge \psi|^{\mathcal{M}} &= |\varphi|^{\mathcal{M}} \cap |\psi|^{\mathcal{M}}. \\
|\neg\varphi|^{\mathcal{M}} &= -|\varphi|^{\mathcal{M}}. \\
|\Box\varphi|^{\mathcal{M}} &= [R]|\varphi|^{\mathcal{M}}. \\
|\forall x\varphi|^{\mathcal{M}} &= \forall_x|\varphi|^{\mathcal{M}}. \\
|\varphi \vee \psi|^{\mathcal{M}} &= |\varphi|^{\mathcal{M}} \cup |\psi|^{\mathcal{M}}. \\
|\varphi \to \psi|^{\mathcal{M}} &= |\varphi|^{\mathcal{M}} \Rightarrow |\psi|^{\mathcal{M}}. \\
|\Diamond\varphi|^{\mathcal{M}} &= \langle R\rangle|\varphi|^{\mathcal{M}}. \\
|\exists x\varphi|^{\mathcal{M}} &= \exists_x|\varphi|^{\mathcal{M}}.
\end{aligned}
\tag{4.1.2}
$$

A premodel $\mathcal{M}$ is called a *model* if

$$|\varphi|^{\mathcal{M}} \in \mathit{PropFun} \text{ for all } \textit{atomic} \text{ formulas } \varphi.$$

THEOREM 4.1.1. *In any model $\mathcal{M}$ on a functional model structure,*

(1) $|\varphi|^{\mathcal{M}} \in \mathit{PropFun}$ *for every formula* φ.
(2) $|\varphi|^{\mathcal{M}} f \in \mathit{Prop}$ *for every formula* φ *and every* $f \in U^{\omega}$.
(3) $\mathfrak{S}$ *is E-admissible*: *every existence set Ea belongs to Prop.*

PROOF. For (1) we show $|\varphi|^{\mathcal{M}} \in \mathit{PropFun}$ by induction on the formation of φ. The case of an atomic φ, other than F, holds by definition of $\mathcal{M}$ as a model. The case that φ is F holds because $\alpha_{\emptyset} \in \mathit{PropFun}$. The inductive cases all follow from the equations just given above, together with the fact that *PropFun* is closed under the relevant operation appearing on the right of each equation.

(2) is immediate from (1), since every member of *PropFun* is a function whose range is included in *Prop*.

For (3), as in the introduction to Chapter 3, picking an f and an x with $fx = a$ gives $Ea = |\mathcal{E}x|^{\mathcal{M}} f$. But $|\mathcal{E}x|^{\mathcal{M}} f \in \mathit{Prop}$ by (2). ⊣

The import of this Theorem is that in order to have every formula define an admissible propositional function it is sufficient to ensure that every *atomic* formula does so. In addition, part (2) of the Theorem states that every truth set of a model is an admissible proposition. So a model on $\mathfrak{S}$ is a model, in our earlier sense, on the model structure underlying $\mathfrak{S}$ (i.e. the structure got by deleting *PropFun*). Hence any soundness result holding for validity in a model on a model structure holds for validity in a model, as we have now defined it, on a functional model structure.

We say that a formula φ is *valid in* $\mathfrak{S}$ if it is valid in every model on $\mathfrak{S}$.

THEOREM 4.1.2. *For any functional model structure $\mathfrak{S}$, the set $\mathrm{L}_{\mathfrak{S}}$ of all formulas valid in $\mathfrak{S}$ is an $\mathcal{E}$-logic.*

PROOF. By the soundness results of Section 1.7 as well as Section 3.1, all of the axioms defining an $\mathcal{E}$-logic are valid in all models on model structures, hence valid in all models on functional model structures, and in particular valid in all the models on $\mathfrak{S}$. Also, all of the rules defining an $\mathcal{E}$-logic, except for Generalisation on Constants, are sound for validity in any model, hence sound for validity in $\mathfrak{S}$.

It remains to prove that $L_{\mathfrak{S}}$ is closed under GC, the only rule that is not sound for validity in a single model. For this we use the soundness proof for GC given in Theorem 1.7.7. That proof will go through unchanged here, provided we can show that if $\mathcal{M}$ is any model on $\mathfrak{S}$, then for all $b \in U$, $\mathcal{M}[b]$ is a model on $\mathfrak{S}$, where $\mathcal{M}[b]$ is the premodel differing from $\mathcal{M}$ only in having $|\mathsf{c}|^{\mathcal{M}[b]} = b$.

So we have to show that $|\varphi|^{\mathcal{M}[b]} \in \mathit{PropFun}$ for any atomic formula φ . This is where we need the updatability of $\mathfrak{S}$. Take a variable x not in φ. By Lemma 1.7.5, for all $g \in U^\omega$,

$$|\varphi|^{\mathcal{M}[b]}g = |\varphi(x/\mathsf{c})|^{\mathcal{M}}g[b/x].$$

Now if α is the function $|\varphi(x/\mathsf{c})|^{\mathcal{M}}$, then $\alpha \in \mathit{PropFun}$ by Theorem 4.1.1 as $\mathcal{M}$ is a model, and so $\alpha[b/x] \in \mathit{PropFun}$ as $\mathfrak{S}$ is updatable. But for any $g \in U^\omega$,

$$\alpha[b/x]g = \alpha(g[b/x]) = |\varphi|^{\mathcal{M}[b]}g.$$

This shows that $|\varphi|^{\mathcal{M}[b]} = \alpha[b/x]$, so $|\varphi|^{\mathcal{M}[b]}$ is admissible as required. (In fact the argument directly shows that $|\varphi|^{\mathcal{M}[b]} \in \mathit{PropFun}$ for all formulas φ, not just the atomic ones.) ⊣

Constraints on the notion of functional model structure are illustrated by the fact that we cannot always turn a model structure into a functional one just by defining *PropFun* to be the collection of all functions of the form $U^\omega \to \mathit{Prop}$. The reason is that this collection need not be closed under $\forall_x$. The function $\forall_x\alpha$ may not take all its values in *Prop*, even when α does.

EXAMPLE 4.1.3. Take the structure introduced in Example 1.5.2 and studied further in Example 1.6.6. This has $W = U = \omega$; $E0 = \emptyset$ and $En = \{n\}$ for $n > 0$; with *Prop* being the Boolean algebra consisting of all the finite subsets of $\{n : n > 0\}$ and their complements in W.

For a given variable x, define $\alpha_x : U^\omega \to \mathit{Prop}$ by

$$\alpha_x f = \begin{cases} \{fx+1\} & \text{if } fx \text{ is odd,} \\ W & \text{otherwise.} \end{cases}$$

We have

$$\forall_x\alpha_x f = \bigsqcap_{n\in\omega} En \Rightarrow \alpha_x(f[n/x]),$$

and since $f[n/x]x = n$,

$$\alpha_x(f[n/x]) = \begin{cases} \{n+1\} & \text{if } n \text{ is odd,} \\ W & \text{otherwise.} \end{cases}$$

Using the facts that $\{n\} \Rightarrow \{n+1\} = -\{n\}$ and $En \Rightarrow W = W$, this yields

$$\forall_x \alpha_x f = \bigsqcap_{n \text{ odd}} -\{n\}.$$

But $\bigcap_{n \text{ odd}} -\{n\}$ is the non-admissible set of all even natural numbers, and so $\bigsqcap_{n \text{ odd}} -\{n\}$ is the non-admissible set $\{2n : n > 0\}$ of all positive even numbers. Thus for all $f \in U^\omega$, $\forall_x \alpha_x f \notin Prop$. ⊣

Now fix an $\mathcal{E}$-logic L. We are going to construct a canonical functional model structure $\mathfrak{S}_{\mathrm{L}}$ for L, refining the model structure $\mathcal{S}_{\mathrm{L}}$. Recalling that $U_{\mathcal{L}}$ is the set of all closed $\mathcal{L}$-terms, and W_{L} is the set of all L-maximal sets of $\mathcal{L}$-formulas, we associate with each formula φ the function $\alpha_\varphi : U_{\mathcal{L}}{}^\omega \to \wp W_{\mathrm{L}}$ defined by

$$\alpha_\varphi f = |\varphi^f|_{\mathrm{L}} = \{w \in W_{\mathrm{L}} : \varphi^f \in w\}.$$

Since φ^f is a sentence, $|\varphi^f|_{\mathrm{L}}$ belongs to the set $Prop_{\mathrm{L}}$ of admissible propositions of $\mathcal{S}_{\mathrm{L}}$, so we have $\alpha_\varphi : U_{\mathcal{L}}{}^\omega \to Prop_{\mathrm{L}}$.

LEMMA 4.1.4. (1) $\alpha_\emptyset = \alpha_{\mathsf{F}}$.
(2) $\alpha_\varphi \cap \alpha_\psi = \alpha_{\varphi \wedge \psi}$.
(3) $-\alpha_\varphi = \alpha_{\neg\varphi}$.
(4) $[R]\alpha_\varphi = \alpha_{\Box\varphi}$.
(5) $\forall_x \alpha_\varphi = \alpha_{\forall x \varphi}$.
(6) $\alpha_\varphi[\tau/x] = \alpha_{\varphi(\tau/x)}$ *for all* $\tau \in U_{\mathcal{L}}$.

PROOF. (1) holds because $|\mathsf{F}^f|_{\mathrm{L}} = |\mathsf{F}|_{\mathrm{L}} = \emptyset$ for all $f \in U_{\mathcal{L}}{}^\omega$.

(2) asserts that for all f, $|\varphi^f|_{\mathrm{L}} \cap |\psi^f|_{\mathrm{L}} = |(\varphi \wedge \psi)^f|_{\mathrm{L}}$. This holds because $(\varphi \wedge \psi)^f = \varphi^f \wedge \psi^f$.

(3) and (4) hold similarly, as $-|\varphi^f|_{\mathrm{L}} = |\neg\varphi^f|_{\mathrm{L}}$ and $[R]|\varphi^f|_{\mathrm{L}} = |\Box\varphi^f|_{\mathrm{L}}$.

For (5), since

$$\forall_x \alpha_\varphi f = \bigsqcap_{\tau \in U_{\mathcal{L}}} (E\tau \Rightarrow \alpha_\varphi f[\tau/x])$$

and $\alpha_\varphi f[\tau/x] = |\varphi^{f[\tau/x]}|_{\mathrm{L}}$, equation (5) asserts that for all $f \in U_{\mathcal{L}}{}^\omega$,

$$\bigsqcap_{\tau \in U_{\mathcal{L}}} (E\tau \Rightarrow |\varphi^{f[\tau/x]}|_{\mathrm{L}}) = |(\forall x \varphi)^f|_{\mathrm{L}},$$

which was proved in Lemma 1.9.3(5).

In (6), since τ is closed, for all $f \in U_{\mathcal{L}}{}^\omega$ we have $\varphi^{f[\tau/x]} = \varphi(\tau/x)^f$ as in Lemma 1.9.3(3). Hence

$$(\alpha_\varphi[\tau/x])f = \alpha_\varphi(f[\tau/x]) = |\varphi^{f[\tau/x]}|_{\mathrm{L}} = |\varphi(\tau/x)^f|_{\mathrm{L}} = \alpha_{\varphi(\tau/x)} f.$$

This shows that the updating $\alpha_\varphi[\tau/x]$ is just the function $\alpha_{\varphi(\tau/x)}$. ⊣

COROLLARY 4.1.5. (1) $\alpha_\varphi \cup \alpha_\psi = \alpha_{\varphi\vee\psi}$.
(2) $\alpha_\varphi \Rightarrow \alpha_\psi = \alpha_{\varphi\to\psi}$.
(3) $\langle R\rangle\alpha_\varphi = \alpha_{\Diamond\varphi}$.
(4) $\exists_x\alpha_\varphi = \alpha_{\exists x\varphi}$.
(5) $\alpha_\varphi \subseteq \alpha_\psi$ *iff* $\vdash_{\mathrm{L}} \varphi \to \psi$.

PROOF. (1)–(4) are straightforward. For (5), we reason that

$\alpha_\varphi \subseteq \alpha_\psi$ iff $\alpha_\varphi \Rightarrow \alpha_\psi = \alpha_{W_{\mathrm{L}}}$ by Boolean algebra,
iff for all $f \in U_{\mathcal{L}}{}^\omega$, $\alpha_{\varphi\to\psi} f = W_{\mathrm{L}}$, by part (2),
iff for all $f \in U_{\mathcal{L}}{}^\omega$, $|(\varphi \to \psi)^f|_{\mathrm{L}} = W_{\mathrm{L}}$,
iff for all $f \in U_{\mathcal{L}}{}^\omega$, $\vdash_{\mathrm{L}} (\varphi \to \psi)^f$, by properties of L-maximal sets.

But $\vdash_{\mathrm{L}} \varphi \to \psi$ iff for all $f \in U_{\mathcal{L}}{}^\omega$, $\vdash_{\mathrm{L}} (\varphi \to \psi)^f$, by the argument of the proof of Theorem 1.9.6. Indeed, if $\vdash_{\mathrm{L}} \varphi \to \psi$, then for all $f \in U_{\mathcal{L}}{}^\omega$, $\vdash_{\mathrm{L}} (\varphi \to \psi)^f$ by the rule TI* of Lemma 1.2.3; while if $\nvdash_{\mathrm{L}} \varphi \to \psi$, then $\nvdash_{\mathrm{L}} (\varphi \to \psi)^f$ where f is any member of $U_{\mathcal{L}}{}^\omega$ that assigns a fresh constant to each free variable of $\varphi \to \psi$, by the rule GC*. ⊣

We can now define the structure

$$\mathfrak{S}_{\mathrm{L}} = (W_{\mathrm{L}}, R_{\mathrm{L}}, Prop_{\mathrm{L}}, U_{\mathcal{L}}, D_{\mathrm{L}}, PropFun_{\mathrm{L}}),$$

where $(W_{\mathrm{L}}, R_{\mathrm{L}}, Prop_{\mathrm{L}}, U_{\mathcal{L}}, D_{\mathrm{L}})$ is the canonical model structure $\mathcal{S}_{\mathrm{L}}$ as in Section 1.9 (see also Section 3.1), and

$$PropFun_{\mathrm{L}} = \{\alpha_\varphi : \varphi \text{ is any } \mathcal{L}\text{-formula}\}.$$

By equations (1)–(5) of Lemma 4.1.4, $PropFun_{\mathrm{L}}$contains $\alpha_\emptyset$ and is closed under the operations $\cap$, $-$, $[R]$ and $\forall_x$ for all $x \in \mathsf{InVar}$. By equation (6), $\alpha \in PropFun_{\mathrm{L}}$ implies $\alpha[\tau/x] \in PropFun_{\mathrm{L}}$ for all $\tau \in U_{\mathcal{L}}$, i.e. $\mathfrak{S}_{\mathrm{L}}$ is updatable. Hence $\mathfrak{S}_{\mathrm{L}}$ is a functional model structure.

The *canonical premodel* $\mathfrak{M}_{\mathrm{L}} = (\mathfrak{S}_{\mathrm{L}}, |-|^{\mathfrak{M}_{\mathrm{L}}})$ on $\mathfrak{S}_{\mathrm{L}}$ is defined by taking $|-|^{\mathfrak{M}_{\mathrm{L}}}$ to be the function $|-|^{\mathcal{M}_{\mathrm{L}}}$, i.e

- $|\mathsf{c}|^{\mathfrak{M}_{\mathrm{L}}} = \mathsf{c} \in U_{\mathcal{L}}$, for all constants $\mathsf{c} \in \mathsf{Con}_{\mathcal{L}}$.
- $|F|^{\mathfrak{M}_{\mathrm{L}}}(\tau_1, \dots, \tau_n) = F\tau_1 \cdots \tau_n \in U_{\mathcal{L}}$, for all $\tau_1, \dots, \tau_n \in U_{\mathcal{L}}$.
- $|P|^{\mathfrak{M}_{\mathrm{L}}}(\tau_1, \dots, \tau_n) = |P\tau_1 \cdots \tau_n|_{\mathrm{L}} \in Prop_{\mathrm{L}}$, for all $\tau_1, \dots, \tau_n \in U_{\mathcal{L}}$.

So the difference between $\mathfrak{M}_{\mathrm{L}}$ and $\mathcal{M}_{\mathrm{L}}$ is just that $\mathfrak{M}_{\mathrm{L}}$ is based on $\mathfrak{S}_{\mathrm{L}}$ while $\mathcal{M}_{\mathrm{L}}$ is based on $\mathcal{S}_{\mathrm{L}}$. This difference, the presence or abscence of a specified *PropFun*, is not relevant to the definition of truth sets in a given premodel, and so for every formula φ, the function $|\varphi|^{\mathfrak{M}_{\mathrm{L}}} : U_{\mathcal{L}}{}^\omega \to \wp W_{\mathrm{L}}$ is identical to $|\varphi|^{\mathcal{M}_{\mathrm{L}}}$. Using the Truth Lemma (Theorem 1.9.4) we calculate that for each f,

$$|\varphi|^{\mathfrak{M}_{\mathrm{L}}} f = |\varphi|^{\mathcal{M}_{\mathrm{L}}} f = |\varphi^f|_{\mathrm{L}} = \alpha_\varphi f.$$

Thus $|\varphi|^{\mathfrak{M}_{\mathrm{L}}}$ is just the function $\alpha_\varphi \in PropFun_{\mathrm{L}}$. This shows that every propositional function $|\varphi|^{\mathfrak{M}_{\mathrm{L}}}$ is admissible in $\mathfrak{S}_{\mathrm{L}}$, and so $\mathfrak{M}_{\mathrm{L}}$ is a *model* on $\mathfrak{S}_L$. Given that $\mathcal{M}_L$ characterises L (Theorem 1.9.6), we can draw the following conclusions.

THEOREM 4.1.6. *For any $\mathcal{E}$-logic* L,

(1) L *is characterised by the model* $\mathfrak{M}_L$ *on the functional model structure* $\mathfrak{S}_L$.
(2) *If* L *is characterised by the model structure* $\mathcal{S}_L$, *then it is characterised by* $\mathfrak{S}_L$.

PROOF. (1): Since $\mathcal{M}_L$ characterises L (Theorem 1.9.6), and in general $|\varphi|^{\mathfrak{M}_L} = |\varphi|^{\mathcal{M}_L}$ as above, we have $\mathfrak{M}_L \models \varphi$ iff $\mathcal{M}_L \models \varphi$ iff $\vdash_L \varphi$.

(2): Suppose $\mathcal{S}_L$ characterises L. If $\vdash_L \varphi$, then φ is valid in $\mathcal{S}_L$, hence valid in $\mathfrak{S}_L$ as every model on $\mathfrak{S}_L$ is a model on $\mathcal{S}_L$. Conversely, if φ is valid in $\mathfrak{S}_L$, then it is valid in $\mathfrak{M}_L$, hence $\vdash_L \varphi$ by (1). Thus $\mathfrak{S}_L$ characterises L. ⊣

We can equally well define this $\mathfrak{S}_L$ construction for logics in languages without $\mathcal{E}$, using the original definition of $\mathcal{S}_L$ from Section 1.9, in which

$$E\tau = \{w \in W_L : \tau \in D_L(w)\} = \{w \in W_L : \mathrm{UI}(\tau) \subseteq w\},$$

without having $E\tau = |\mathcal{E}\tau|_L$. Then Theorem 4.1.6 holds also for $\mathcal{E}$-free logics.

The converse of part (2) of Theorem 4.1.6 is not true in general. The logic $L_{\mathfrak{S}_L}$ characterised by $\mathfrak{S}_L$ includes the logic $L_{\mathcal{S}_L}$ characterised by $\mathcal{S}_L$, since every model on $\mathfrak{S}_L$ is a model on $\mathcal{S}_L$ so validity in $\mathcal{S}_L$ implies validity in $\mathfrak{S}_L$. But $\mathcal{S}_L$ may have some model $\mathcal{M}$ that is not a model on $\mathfrak{S}_L$, since some $|\varphi|^{\mathcal{M}}$ may not belong to $PropFun_L$. Such a model on $\mathcal{S}_L$ may falsify L while no model on $\mathfrak{S}_L$ does, implying that

$$L_{\mathcal{S}_L} \subsetneq L_{\mathfrak{S}_L} = L.$$

To show that there can be an $\mathcal{M}$ that a model on $\mathcal{S}_L$ but not on $\mathfrak{S}_L$ we must show that we can have every truth set $|\varphi|^{\mathcal{M}} f$ belonging $Prop_L$ without requiring that the propositional function $|\varphi|^{\mathcal{M}}$ itself always belongs to $PropFun_L$.

EXAMPLE 4.1.7. $L = \mathrm{QEK} + \neg\Box\mathsf{F} + \mathrm{UI} + (\exists x\Box\varphi \to \forall x\Diamond\varphi)$.

This example exhibits the behaviour just described. L is the smallest logic that contains the sentence $\neg\Box\mathsf{F}$ and includes the Universal Instantiation scheme UI and the scheme $\exists x\Box\varphi \to \forall x\Diamond\varphi$. The inclusion of UI makes the existence predicate redundant (see the last paragraph of Section 3.1), and we could just as well take this as a logic in the $\mathcal{E}$-free language. L is a consistent logic, since it is validated by a full model structure having a single world that is R-related to itself, and one universal domain containing a single element. Therefore W_L, the set of maximally L-consistent sets of formulas, is non-empty.

The scheme $\exists x\Box\varphi \to \forall x\Diamond\varphi$ may have some philosophical interest (if only for its implausibility!), but is introduced here to give a simple example, involving both quantifiers and modalities, that illustrates the differences between validity in $\mathcal{S}_L$ and in $\mathfrak{S}_L$. We will show that L is valid in $\mathfrak{S}_L$, but not in $\mathcal{S}_L$.

First, the presence of $\neg\Box\mathsf{F}$ in L ensures that $[R_{\mathrm{L}}]\emptyset = \emptyset$, which in turn ensures that $\neg\Box\mathsf{F}$ is valid in both $\mathfrak{S}_{\mathrm{L}}$ and $\mathcal{S}_{\mathrm{L}}$. The presence of UI ensures that $\mathcal{S}_{\mathrm{L}}$ has one universal domain (Lemma 2.4.2), which in turn ensures that UI is valid in both $\mathfrak{S}_{\mathrm{L}}$ and $\mathcal{S}_{\mathrm{L}}$. Then $E\tau = W_{\mathrm{L}}$ for all $\tau \in U_{\mathcal{L}}{}^{\omega}$, which means that the description of any propositional function of the form $\forall_x\alpha : U_{\mathcal{L}}{}^{\omega} \to \wp W_{\mathrm{L}}$ simplifies to

$$(\forall_x\alpha)f = \bigsqcap_{\tau\in U_{\mathcal{L}}} \alpha f[\tau/x]. \tag{4.1.3}$$

Now if $\mathcal{M}$ is any model on the functional structure $\mathfrak{S}_{\mathrm{L}}$, then for each formula φ we have $|\varphi|^{\mathcal{M}} \in PropFun_{\mathrm{L}}$, so there is some ψ such that $|\varphi|^{\mathcal{M}} = \alpha_\psi = |\psi|^{\mathfrak{M}_{\mathrm{L}}}$. Then using equations from (4.1.2) we see that

$$\begin{aligned} |\exists x\Box\varphi \to \forall x\Diamond\varphi|^{\mathcal{M}} &= \exists_x[R_{\mathrm{L}}]|\varphi|^{\mathcal{M}} \Rightarrow \forall_x\langle R_{\mathrm{L}}\rangle|\varphi|^{\mathcal{M}} \\ &= \exists_x[R_{\mathrm{L}}]|\psi|^{\mathfrak{M}_{\mathrm{L}}} \Rightarrow \forall_x\langle R_{\mathrm{L}}\rangle|\psi|^{\mathfrak{M}_{\mathrm{L}}} \\ &= |\exists x\Box\psi \to \forall x\Diamond\psi|^{\mathfrak{M}_{\mathrm{L}}}. \end{aligned}$$

But $\exists x\Box\psi \to \forall x\Diamond\psi$ is an L-theorem, by definition, hence is valid in $\mathfrak{M}_{\mathrm{L}}$ by Theorem 4.1.6(1). It follows that $\exists x\Box\varphi \to \forall x\Diamond\varphi$ is valid in $\mathcal{M}$.

This shows that the scheme $\exists x\Box\varphi \to \forall x\Diamond\varphi$ is valid in all models on $\mathfrak{S}_{\mathrm{L}}$. Altogether, L is validated by $\mathfrak{S}_{\mathrm{L}}$, and indeed $\mathrm{L} = \mathrm{L}_{\mathfrak{S}_{\mathrm{L}}}$.

It remains to show that L is not valid in $\mathcal{S}_{\mathrm{L}}$, and so $\mathrm{L}_{\mathcal{S}_{\mathrm{L}}} \subsetneq \mathrm{L}_{\mathfrak{S}_{\mathrm{L}}}$. We take a monadic predicate symbol P and construct a model $\mathcal{M}$ on $\mathcal{S}_{\mathrm{L}}$ that falsifies $\exists x\Box Px \to \forall x\Diamond Px$. To define $|P|^{\mathcal{M}} : U_{\mathcal{L}} \to Prop_{\mathrm{L}}$ we fix a particular constant c and put, for all $\tau \in U_{\mathcal{L}}$,

$$|P|^{\mathcal{M}}\tau = \begin{cases} \emptyset & \text{if } \tau = \mathsf{c} \\ W_L & \text{if } \tau \neq \mathsf{c}. \end{cases}$$

For any other predicate symbol Q we define $|Q|^{\mathcal{M}}$ to have constant value W_{L}. The interpretation of individual constants and function symbols by $\mathcal{M}$ can be arbitrary.

To explain why $\mathcal{M}$ is a model, we will say that a propositional function $\alpha : U_{\mathcal{L}}{}^{\omega} \to Prop_{\mathrm{L}}$ is 2-*valued* if $\alpha f \in \{W_{\mathrm{L}}, \emptyset\}$ for all $f \in U_{\mathcal{L}}{}^{\omega}$. The constant function $\alpha_\emptyset$ is 2-valued. So are the propositional functions $|\varphi|^{\mathcal{M}}$ for all *atomic* φ, since the functions $|P|^{\mathcal{M}}$ and $|Q|^{\mathcal{M}}$ just defined take all their values in $\{W_{\mathrm{L}}, \emptyset\}$, and also $|\mathcal{E}\tau|^{\mathcal{M}} f = W_{\mathrm{L}}$ in general. We will show that the set of 2-valued functions is closed under the operations $\cap$, $-$, $[R_{\mathrm{L}}]$ and $\forall_x$ used to inductively define $|\varphi|^{\mathcal{M}}$ for all φ. Then every $|\varphi|^{\mathcal{M}}$ is 2-valued, and therefore every truth set $|\varphi|^{\mathcal{M}} f$ belongs to $\{W_{\mathrm{L}}, \emptyset\} \subseteq Prop_{\mathrm{L}}$, hence is admissible in $\mathcal{S}_{\mathrm{L}}$. This means that $\mathcal{M}$ is a model on $\mathcal{S}_{\mathrm{L}}$ as claimed.

$\alpha \cap \beta$ and $-\alpha$ are 2-valued when α and β are, because $\{W_{\mathrm{L}}, \emptyset\}$ is closed under the Boolean set operations $\cap$ and $-$. Similarly, $[R_{\mathrm{L}}]\alpha$ is 2-valued when α is, because $[R_{\mathrm{L}}]W_{\mathrm{L}} = W_{\mathrm{L}}$ and $[R_{\mathrm{L}}]\emptyset = \emptyset$. For a function $\forall_x\varphi$, we use

(4.1.3). If α is 2-valued, then for each $f \in U_{\mathcal{L}}{}^{\omega}$ there are two cases. If $\alpha f[\tau/x] = W_{\mathrm{L}}$ for all $\tau \in$ UI, then $(\forall_x\alpha)f = \bigsqcap\{W_{\mathrm{L}}\} = W_{\mathrm{L}}$. Otherwise, $\alpha f[\tau/x] = \emptyset$ for some τ, so $(\forall_x\alpha)f = \emptyset$ because $(\forall_x\alpha)f \subseteq \alpha f[\tau/x]$. Thus $\forall_x\alpha$ is 2-valued.

That completes the proof that $\mathcal{M}$ is a model. Now for any $f \in U_{\mathrm{L}}{}^{\omega}$ and any $\tau \in U_{\mathcal{L}}$, $|Px|^{\mathcal{M}} f[\tau/x] = |P|^{\mathcal{M}}\tau$, so $|\Diamond Px|^{\mathcal{M}} f[\tau/x] = \langle R_{\mathrm{L}}\rangle|P|^{\mathcal{M}}\tau$. Thus

$$|\forall x\Diamond Px|^{\mathcal{M}} f = \bigsqcap_{\tau\in U_{\mathcal{L}}} \langle R_{\mathrm{L}}\rangle|P|^{\mathcal{M}}\tau = \emptyset,$$

since in particular, $\langle R_{\mathrm{L}}\rangle|P|^{\mathcal{M}}\mathsf{c} = \langle R_{\mathrm{L}}\rangle\emptyset = \emptyset$.

Now taking any f with $fx \neq \mathsf{c}$, we have $|Px|^{\mathcal{M}} f = |P|^{\mathcal{M}} fx = W_{\mathrm{L}}$, so $|\Box Px|^{\mathcal{M}} f = [R_{\mathrm{L}}]W_{\mathrm{L}} = W_{\mathrm{L}}$. But $|\Box Px|^{\mathcal{M}} f \subseteq |\exists x\Box Px|^{\mathcal{M}} f$, so this gives $|\exists x\Box Px|^{\mathcal{M}} f = W_{\mathrm{L}}$. Therefore

$$|\exists x\Box Px \to \forall x\Diamond Px|^{\mathcal{M}} f = W_{\mathrm{L}} \Rightarrow \emptyset = \emptyset \neq W_{\mathrm{L}},$$

so $\exists x\Box Px \to \forall x\Diamond Px$ is falsified by $\mathcal{M}$, hence is not valid in $\mathcal{S}_{\mathrm{L}}$.

It follows that $\mathcal{M}$ is not a model on $\mathfrak{S}_{\mathrm{L}}$, and indeed the propositional function $|Px|^{\mathcal{M}}$ fails to belong to $PropFun_{\mathrm{L}}$, while all of its values $|Px|^{\mathcal{M}} f$ do belong to $Prop_{\mathrm{L}}$. ⊣

Which logics L are in fact characterised by their canonical functional model structure $\mathfrak{S}_{\mathrm{L}}$? We are going to answer this question in terms of closure of L under the rule of substitution for predicate letters (see Theorem 4.5.1). First we need to develop the complex theory of this rule.

4.2. Predicate Substitution: Notation and Terminology

Boldface symbols $\boldsymbol{x}, \boldsymbol{a}, \boldsymbol{\tau}, \boldsymbol{\pi}$ etc. will be used to denote finite tuples, or lists, of terms or individuals, presented in the form $\boldsymbol{\tau} = \tau_1, \dots, \tau_n$. Two tuples $\boldsymbol{\tau}, \boldsymbol{\sigma}$ will be called *disjoint* if they have no members in common: i.e. $\tau_i \neq \sigma_j$ in general. A tuple is *disjoint* from a formula if no member of the tuple occurs in the formula. Tuples may contain repetitions, but when a list of individual variables is being substituted for, it will be assumed that all member of the list are distinct from each other. An indexed collection $\{\boldsymbol{\sigma}_P : P \in \Pi\}$ of tuples will be called *pairwise disjoint* if $\boldsymbol{\sigma}_P$ is disjoint from $\boldsymbol{\sigma}_{P'}$ whenever their indices $P, P' \in \Pi$ are distinct. Concatenation of tuples $\boldsymbol{\tau}$ and $\boldsymbol{\sigma}$ will be written $\boldsymbol{\tau\sigma}$.

We now introduce several items of definition and notation, concerning conditions under which various entities are freely substitutable for others. The first two items are familiar.

- Term τ is *free for individual variable x in φ* if no free occurrence of x in φ is within the scope of $\forall y$ where y is any variable occurring in τ.
- If τ is free for x in φ, then $\varphi(\tau/x)$ is the formula obtained by uniform substitution of τ in place of the free occurrences of x in φ.

- If $\boldsymbol{\tau} = \tau_1, \ldots, \tau_n$ and $\boldsymbol{x} = x_1, \ldots, x_n$ are n-tuples of terms and individual variables, respectively, then $\boldsymbol{\tau}$ *is free for* $\boldsymbol{x}$ *in* φ if τ_i is free for x_i in φ for all $i \leq n$. If this holds, then $\varphi(\boldsymbol{\tau}/\boldsymbol{x})$ is the result of simultaneous substitution, for all $i \leq n$, of τ_i for free x_i in φ.
- We may write $(\boldsymbol{\tau}/\boldsymbol{x})$ for the *substitution operator* that assigns $\varphi(\boldsymbol{\tau}/\boldsymbol{x})$ to φ. This is defined at φ just when $\boldsymbol{\tau}$ is free for $\boldsymbol{x}$ in φ.
- An atomic formula may be viewed as having the form $P\boldsymbol{\tau}$ with P a predicate letter, of some finite arity δP, and $\boldsymbol{\tau}$ a δP-tuple of terms. Such a formula will be called an *instance of* P, or a P*-instance*.
- Let $\boldsymbol{x}$ be a δP-tuple of variables. A formula ψ is *free for* P *in* φ *via* $\boldsymbol{x}$ when:

  (S1) each P-instance $P\boldsymbol{\tau}$ occurring in φ has $\boldsymbol{\tau}$ free for $\boldsymbol{x}$ in ψ (so that $\psi(\boldsymbol{\tau}/\boldsymbol{x})$ is defined); and

  (S2) if y is any free variable of ψ that is not in the list $\boldsymbol{x}$, then y does not occur bound in φ.
- We will write $(\psi/P\boldsymbol{x})$ for a substitution operator that is defined at φ just when ψ is free for P in φ via $\boldsymbol{x}$. If this holds we will also say that $(\psi/P\boldsymbol{x})$ *is free at* φ.

  If $(\psi/P\boldsymbol{x})$ is free at φ, then $\varphi(\psi/P\boldsymbol{x})$ is defined to be the formula obtained by replacing each P-instance $P\boldsymbol{\tau}$ occurring in φ by $\psi(\boldsymbol{\tau}/\boldsymbol{x})$. $\varphi(\psi/P\boldsymbol{x})$ will be said to be obtained by *substitution of* ψ *for* P *in* φ *via* $\boldsymbol{x}$.
- Define a *parameter of* $(\psi/P\boldsymbol{x})$ to be any free variable of ψ that is not in the list $\boldsymbol{x}$. Thus (S2) above can be expressed as "no parameter occurs bound in φ". An occurrence of a parameter in ψ is not affected by the substitution $(\psi/P\boldsymbol{x})$. (S2) ensures that it does not become become bound in $\psi(\boldsymbol{\tau}/\boldsymbol{x})$ within $\varphi(\psi/P\boldsymbol{x})$.

  We will sometimes write $\boldsymbol{\pi}_P$ for a listing of the parameters of $(\psi/P\boldsymbol{x})$. This need not be of the same length as $\boldsymbol{x}$, and may well be empty. If it is empty, we call $(\psi/P\boldsymbol{x})$ a *parameterless* substitution.
- Let Π be a finite set of predicate letters that indexes collections $\{\psi_P : P \in \Pi\}$ of formulas and $\{\boldsymbol{x}_P : P \in \Pi\}$ of δP-tuples of variables. The notation $(\psi_P/P\boldsymbol{x}_P : P \in \Pi)$ denotes a substitution operator that is defined at a formula φ just when $(\psi_P/P\boldsymbol{x}_P)$ is free at φ for all $P \in \Pi$. Then

  $$\varphi(\psi_P/P\boldsymbol{x}_P : P \in \Pi)$$

  is the formula resulting from simultaneous substitution, for all $P \in \Pi$, of ψ_P for P in φ via $\boldsymbol{x}_P$.

  In this situation we say that the substitution $(\psi_P/P\boldsymbol{x}_P : P \in \Pi)$ is *free at* φ, and that $\varphi(\psi_P/P\boldsymbol{x}_P : P \in \Pi)$ is a *free substitution instance of* φ.

EXAMPLE 4.2.1. Let P be binary and $\boldsymbol{x} = x_1, x_2$. Let φ be $\forall yPyy$ and ψ be $\forall zPx_1z$. The only P-instance in φ is $P\boldsymbol{\tau}$ with $\boldsymbol{\tau} = y, y$. This $\boldsymbol{\tau}$ free for $\boldsymbol{x}$ in ψ, and $\psi(\boldsymbol{\tau}/\boldsymbol{x})$ is $\forall zPyz$. Then $(\psi/P\boldsymbol{x})$ is free at φ, and $\varphi(\psi/P\boldsymbol{x})$ is $\forall y\forall zPyz$.

For an example with both x_1 and x_2 free in the substituting formula, let ψ' be $\forall z(Px_1z \wedge Px_2z)$. Then $\psi'(\boldsymbol{\tau}/\boldsymbol{x})$ is $\forall z(Pyz \wedge Pyz)$, and $\varphi(\psi'/P\boldsymbol{x})$ is $\forall y\forall z(Pyz \wedge Pyz)$, equivalent to $\varphi(\psi/P\boldsymbol{x})$.

Of course there are models $\mathcal{M}$ in which $\forall yPyy$ is valid but $\forall y\forall zPyz$ is not, so the logic $\{\varphi : \mathcal{M} \models \varphi\}$ is not closed under free substitution instances. As mentioned in the introduction to this chapter, in a logic that did have that closure property, the theorems would represent "universal laws", expressing properties that hold of all predicates, i.e. hold no matter what interpretation is given to the predicate letters. A logic in which $\forall yPyy$ is an axiom or theorem would be viewed as a *theory* about a particular kind of predicate, say the theory of reflexive predicates. ⊣

Now if $(\psi/P\boldsymbol{x})$ is free at φ, and θ is any subformula of φ, then $(\psi/P\boldsymbol{x})$ is also free at θ. This follows readily from the fact that each P-instance in θ is also a P-instance in φ. Thus we get

LEMMA 4.2.2. *For a given operator* $(\psi_P/P\boldsymbol{x}_P : P \in \Pi)$, *the set*

$$\{\varphi : (\psi_P/P\boldsymbol{x}_P : P \in \Pi) \text{ is free at } \varphi\}$$

is closed under subformulas. ⊣

Next we extend the notation for updating variable-assignments to provide for simultaneous updating of lists of variables. Let $\boldsymbol{x} = x_1, \dots, x_n$ be a tuple of variables without repetitions: $x_i \neq x_j$ whenever $i \neq j$.

Given $f \in U^\omega$, and $\boldsymbol{a} \in U^n$, let $f[\boldsymbol{a}/\boldsymbol{x}]$ be the result of updating f by simultaneously assigning x_i the value a_i for each $i \leq n$. Formally, for each variable y,

$$f[\boldsymbol{a}/\boldsymbol{x}](y) = \begin{cases} a_i & \text{if } y = x_i, \\ f(y) & \text{if } y \neq x_i \text{ for all } i \leq n. \end{cases}$$

Since the x_i's are distinct, the simultaneous update $f[\boldsymbol{a}/\boldsymbol{x}]$ can be obtained by the *sequential* updating of the x_i by a_i:

$$f[\boldsymbol{a}/\boldsymbol{x}] = f[a_1/x_1][a_2/x_2]\cdots[a_n/x_n], \tag{4.2.1}$$

and these updates $[a_i/x_i]$ can be arbitrarily permuted without changing the result.

Using concatenations $\boldsymbol{xy}$ of *disjoint* tuples $\boldsymbol{x}, \boldsymbol{y}$ of variables we can form updates $f[\boldsymbol{ab}/\boldsymbol{xy}]$, which we may also write as $f[\boldsymbol{a}/\boldsymbol{x}, \boldsymbol{b}/\boldsymbol{y}]$ or $f[\boldsymbol{b}/\boldsymbol{y}, \boldsymbol{a}/\boldsymbol{x}]$, since the order of updating of distinct variables is immaterial.

For a tuple $\boldsymbol{\tau} = \tau_1, \dots, \tau_n$ of terms, in a given premodel $\mathcal{M}$ we can define

$$|\boldsymbol{\tau}|^{\mathcal{M}} f = |\tau_1|^{\mathcal{M}} f, \dots, |\tau_n|^{\mathcal{M}} f \in U^n.$$

This provides a convenient notation for the semantics of atomic formulas, as we then have

$$|P\boldsymbol{\tau}|^{\mathcal{M}} f = |P|^{\mathcal{M}} |\boldsymbol{\tau}|^{\mathcal{M}} f.$$

Simultaneous updates of the form

$$f[\,|\boldsymbol{\tau}|^{\mathcal{M}} f/\boldsymbol{x}] = f[\,|\tau_1|^{\mathcal{M}} f/x_1, \dots, |\tau_n|^{\mathcal{M}} f/x_n]$$

play an important role in the evaluation of simultaneous substitution instances $\varphi(\boldsymbol{\tau}/\boldsymbol{x})$, as we explain in the next Lemma. But first note that if $\boldsymbol{\tau}$ is a list $x_1, \dots,$ x_n of *variables*, then $|\boldsymbol{\tau}|^{\mathcal{M}} f$ is the list $fx_1, \dots, fx_n$ of individuals, depending only on f and not $\mathcal{M}$. In that case we can write $|\boldsymbol{\tau}|^{\mathcal{M}} f$ just as $\boldsymbol{\tau} f$.

To view the formula $\varphi(\boldsymbol{\tau}/\boldsymbol{x})$ as also being obtained by a sequence of single-variable substitutions

$$\varphi(\tau_1/x_1)(\tau_2/x_2)\cdots(\tau_n/x_n)$$

we would need to assume that $\boldsymbol{\tau}$ and $\boldsymbol{x}$ are disjoint, i.e. no x_i occurs in any τ_j. Now from the Substitution Lemma 1.6.2 we have the result

$$|\varphi(\tau/x)|^{\mathcal{M}} f = |\varphi|^{\mathcal{M}} f[|\tau|^{\mathcal{M}} f/x],$$

for *single-variable* substitutions. But the proof of this can be adjusted to show the same result for simultaneous substitutions, as follows, without requiring $\boldsymbol{\tau}$ and $\boldsymbol{x}$ to be disjoint.

LEMMA 4.2.3 (Simultaneous Substitution). *If $\boldsymbol{\tau}$ is free for $\boldsymbol{x}$ in φ, then in any premodel $\mathcal{M}$,*

$$|\varphi(\boldsymbol{\tau}/\boldsymbol{x})|^{\mathcal{M}} f = |\varphi|^{\mathcal{M}} f[\,|\boldsymbol{\tau}|^{\mathcal{M}} f/\boldsymbol{x}].$$

PROOF. By induction on φ. First we need to know that for any tuple $\boldsymbol{\sigma} = \sigma_1, \dots, \sigma_n$ of terms, the tuple $\boldsymbol{\sigma}(\boldsymbol{\tau}/\boldsymbol{x}) = \sigma_1(\boldsymbol{\tau}/\boldsymbol{x}), \dots, \sigma_n(\boldsymbol{\tau}/\boldsymbol{x})$ has

$$|\boldsymbol{\sigma}(\boldsymbol{\tau}/\boldsymbol{x})|^{\mathcal{M}} f = |\boldsymbol{\sigma}|^{\mathcal{M}} f[\,|\boldsymbol{\tau}|^{\mathcal{M}} f/\boldsymbol{x}].$$

This is a standard fact from non-modal first-order logic, requiring nothing new here. We use it for the atomic case $\varphi = P\boldsymbol{\sigma}$, shown by

$$|(P\boldsymbol{\sigma})(\boldsymbol{\tau}/\boldsymbol{x})|^{\mathcal{M}} f = |P(\boldsymbol{\sigma}(\boldsymbol{\tau}/\boldsymbol{x}))|^{\mathcal{M}} f = |P|^{\mathcal{M}} |\boldsymbol{\sigma}(\boldsymbol{\tau}/\boldsymbol{x})|^{\mathcal{M}} f,$$

which from above is $|P|^{\mathcal{M}} |\boldsymbol{\sigma}|^{\mathcal{M}} f[\,|\boldsymbol{\tau}|^{\mathcal{M}} f/\boldsymbol{x}] = |P\boldsymbol{\sigma}|^{\mathcal{M}} f[\,|\boldsymbol{\tau}|^{\mathcal{M}} f/\boldsymbol{x}]$.

The inductive cases of the propositional connectives are straightforward, using $(\neg\varphi)(\boldsymbol{\tau}/\boldsymbol{x}) = \neg(\varphi(\boldsymbol{\tau}/\boldsymbol{x}))$ etc.

For the quantifier case, suppose φ is $\forall y\psi$, with $\boldsymbol{\tau}$ free for $\boldsymbol{x}$ in φ, and assume the result inductively for ψ. Let $\boldsymbol{x}'$ be the sublist of $\boldsymbol{x}$ consisting of those members that actually occur free in φ, and let $\boldsymbol{\tau}'$ be the corresponding sublist of $\boldsymbol{\tau}$. Then $\varphi(\boldsymbol{\tau}/\boldsymbol{x}) = \varphi(\boldsymbol{\tau}'/\boldsymbol{x}')$ and

$$|\varphi|^{\mathcal{M}} f[\,|\boldsymbol{\tau}|^{\mathcal{M}} f/\boldsymbol{x}] = |\varphi|^{\mathcal{M}} f[\,|\boldsymbol{\tau}'|^{\mathcal{M}} f/\boldsymbol{x}'],$$

since the valuations $f[\,|\boldsymbol{\tau}|^{\mathcal{M}} f/\boldsymbol{x}]$ and $f[\,|\boldsymbol{\tau}'|^{\mathcal{M}} f/\boldsymbol{x}']$ agree on all free variables of φ. So it is enough to prove the result for $\boldsymbol{\tau}'$ and $\boldsymbol{x}'$.

Thus we might as well assume every member of $\boldsymbol{x}$ occurs free in $\forall y\psi$. It follows that y does not occur in the list $\boldsymbol{x}$ and $\varphi(\boldsymbol{\tau}/\boldsymbol{x}) = \forall y(\psi(\boldsymbol{\tau}/\boldsymbol{x}))$, with every member of $\boldsymbol{x}$ occurring in ψ. Since $\boldsymbol{\tau}$ is free for $\boldsymbol{x}$ in $\forall y\psi$, it must also be free for $\boldsymbol{x}$ in ψ, and y does not occur in any member of $\boldsymbol{\tau}$. Thus any f agrees with any of its y-updates $f[a/y]$ on the variables of $\boldsymbol{\tau}$, so $|\boldsymbol{\tau}|^{\mathcal{M}}f[a/y] = |\boldsymbol{\tau}|^{\mathcal{M}}f$. Then we have

$$\begin{aligned}
&|(\forall y\psi)(\boldsymbol{\tau}/\boldsymbol{x})|^{\mathcal{M}}f \\
&= |\forall y(\psi(\boldsymbol{\tau}/\boldsymbol{x}))|^{\mathcal{M}}f \\
&= \bigsqcap_{a\in U} Ea \Rightarrow |\psi(\boldsymbol{\tau}/\boldsymbol{x})|^{\mathcal{M}}f[a/y] && \text{by the semantics of } \forall, \\
&= \bigsqcap_{a\in U} Ea \Rightarrow |\psi|^{\mathcal{M}}f[a/y][|\boldsymbol{\tau}|^{\mathcal{M}}f[a/y]/\boldsymbol{x}] && \text{by hypothesis on } \psi, \\
&= \bigsqcap_{a\in U} Ea \Rightarrow |\psi|^{\mathcal{M}}f[a/y][|\boldsymbol{\tau}|^{\mathcal{M}}f/\boldsymbol{x}], && \text{as } |\boldsymbol{\tau}|^{\mathcal{M}}f[a/y] = |\boldsymbol{\tau}|^{\mathcal{M}}f, \\
&= \bigsqcap_{a\in U} Ea \Rightarrow |\psi|^{\mathcal{M}}f[|\boldsymbol{\tau}|^{\mathcal{M}}f/\boldsymbol{x}][a/y], && \text{as } y \text{ is disjoint from } \boldsymbol{x}, \\
&= |\forall y\psi|^{\mathcal{M}}f[|\boldsymbol{\tau}|^{\mathcal{M}}f/\boldsymbol{x}] && \text{by the semantics of } \forall,
\end{aligned}$$

so the result holds for $\forall y\psi$. ⊣

The simultaneous updating $f \mapsto f[\boldsymbol{a}/\boldsymbol{x}]$ of variable assignments will now be used to lift the updating of propositional functions to lists of variables. The *simultaneous updating* $\alpha[\boldsymbol{a}/\boldsymbol{x}] : U^{\omega} \to \wp W$ of a propositional function $\alpha : U^{\omega} \to \wp W$ is defined by putting

$$\alpha[\boldsymbol{a}/\boldsymbol{x}]f = \alpha(f[\boldsymbol{a}/\boldsymbol{x}]) \tag{4.2.2}$$

for all $f \in U^{\omega}$. This includes the case of the single-variable updatings $\alpha[a/x]$.

LEMMA 4.2.4. *Every functional model structure is closed under simultaneous updating, i.e.* $\alpha \in$ *PropFun implies* $\alpha[\boldsymbol{a}/\boldsymbol{x}] \in$ *PropFun.*

PROOF. Let $\boldsymbol{a} = a_1, \dots, a_n$ and $\boldsymbol{x} = x_1, \dots, x_n$, with the x_i's distinct. Now by (4.2.1) we have

$$f[\boldsymbol{a}/\boldsymbol{x}] = f[a_1/x_1][a_2/x_2]\cdots[a_n/x_n],$$

showing that $f[\boldsymbol{a}/\boldsymbol{x}]$ can be obtain by a finite sequence of single-variable updatings. Hence by repeated use of the single-variable case of (4.2.2),

$$\begin{aligned}
\big(\alpha[a_n/x_n]\cdots[a_1/x_1]\big)f &= \big(\alpha[a_n/x_n]\cdots[a_2/x_2]\big)\big(f[a_1/x_1]\big) \\
&= \big(\alpha[a_n/x_n]\cdots[a_3/x_3]\big)\big(f[a_1/x_1][a_2/x_2]\big) \\
&\ \ \vdots \\
&= \alpha\big(f[a_1/x_1][a_2/x_2]\cdots[a_n/x_n]\big) \\
&= \alpha(f[\boldsymbol{a}/\boldsymbol{x}]) = \alpha[\boldsymbol{a}/\boldsymbol{x}]f.
\end{aligned}$$

Thus $\alpha[\boldsymbol{a}/\boldsymbol{x}]$ is equal to $\alpha[a_n/x_n][a_{n-1}/x_{n-1}]\cdots[a_1/x_1]$, got from α by a finite sequence of single-variable updatings. But *PropFun* is closed under single-variable updatings in any functional model structure, hence it is also closed under simultaneous updatings. ⊣

4.3. The Anatomy of Predicate Substitution

It is possible to make an appropriate definition of $\varphi(\psi_P/\boldsymbol{x}_P : P \in \Pi)$ even when $(\psi_P/\boldsymbol{x}_P : P \in \Pi)$ is not free at φ. This takes some explaining. The explanation is given in this section, along with an analysis of several types of substitution operator. It leads to the conclusion that for a logic to be closed under simultaneous substitutions in general, it is enough for it to be closed under substitutions that have strong "no conflict" restrictions on their parameters. Moreover, under the assumption that the signature has infinitely many individual constants, it is enough for the logic to be closed under substitutions that have no parameters at all.

4.3.1. Changing Bound Variables. We elaborate on the brief discussion in Section 1.2 of relettering of bound variables. Each formula of the form

$$\forall x\varphi \leftrightarrow \forall y\varphi(y/x),$$

with y not in $\forall x\varphi$ but free for x in φ, is derivable in any quantified modal logic (Lemma 1.2.1(4)). In particular it is derivable in the logic $\mathrm{L}_{\mathcal{S}}$ characterised by any model structure $\mathcal{S}$, and so $|\forall x\varphi|^{\mathcal{M}} = |\forall y\varphi(y/x)|^{\mathcal{M}}$ in any model $\mathcal{M}$ on $\mathcal{S}$. This is the essential scheme that allows rewriting of bound variables.

Recall that formulas ψ and ψ' are be called *congruent* if there is a bijection $x_i \mapsto x'_i$ between their sets of bound variables such that ψ and ψ' are identical except that ψ has bound ocurrences of x_i exactly where ψ' has bound ocurrences of x'_i and vice versa. Extending Lemma 1.2.2, we have

LEMMA 4.3.1. *If ψ and ψ' are congruent, then the formula $\psi \leftrightarrow \psi'$ is derivable in any quantified modal logic, hence $|\varphi|^{\mathcal{M}} = |\varphi'|^{\mathcal{M}}$ in any model $\mathcal{M}$ on any model structure.* ⊣

By this result we can rewrite the bound variables of any formula to suit without changing its semantics.

4.3.2. Sufficient Freeness. Recall the conditions for a substitution $(\psi_P/P\boldsymbol{x}_P)$ to be free at φ:

(S1) each P-instance $P\boldsymbol{\tau}$ occurring in φ has $\boldsymbol{\tau}$ free for $\boldsymbol{x}$ in ψ ;
(S2) no parameter of $(\psi_P/P\boldsymbol{x}_P)$ is bound in φ.

These are the conditions given in Church's first edition [1944, pp. 57–58]. In the second edition [1956, pp. 192–193], and in [Kleene 1952, p. 156], (S2) is replaced by the weaker

(S2') if y is any parameter of $(\psi_P/P\boldsymbol{x}_P)$, then no P-instance occurs in φ within the scope of $\forall y$.

We will say $(\psi_P/P\boldsymbol{x}_P)$ is *sufficiently free* at φ if (S1) and (S2') hold, and that $(\psi_P/P\boldsymbol{x}_P : P \in \Pi)$ is sufficiently free at φ if each $(\psi_P/P\boldsymbol{x}_P)$ is. Church [1956, p. 193] gives a proof that first-order (non-modal) logic is closed under the rule of sufficiently free substitutions for a single predicate letter. Kleene [1952,

Theorem 15] proves closure under simultaneous sufficiently free substitutions for finitely many predicate letters.

Since (S2) implies (S2'), every substitution free at φ is sufficiently free at φ, so sufficiently free substitution is at least as strong a rule of inference as free substitution. But in fact the two are equivalent. This can be seen with the help of the following result of Kleene [1952, Lemma 16b].

LEMMA 4.3.2. *Suppose that* $(\psi_P/P\boldsymbol{x}_P : P \in \Pi)$ *is sufficiently free at* φ *and* $(\psi'_P/P\boldsymbol{x}_P : P \in \Pi)$ *is sufficiently free at* φ'. *If* φ *is congruent to* φ', *and each* ψ_P *is congruent to* ψ'_P, *then* $\varphi(\psi_P/P\boldsymbol{x}_P : P \in \Pi)$ *is congruent to* $\varphi'(\psi'_P/P\boldsymbol{x}_P : P \in \Pi)$. ⊣

This statement is purely about the syntactic structure of formulas and does not depend on any proof theory or semantics. It applies just as well to modal formulas as to non-modal ones.

We say that a logic L is *closed under* a certain type of substitution if, whenever $\vdash_L \varphi$, and $(\psi_P/P\boldsymbol{x}_P : P \in \Pi)$ is that type of substitution in relation to φ, then $\vdash_L \varphi(\psi_P/P\boldsymbol{x}_P : P \in \Pi)$.

THEOREM 4.3.3. *A logic is closed under free predicate substitutions if, and only if, it is closed under sufficiently free predicate substitutions.*

PROOF. If L is closed under sufficiently free predicate substitutions, then it is closed under free predicate substitutions for the reason just explained: every free substitution is sufficiently free.

Conversely, suppose L is closed under free predicate substitutions. Then we show that it is also closed under sufficiently free predicate substitutions, as follows. Let $(\psi_P/P\boldsymbol{x}_P : P \in \Pi)$ be *sufficiently* free at φ, with $\vdash_L \varphi$. We want to show $\vdash_L \varphi(\psi_P/P\boldsymbol{x}_P : P \in \Pi)$. Replace all the bound variables of φ by fresh (bound) variables that do not occur at all in any of the formulas ψ_P for all $P \in \Pi$. The resulting formula φ' is congruent to φ, so $\vdash_L \varphi'$ by Lemma 4.3.1. But now $(\psi_P/P\boldsymbol{x}_P : P \in \Pi)$ is free at φ': since no parameter of any ψ_P is a bound variable of φ', (S2) is satisfied. For (S1), the only issue that arises is when there is an atomic formula $P\tau$ in φ, hence with τ free for $\boldsymbol{x}_P$ in ψ_P, that gets re-written as $P\tau'$ in φ' when some of the variables in τ are replaced by the fresh variables that are bound in φ'. Since none of these fresh variables occur in ψ_P at all, and τ is free for $\boldsymbol{x}_P$ in ψ_P, it follows that τ' is also free for $\boldsymbol{x}_P$ in ψ_P.

From the assumed substitution-closure of L, we draw the conclusion that $\vdash_L \varphi'(\psi_P/P\boldsymbol{x}_P : P \in \Pi)$. But this formula is congruent to $\varphi(\psi_P/P\boldsymbol{x}_P : P \in \Pi)$ by Lemma 4.3.2, so the latter is an L-theorem as desired. ⊣

4.3.3. Defining $\varphi(\psi_P/P\boldsymbol{x}_P : P \in \Pi)$ in General. The above kind of reasoning based on Lemma 4.3.2 can be used to define $\varphi(\psi_P/P\boldsymbol{x}_P : P \in \Pi)$ in the case that $(\psi_P/P\boldsymbol{x}_P : P \in \Pi)$ is not free at φ.

Take a φ' congruent to φ whose bound variables do not occur in ψ_P for all $P \in \Pi$. Then for each P, take a ψ'_P congruent to ψ_P whose bound variables do not occur in φ'. The resulting substitution operator $(\psi'_P/P\boldsymbol{x}_P : P \in \Pi)$ is seen to be free at φ', as follows. If y is a parameter of some $(\psi'_P/P\boldsymbol{x}_P)$, then y is free in ψ'_P and hence occurs (free) in ψ_P because ψ'_P and ψ_P differ only on their bound variables, so y is not a bound variable of φ', giving (S2). For (S1), if a P-instance $P\tau$ occurs in φ', then the variables of τ are not bound in ψ'_P, as the bound variables of ψ'_P do not occur in φ', so τ is free for $\boldsymbol{x}_P$ in ψ'_P.

We now define $\varphi(\psi_P/P\boldsymbol{x}_P : P \in \Pi)$ to be $\varphi'(\psi'_P/P\boldsymbol{x}_P : P \in \Pi)$. This does not define $\varphi(\psi_P/P\boldsymbol{x}_P : P \in \Pi)$ uniquely, because of the choice in selection of φ' and ψ'_P, but it does define it uniquely up to congruence, by Lemma 4.3.2, and that determines its semantics uniquely in any model.

THEOREM 4.3.4. *A logic is closed under free predicate substitutions if, and only if, it is closed under predicate substitutions in general.*

PROOF. Suppose L is closed under free predicate substitutions. Take any operator $(\psi_P/P\boldsymbol{x}_P : P \in \Pi)$ and any formula φ such that $\vdash_L \varphi$. If the operator is free at φ, then $\vdash_L \varphi(\psi_P/P\boldsymbol{x}_P : P \in \Pi)$ by supposition.

Otherwise take φ' and $(\psi'_P/P\boldsymbol{x}_P : P \in \Pi)$ as above with $\varphi(\psi_P/P\boldsymbol{x}_P : P \in \Pi)$ defined to be $\varphi'(\psi'_P/P\boldsymbol{x}_P : P \in \Pi)$. Then $\vdash_L \varphi'$ as φ' is congruent to φ, hence $\vdash_L \varphi'(\psi'_P/P\boldsymbol{x}_P : P \in \Pi)$ as $(\psi'_P/P\boldsymbol{x}_P : P \in \Pi)$ is free at φ' and L is closed under free predicate substitutions. So again $\vdash_L \varphi(\psi_P/P\boldsymbol{x}_P : P \in \Pi)$ holds. ⊣

4.3.4. Strongly Free Substitution. By strengthening (S2) instead of weakening it, we obtain a more restricted notion of substitution, hence a potentially more restricted inference rule. If that rule turns out to be equivalent to substitution in general, then we gain the advantage of being able to use rules satisfying restrictions that may make them easier to work with. From that point of view it makes sense to impose the strongest restrictions we can.

For instance, $(\psi_P/P\boldsymbol{x}_P : P \in \Pi)$ will be called *strongly free at* φ if

(i) (S1) holds for every $P \in \Pi$,
(ii) no parameter of any of the operators $(\psi_P/P\boldsymbol{x}_P)$ occurs in φ,
(iii) no two operators $(\psi_P/P\boldsymbol{x}_P)$ share any parameter, and
(iv) no parameter occurs in any of the $\boldsymbol{x}$-lists.

For each $P \in \Pi$, let $\boldsymbol{\pi}_P$ be a list of the parameters of $(\psi_P/P\boldsymbol{x}_P)$. Thus $\boldsymbol{\pi}_P$ is disjoint from $\boldsymbol{x}_P$ by definition of the notion of parameter. Condition (ii) says that each $\boldsymbol{\pi}_P$ is disjoint from φ, (iii) says that the members of $\{\boldsymbol{\pi}_P : P \in \Pi\}$ are pairwise disjoint, and (iv) says that $\boldsymbol{\pi}_P$ is disjoint, not just from $\boldsymbol{x}_P$, but from $\boldsymbol{x}_Q$ for all $Q \in \Pi$. Conditions (iii) and (iv) are thus independent of φ, so are intrinsic features of the operator $(\psi_P/P\boldsymbol{x}_P : P \in \Pi)$.

If no parameter occurs in φ, as in (ii), then certainly no parameter occurs bound in φ, i.e. (S2) holds. So every substitution that is strongly free at φ

must be free at φ. The converse is not true, but the two types of substitution are equivalent as inference rules:

THEOREM 4.3.5. *A logic is closed under free predicate substitutions if, and only if, it is closed under strongly free predicate substitutions.*

PROOF. Closure under free substitutions implies closure under strongly free ones, as every strongly free substitution is free.

Conversely, suppose L is closed under strongly free substitutions. Assume that $\vdash_{\mathrm{L}} \varphi$, and suppose $(\psi_P/P\boldsymbol{x}_P : P \in \Pi)$ is free at φ, with associated parameter lists $\{\boldsymbol{\pi}_P : P \in \Pi\}$. To show that $\vdash_{\mathrm{L}} \varphi(\psi_P/P\boldsymbol{x}_P : P \in \Pi)$ we construct a substitution operator $(\psi'_P/P\boldsymbol{x}_P : P \in \Pi)$ that is strongly free at φ, such that $\varphi(\psi_P/P\boldsymbol{x}_P : P \in \Pi)$ is derivable from $\varphi(\psi'_P/P\boldsymbol{x}_P : P \in \Pi)$ by Term Instantiation.

For each $P \in \Pi$, choose a fresh list $\boldsymbol{\pi}'_P$ of variables, of the same length as $\boldsymbol{\pi}_P$, such that each $\boldsymbol{\pi}'_P$ is disjoint from φ and from $\boldsymbol{\pi}_Q$ and $\boldsymbol{x}_Q$ and ψ_Q for every $Q \in \Pi$ (including $Q = P$); and such that $\{\boldsymbol{\pi}'_P : P \in \Pi\}$ is pairwise disjoint. Since there are infinitely many variables, a selection as described is possible. Let ψ'_P be $\psi_P(\boldsymbol{\pi}'_P/\boldsymbol{\pi}_P)$. Then the free variables of ψ'_P that are not in $\boldsymbol{x}_P$ are precisely the members of $\boldsymbol{\pi}'_P$, so $\boldsymbol{\pi}'_P$ is the parameter list of $(\psi'_P/P\boldsymbol{x}_P)$.

Now by supposition, (S1) holds for each P. Thus if $P\boldsymbol{\tau}$ occurs in φ, then $\boldsymbol{\tau}$ is free for $\boldsymbol{x}_P$ in ψ_P. But since $\boldsymbol{\tau}$ occurs in φ, which is disjoint from $\boldsymbol{\pi}'_P$, $\boldsymbol{\tau}$ is disjoint from $\boldsymbol{\pi}'_P$, so $\boldsymbol{\tau}$ is free for $\boldsymbol{x}_P$ in $\psi_P(\boldsymbol{\pi}'_P/\boldsymbol{\pi}_P) = \psi'_P$. Thus (S1) holds with ψ'_P in place of ψ_P. The choice of the $\boldsymbol{\pi}'_P$'s then ensures that $(\psi'_P/P\boldsymbol{x}_P : P \in \Pi)$ is strongly free at φ.

Let $\varphi' = \varphi(\psi'_P/P\boldsymbol{x}_P : P \in \Pi)$. Then $\vdash_{\mathrm{L}} \varphi'$ by closure of L under strongly free substitutions. Now we ask: whereabouts in φ' can a variable π' from $\boldsymbol{\pi}'_P$ occur? Answer:

> Since $\boldsymbol{\pi}'_P$ is disjoint from φ, from every ψ_Q, and from every $\boldsymbol{\pi}'_Q$ with $Q \neq P$, such a π' can only occur in φ' within a subformula $\psi'_P(\boldsymbol{\tau}/\boldsymbol{x}_P)$ that is substituted for some instance $P\boldsymbol{\tau}$ in φ when $(\psi'_P/P\boldsymbol{x}_P)$ is applied to φ in forming φ'.

This means that if we apply the substitution $(\boldsymbol{\pi}_P/\boldsymbol{\pi}'_P)$ to φ', it will only affect subformulas of φ' of the form $\psi'_P(\boldsymbol{\tau}/\boldsymbol{x}_P)$ with $P\boldsymbol{\tau}$ in φ. But

$$\psi'_P(\boldsymbol{\tau}/\boldsymbol{x}_P)(\boldsymbol{\pi}_P/\boldsymbol{\pi}'_P) = \psi_P(\boldsymbol{\pi}'_P/\boldsymbol{\pi}_P)(\boldsymbol{\tau}/\boldsymbol{x}_P)(\boldsymbol{\pi}_P/\boldsymbol{\pi}'_P) = \psi_P(\boldsymbol{\tau}/\boldsymbol{x}_P).$$

The first of these two equalities holds just by definition of ψ'_P. For the second, since $\boldsymbol{\pi}_P$ is disjoint from $\boldsymbol{x}_P$ the effect of applying $(\boldsymbol{\pi}'_P/\boldsymbol{\pi}_P)$ and then $(\boldsymbol{\tau}/\boldsymbol{x}_P)$ to ψ_P is to replace each member of $\boldsymbol{\pi}_P$ in ψ_P by the corresponding member of $\boldsymbol{\pi}'_P$, and then replace all members of $\boldsymbol{x}_P$ by the corresponding member of $\boldsymbol{\tau}$, with the second step not touching the $\boldsymbol{\pi}'_P$'s newly introduced by the first step, because $\boldsymbol{\pi}'_P$ was chosen to be disjoint from $\boldsymbol{x}_P$. Now there are no $\boldsymbol{\pi}'_P$'s in $\boldsymbol{\tau}$: the two lists are disjoint as explained above. So if we then take the third step of applying $(\boldsymbol{\pi}_P/\boldsymbol{\pi}'_P)$, this will just replace the $\boldsymbol{\pi}'_P$'s introduced in the first step by their original counterparts in π_P, and leave the τ's introduced in the second step unchanged. The overall effect of the three steps is the same as applying (τ/x_P) to ψ_P.

Thus we see that the substitution (π_P/π'_P) converts $\psi'_P(\tau/x_P)$ into $\psi_P(\tau/x_P)$, which is the formula that is substituted for $P\tau$ in φ by the operator (ψ_P/Px_P). So by performing the substitutions (π_P/π'_P) for all $P \in \Pi$ on φ', which can be done in any order as the π'_P are pairwise disjoint and disjoint from every π_Q, the overall effect is to remove all the π'_P-parameters from φ' and to convert φ' into the same formula that would arise from φ by simultaneously applying all the substitutions (ψ_P/Px_P).

In other words, if $\Pi = \{P_1, \dots, P_n\}$, then

$$\varphi'(\pi_{P_1}/\pi'_{P_1}) \cdots (\pi_{P_n}/\pi'_{P_n}) = \varphi(\psi_P/Px_P : P \in \Pi).$$

Since $\vdash_L \varphi'$, we get $\vdash_L \varphi'(\pi_{P_1}/\pi'_{P_1})$ by the simultaneous Term Instantiation rule TI* of Lemma 1.2.3. Repeated application of TI* in this way leads to $\vdash_L \varphi'(\pi_{P_1}/\pi'_{P_1}) \cdots (\pi_{P_n}/\pi'_{P_n})$, and therefore $\vdash_L \varphi(\psi_P/Px_P : P \in \Pi)$.

That completes the proof that L is closed under free substitutions. ⊣

The ability to confine our attention to strongly free substitutions will be very helpful when it comes to proving the soundness of predicate letter substitution in Section 4.4. There we will also need the fact that preservation of freeness by subformulas (Lemma 4.2.2) extends to strong freeness:

LEMMA 4.3.6. *For a given operator* $(\psi_P/Px_P : P \in \Pi)$, *the set*

$$\{\varphi : (\psi_P/Px_P : P \in \Pi) \text{ is strongly free at } \varphi\}$$

is closed under subformulas. ⊣

4.3.5. Parameterless Substitution. We say that a substitution operator $(\psi_P/Px_P : P \in \Pi)$ is *parameterless* if, for each $P \in \Pi$, the operator (ψ_P/Px_P) has no parameters, i.e. every free variable of ψ_P belongs to the list x_P. This implies the condition (S2) for any φ: if there are no parameters, then there are no parameters bound in φ. Thus a parameterless substitution is free at φ iff it satisfies (S1) for every $P \in \Pi$.

THEOREM 4.3.7. *For a signature with infinitely many individual constants, a logic is closed under free predicate substitutions if, and only if, it is closed under parameterless free predicate substitutions.*

PROOF. Closure under free substitutions implies closure under the parameterless ones in particular.

Conversely, suppose L is closed under parameterless free substitutions. To show that it is closed under all free substitutions, it suffices by Theorem 4.3.5 to show that it is closed under strongly free substitutions.

So, let $\vdash_L \varphi$, and suppose $(\psi_P/Px_P : P \in \Pi)$ is strongly free at φ, with associated parameter lists $\{\pi_P : P \in \Pi\}$. We turn this into a parameterless operator by replacing the parameters π_P by new individual constants, of which we have an infinite supply. This is exactly parallel to what we did in Theorem 4.3.5 in replacing the $\boldsymbol{\pi}_P$'s by the new variables $\boldsymbol{\pi}'_P$.

For each $P \in \Pi$, choose a fresh list $\mathbf{c}_P$ of constants, of the same length as $\boldsymbol{\pi}_P$, such that each $\mathbf{c}_P$ is disjoint from φ and from ψ_Q for every $Q \in \Pi$; and $\{\mathbf{c}_P : P \in \Pi\}$ is pairwise disjoint. Let ψ'_P be $\psi_P(\mathbf{c}_P/\boldsymbol{\pi}_P)$, replacing all the parameters of ψ_P by constants. Thus any free variables of ψ'_P must belong to $\boldsymbol{x}_P$, and so $(\psi'_P/P\boldsymbol{x}_P)$ is a parameterless substitution operator.

Notice also that ψ'_P contains no free occurrence of any member of any $\boldsymbol{\pi}_Q$, since $\boldsymbol{\pi}_Q$ is disjoint from $\boldsymbol{x}_P$ by condition (iv) of strong freeness.

Now reasoning just as in Theorem 4.3.5, but with $\mathbf{c}_P$ in place of $\boldsymbol{\pi}'_P$, we find that $(\psi'_P/P\boldsymbol{x}_P : P \in \Pi)$ is free at φ, and that if $\Pi = \{P_1, \ldots, P_n\}$ and $\varphi' = \varphi(\psi'_P/P\boldsymbol{x}_P : P \in \Pi)$, then

$$\varphi'(\boldsymbol{\pi}_{P_1}/\mathbf{c}_{P_1}) \cdots (\boldsymbol{\pi}_{P_n}/\mathbf{c}_{P_n}) = \varphi(\psi_P/P\boldsymbol{x}_P : P \in \Pi).$$

Now $\vdash_{\mathrm{L}} \varphi'$ by closure of L under parameterless free substitutions. Since $(\psi_P/P\boldsymbol{x}_P : P \in \Pi)$ is strongly free at φ, every $\boldsymbol{\pi}_Q$ is disjoint from φ'. Indeed $\boldsymbol{\pi}_Q$ is disjoint from φ, so a member of $\boldsymbol{\pi}_Q$ could only occur in φ' within a subformula $\psi'_P(\boldsymbol{\tau}/\boldsymbol{x}_P)$ that is substituted for some instance $P\boldsymbol{\tau}$ in φ. But we already noted that $\boldsymbol{\pi}_Q$ is disjoint from ψ'_P, and it is disjoint from $\boldsymbol{\tau}$ since $\boldsymbol{\tau}$ occurs in φ, so it is disjoint from $\psi'_P(\boldsymbol{\tau}/\boldsymbol{x}_P)$.

In particular, $\boldsymbol{\pi}_{P_1}$ is disjoint from φ', so then $\vdash_{\mathrm{L}} \varphi'(\boldsymbol{\pi}_{P_1}/\mathbf{c}_{P_1})$ by the rule Sub* of Lemma 1.2.3. But $\boldsymbol{\pi}_{P_2}$ is disjoint from φ' and from $\boldsymbol{\pi}_{P_1}$, hence from $\varphi'(\boldsymbol{\pi}_{P_1}/\mathbf{c}_{P_1})$, so $\vdash_{\mathrm{L}} \varphi'(\boldsymbol{\pi}_{P_1}/\mathbf{c}_{P_1})(\boldsymbol{\pi}_{P_2}/\mathbf{c}_{P_2})$ by Sub*. Repeated application of Sub* in this way leads to $\vdash_{\mathrm{L}} \varphi'(\boldsymbol{\pi}_{P_1}/\mathbf{c}_{P_1}) \cdots (\boldsymbol{\pi}_{P_n}/\mathbf{c}_{P_n})$, hence $\vdash_{\mathrm{L}} \varphi(\psi_P/P\boldsymbol{x}_P : P \in \Pi)$, showing that L is indeed closed under strongly free substitutions. ⊣

To summarise all the results of this section, we have shown that the rules of

- strongly free substitution,
- free substitution,
- sufficiently free substitution, and
- general substitution

for predicate letters are all deductively equivalent over any quantified modal logic; and that if the signature contains infinitely many individual constants, then they are all deductively equivalent to

- parameterless free substitution.

4.4. Soundness of Predicate Substitution

The rule of substitution for predicate letters is

$$\frac{\varphi}{\varphi(\psi_P/P\boldsymbol{x}_P : P \in \Pi)},$$

i.e. *from* φ *infer any substitution instance* $\varphi(\psi_P/P\boldsymbol{x}_P : P \in \Pi)$.

We are now going to prove the soundness of this rule for validity in any model structure $\mathcal{S}$. This means that the logic $\mathrm{L}_{\mathcal{S}} = \{\varphi : \mathcal{S} \models \varphi\}$ characterised by $\mathcal{S}$ is closed under this rule. The proof will also establish soundness of the rule for validity in any functional model structure $\mathfrak{S}$.

From what we saw in the previous section, it suffices to show that logics of the form $\mathrm{L}_{\mathcal{S}}$ and $\mathrm{L}_{\mathfrak{S}}$ are closed under *strongly free* substitution. In fact if we were to invoke our general assumption that the background signature $\mathcal{L}$ has infinitely many constants, it would suffice to show that such logics are closed under parameterless free substitution. But the proof we give works for any signature.

We start with an operator $(\psi_P/P\boldsymbol{x}_P : P \in \Pi)$ that satisfies the intrinsic strong freeness properties (iii) and (iv) of Section 4.3.4. If $\boldsymbol{\pi}_P$ is the parameter-list for $(\psi_P/P\boldsymbol{x}_P)$, let $\boldsymbol{\pi}$ be a concatenation of all the members of the collection $\{\boldsymbol{\pi}_P : P \in \Pi\}$, which is pairwise disjoint by (iii). Then $\boldsymbol{\pi}$ is disjoint from $\boldsymbol{x}_P$ for all P by (iv). Let n be the length of $\boldsymbol{\pi}$.

For each formula θ put

$$\theta^{\Pi} = \theta(\psi_P/P\boldsymbol{x}_P : P \in \Pi).$$

Suppose that $\mathcal{S} \models \varphi$ and our operator $(\psi_P/P\boldsymbol{x}_P : P \in \Pi)$ is strongly free at φ, i.e. properties (i) and (ii) of Section 4.3.4 also hold for this φ. We want to show $\mathcal{S} \models \varphi^{\Pi}$, i.e. φ^{Π} is valid in all models on $\mathcal{S}$. So fix a model $\mathcal{M}$ on $\mathcal{S}$. It suffices to take an arbitrary $\boldsymbol{b} \in U^n$, and show that

$$|\varphi^{\Pi}|^{\mathcal{M}} f[\boldsymbol{b}/\boldsymbol{\pi}] = W \quad \text{for any } f \in U^{\omega}. \tag{4.4.1}$$

For then, if $f \in U^{\omega}$, we can take $\boldsymbol{b} = |\boldsymbol{\pi}|^{\mathcal{M}} f$, which we write more briefly as $\boldsymbol{\pi} f$, since this depends only on f because $\boldsymbol{\pi}$ consists only of variables. From (4.4.1) we conclude that $|\varphi^{\Pi}|^{\mathcal{M}} f[\boldsymbol{\pi} f/\boldsymbol{\pi}] = W$. But $f[\boldsymbol{\pi} f/\boldsymbol{\pi}] = f$, so this shows that $|\varphi^{\Pi}|^{\mathcal{M}} f = W$ for any f, hence φ^{Π} is valid in $\mathcal{M}$ as desired.

To prove (4.4.1), given $\mathcal{M}$ and an arbitrary $\boldsymbol{b}$, define a premodel $\mathcal{M}^{\boldsymbol{b}}$ whose interpretation of $P \in \Pi$ will reflect the substitution of ψ_P for P. $\mathcal{M}^{\boldsymbol{b}}$ is identical to $\mathcal{M}$ on the interpretation of individual constants, function symbols, and predicate symbols that do not belong to Π. For the interpretation of $P \in \Pi$, note that the lists $\boldsymbol{x}_P$ and $\boldsymbol{\pi}$ contain all the free variables of ψ_P between them, since any free variable of ψ_P that is not in $\boldsymbol{x}_P$ is, by definition, a parameter in $\boldsymbol{\pi}_P$, hence in $\boldsymbol{\pi}$. Thus $|\psi_P|^{\mathcal{M}}$ depends only on the values of the variables in the disjoint lists $\boldsymbol{x}_P$ and $\boldsymbol{\pi}$. Hence for each δP-tuple $\boldsymbol{a}$ of individuals from U we can define

$$|P|^{\mathcal{M}^{\boldsymbol{b}}} \boldsymbol{a} = |\psi_P|^{\mathcal{M}} f[\boldsymbol{a}/\boldsymbol{x}_P, \boldsymbol{b}/\boldsymbol{\pi}], \tag{4.4.2}$$

where f is any member of U^{ω}. This definition does not depend on the choice of f.

Note that, because $\mathcal{M}$ and $\mathcal{M}^{\boldsymbol{b}}$ are identical on constants and function symbols, we have $|\boldsymbol{\tau}|^{\mathcal{M}} f = |\boldsymbol{\tau}|^{\mathcal{M}^{\boldsymbol{b}}} f$ for any term-tuple $\boldsymbol{\tau}$ and any variable-assignment f.

LEMMA 4.4.1. *For any formula θ, if $(\psi_P/P\boldsymbol{x}_P : P \in \Pi)$ is strongly free at θ, then for all $f \in U^\omega$,*

$$|\theta|^{\mathcal{M}^{\boldsymbol{b}}} f = |\theta^\Pi|^{\mathcal{M}} f[\boldsymbol{b}/\boldsymbol{\pi}].$$

PROOF. Repeated use will be made of the fact that if $(\psi_P/P\boldsymbol{x}_P : P \in \Pi)$ is strongly free at θ, then it is strongly free at every subformula of θ (Lemma 4.3.6).

We proceed by induction on the formation of θ. There are three base cases for different kinds of atomic formula, and then the inductive cases of the connectives and quantifiers. In each case, the assumption that $(\psi_P/P\boldsymbol{x}_P : P \in \Pi)$ is strongly free at θ implies that the parameter list $\boldsymbol{\pi}$ is disjoint from θ, by the strong freeness condition (ii).

- If θ is $\mathcal{E}\boldsymbol{\tau}$ for some $\boldsymbol{\tau}$, then $\theta^\Pi = \theta$. For all $f \in U^\omega$, using the above observation that $|\boldsymbol{\tau}|^{\mathcal{M}} f = |\boldsymbol{\tau}|^{\mathcal{M}^{\boldsymbol{b}}} f$,

  $$|\theta|^{\mathcal{M}^{\boldsymbol{b}}} f = |\mathcal{E}\boldsymbol{\tau}|^{\mathcal{M}^{\boldsymbol{b}}} f = E(|\boldsymbol{\tau}|^{\mathcal{M}^{\boldsymbol{b}}} f) = E(|\boldsymbol{\tau}|^{\mathcal{M}} f) = |\mathcal{E}\boldsymbol{\tau}|^{\mathcal{M}} f = |\theta|^{\mathcal{M}} f.$$

  Now $\boldsymbol{\pi}$ is disjoint from θ, so no variable of the atomic θ occurs in $\boldsymbol{\pi}$. Hence f and $f[\boldsymbol{b}/\boldsymbol{\pi}]$ agree on the variables of θ, so $|\theta|^{\mathcal{M}} f = |\theta|^{\mathcal{M}} f[\boldsymbol{b}/\boldsymbol{\pi}]$ by Lemma 1.6.1. Since $\theta^\Pi = \theta$ in this case, we have altogether shown $|\theta|^{\mathcal{M}^{\boldsymbol{b}}} f = |\theta^\Pi|^{\mathcal{M}} f[\boldsymbol{b}/\boldsymbol{\pi}]$ as required.
- If θ is $P\boldsymbol{\tau}$ with $P \notin \Pi$, then again $\theta^\Pi = \theta$. Also $|P|^{\mathcal{M}^{\boldsymbol{b}}} = |P|^{\mathcal{M}}$ by definition of $\mathcal{M}^{\boldsymbol{b}}$, so

  $$|P\boldsymbol{\tau}|^{\mathcal{M}^{\boldsymbol{b}}} f = |P|^{\mathcal{M}^{\boldsymbol{b}}}(|\boldsymbol{\tau}|^{\mathcal{M}^{\boldsymbol{b}}} f) = |P|^{\mathcal{M}}(|\boldsymbol{\tau}|^{\mathcal{M}} f) = |P\boldsymbol{\tau}|^{\mathcal{M}} f.$$

  Hence again $|\theta|^{\mathcal{M}^{\boldsymbol{b}}} f = |\theta|^{\mathcal{M}} f$. Using the disjointness of $\boldsymbol{\pi}$ from θ, the result $|\theta|^{\mathcal{M}^{\boldsymbol{b}}} f = |\theta^\Pi|^{\mathcal{M}} f[\boldsymbol{b}/\boldsymbol{\pi}]$ then follows exactly as in the previous case.
- If θ is $P\boldsymbol{\tau}$ with $P \in \Pi$, then $\boldsymbol{\tau}$ is free for $\boldsymbol{x}_P$ in θ by condition (i) of strong freeness at θ, i.e. (S1), and θ^Π is $\psi_P(\boldsymbol{\tau}/\boldsymbol{x}_P)$. Also, as $\boldsymbol{\pi}$ is disjoint from $P\boldsymbol{\tau}$, $|\boldsymbol{\tau}|^{\mathcal{M}} f = |\boldsymbol{\tau}|^{\mathcal{M}} f[\boldsymbol{b}/\boldsymbol{\pi}]$. Now

$$\begin{aligned}
|P\boldsymbol{\tau}|^{\mathcal{M}^{\boldsymbol{b}}} f &= |P|^{\mathcal{M}^{\boldsymbol{b}}} |\boldsymbol{\tau}|^{\mathcal{M}^{\boldsymbol{b}}} f, \text{ by the semantics of } P\boldsymbol{\tau},\\
&= |P|^{\mathcal{M}^{\boldsymbol{b}}} |\boldsymbol{\tau}|^{\mathcal{M}} f, \text{ since } |\boldsymbol{\tau}|^{\mathcal{M}} f = |\boldsymbol{\tau}|^{\mathcal{M}^{\boldsymbol{b}}} f \text{ as above},\\
&= |\psi_P|^{\mathcal{M}} f[\,|\boldsymbol{\tau}|^{\mathcal{M}} f/\boldsymbol{x}_P, \boldsymbol{b}/\boldsymbol{\pi}], \text{ by definition (4.4.2) of } |P|^{\mathcal{M}^{\boldsymbol{b}}},\\
&= |\psi_P|^{\mathcal{M}} f[\boldsymbol{b}/\boldsymbol{\pi}][\,|\boldsymbol{\tau}|^{\mathcal{M}} f/\boldsymbol{x}_P], \text{ as } \boldsymbol{x}_P \text{ and } \boldsymbol{\pi} \text{ are disjoint by (iv)},\\
&= |\psi_P|^{\mathcal{M}} f[\boldsymbol{b}/\boldsymbol{\pi}][\,|\boldsymbol{\tau}|^{\mathcal{M}} f[\boldsymbol{b}/\boldsymbol{\pi}]/\boldsymbol{x}_P], \text{ as } |\boldsymbol{\tau}|^{\mathcal{M}} f = |\boldsymbol{\tau}|^{\mathcal{M}} f[\boldsymbol{b}/\boldsymbol{\pi}],\\
&= |\psi(\boldsymbol{\tau}/\boldsymbol{x}_P)|^{\mathcal{M}} f[\boldsymbol{b}/\boldsymbol{\pi}], \text{ by Lemma 4.2.3.}
\end{aligned}$$

  So the result holds in this case.

- If the result holds for θ, and $(\psi_P/P\boldsymbol{x}_P : P \in \Pi)$ is free at $\neg\theta$, then

$$\begin{aligned}|\neg\theta|^{\mathcal{M}^{\boldsymbol{b}}} f &= W - |\theta|^{\mathcal{M}^{\boldsymbol{b}}} f \\ &= W - |\theta^{\Pi}|^{\mathcal{M}} f[\boldsymbol{b}/\boldsymbol{\pi}], \quad \text{by induction hypothesis,} \\ &= |\neg(\theta^{\Pi})|^{\mathcal{M}} f[\boldsymbol{b}/\boldsymbol{\pi}],\end{aligned}$$

  so the result holds for $\neg\theta$ as $\neg(\theta^{\Pi}) = (\neg\theta)^{\Pi}$.
- If the result holds for θ_1 and θ_2, and $(\psi_P/P\boldsymbol{x}_P : P \in \Pi)$ is free at $\theta_1 \wedge \theta_2$, then

$$\begin{aligned}|\theta_1 \wedge \theta_2|^{\mathcal{M}^{\boldsymbol{b}}} f &= |\theta_1|^{\mathcal{M}^{\boldsymbol{b}}} f \cap |\theta_2|^{\mathcal{M}^{\boldsymbol{b}}} f \\ &= |\theta_1^{\Pi}|^{\mathcal{M}} f[\boldsymbol{b}/\boldsymbol{\pi}] \cap |\theta_2^{\Pi}|^{\mathcal{M}} f[\boldsymbol{b}/\boldsymbol{\pi}], \quad \text{by induction hypothesis,} \\ &= |\theta_1^{\Pi} \wedge \theta_2^{\Pi}|^{\mathcal{M}} f[\boldsymbol{b}/\boldsymbol{\pi}],\end{aligned}$$

  so the result holds for $\theta_1 \wedge \theta_2$ as $\theta_1^{\Pi} \wedge \theta_2^{\Pi} = (\theta_1 \wedge \theta_2)^{\Pi}$.
- If the result holds for θ, and $(\psi_P/P\boldsymbol{x}_P : P \in \Pi)$ is free at $\Box\theta$, then

$$|\Box\theta|^{\mathcal{M}^{\boldsymbol{b}}} f = [R]|\theta|^{\mathcal{M}^{\boldsymbol{b}}} f = [R]|\theta^{\Pi}|^{\mathcal{M}} f[\boldsymbol{b}/\boldsymbol{\pi}] = |\Box(\theta^{\Pi})|^{\mathcal{M}} f[\boldsymbol{b}/\boldsymbol{\pi}],$$

  so the result holds for $\Box\theta$ as $\Box(\theta^{\Pi}) = (\Box\theta)^{\Pi}$.
- If the result holds for θ, and $(\psi_P/P\boldsymbol{x}_P : P \in \Pi)$ is free at $\forall y\theta$, then no parameter occurs in $\forall y\theta$ by condition (ii) of strong freeness, so y is not in the list $\boldsymbol{\pi}$. Thus

$$\begin{aligned}|\forall y\theta|^{\mathcal{M}^{\boldsymbol{b}}} f &= \prod_{a\in U} Ea \Rightarrow |\theta|^{\mathcal{M}^{\boldsymbol{b}}} f[a/y] \\ &= \prod_{a\in U} Ea \Rightarrow |\theta^{\Pi}|^{\mathcal{M}} f[a/y][\boldsymbol{b}/\boldsymbol{\pi}], \quad \text{by induction hypothesis,} \\ &= \prod_{a\in U} Ea \Rightarrow |\theta^{\Pi}|^{\mathcal{M}} f[\boldsymbol{b}/\boldsymbol{\pi}][a/y], \quad \text{as } y \text{ is not in } \boldsymbol{\pi}, \\ &= |\forall y(\theta^{\Pi})|^{\mathcal{M}} f[\boldsymbol{b}/\boldsymbol{\pi}],\end{aligned}$$

  so the result holds for $\forall y\theta$ as $\forall y(\theta^{\Pi}) = (\forall y\theta)^{\Pi}$.

This completes the inductive proof. ⊣

LEMMA 4.4.2. *For each formula θ there is a formula θ^* and an n-tuple $\boldsymbol{\pi}^*$ of variables such that the propositional function $|\theta|^{\mathcal{M}^{\boldsymbol{b}}}$ is equal to the simultaneous updating $|\theta^*|^{\mathcal{M}}[\boldsymbol{b}/\boldsymbol{\pi}^*]$ of the function $|\theta^*|^{\mathcal{M}}$.*

PROOF. Recall from (4.2.2) that $\big(|\theta^*|^{\mathcal{M}}[\boldsymbol{b}/\boldsymbol{\pi}^*]\big)f$ is defined to be the updating $|\theta^*|^{\mathcal{M}} f[\boldsymbol{b}/\boldsymbol{\pi}^*]$, so we have to show that

$$|\theta|^{\mathcal{M}^{\boldsymbol{b}}} f = |\theta^*|^{\mathcal{M}} f[\boldsymbol{b}/\boldsymbol{\pi}^*] \tag{4.4.3}$$

for every $f \in U^{\omega}$. We distinguish three cases.

First, suppose that $(\psi_P/P\boldsymbol{x}_P : P \in \Pi)$ is strongly free at θ. Put $\theta^* = \theta^{\Pi}$ and $\boldsymbol{\pi}^* = \boldsymbol{\pi}$. Then Lemma 4.4.1 gives (4.4.3).

For the second case, suppose that $(\psi_P/P\boldsymbol{x}_P : P \in \Pi)$ is free at θ, but not strongly free. Then, as in the proof of Theorem 4.3.5, we construct another substitution operator $(\psi'_P/P\boldsymbol{x}_P : P \in \Pi)$ that is strongly free at θ. We choose, for each $P \in \Pi$, a fresh list $\boldsymbol{\pi}'_P$ of variables, of the same length as $\boldsymbol{\pi}_P$, such that each $\boldsymbol{\pi}'_P$ is disjoint from θ and from $\boldsymbol{\pi}_Q$ and $\boldsymbol{x}_Q$ and ψ_Q for every $Q \in \Pi$; and such that $\{\boldsymbol{\pi}'_P : P \in \Pi\}$ is pairwise disjoint. ψ'_P is defined to be $\psi_P(\boldsymbol{\pi}'_P/\boldsymbol{\pi}_P)$. Then $\boldsymbol{\pi}'_P$ is the parameter list of $(\psi'_P/P\boldsymbol{x}_P)$. Let $\boldsymbol{\pi}'$ be a concatenation of $\boldsymbol{\pi}'_P$'s for all $P \in \Pi$.

Now we repeat the construction and proof of Lemma 4.4.1, using the new operator $(\psi'_P/P\boldsymbol{x}_P : P \in \Pi)$ in place of $(\psi_P/P\boldsymbol{x}_P : P \in \Pi)$, and its parameter list $\boldsymbol{\pi}'$ in place of $\boldsymbol{\pi}$. This involves defining a premodel $\mathcal{M}^{\boldsymbol{b}'}$ that agrees with $\mathcal{M}$ except on members of Π, where it has

$$|P|^{\mathcal{M}^{\boldsymbol{b}'}}\boldsymbol{a} = |\psi'_P|^{\mathcal{M}} f[\boldsymbol{a}/\boldsymbol{x}_P, \boldsymbol{b}/\boldsymbol{\pi}'] \tag{4.4.4}$$

for arbitrary f, in place of (4.4.2). Since the new operator is strongly free at θ, by the proof of Lemma 4.4.1 we get

$$|\theta|^{\mathcal{M}^{\boldsymbol{b}'}} f = |\theta(\psi'_P/P\boldsymbol{x}_P : P \in \Pi)|^{\mathcal{M}} f[\boldsymbol{b}/\boldsymbol{\pi}']$$

in general. But it turns out that $\mathcal{M}^{\boldsymbol{b}'}$ is identical to $\mathcal{M}^{\boldsymbol{b}}$, so putting $\theta^* = \theta(\psi'_P/P\boldsymbol{x}_P : P \in \Pi)$ and $\boldsymbol{\pi}^* = \boldsymbol{\pi}'$ gives the desired result (4.4.3) in this case.

To see why $\mathcal{M}^{\boldsymbol{b}'} = \mathcal{M}^{\boldsymbol{b}}$, observe that for $P \in \Pi$, by definition of ψ'_P we have

$$|P|^{\mathcal{M}^{\boldsymbol{b}'}}\boldsymbol{a} = |\psi_P(\boldsymbol{\pi}'_P/\boldsymbol{\pi}_P)|^{\mathcal{M}} f[\boldsymbol{a}/\boldsymbol{x}_P, \boldsymbol{b}/\boldsymbol{\pi}'].$$

If $g = f[\boldsymbol{a}/\boldsymbol{x}_P, \boldsymbol{b}/\boldsymbol{\pi}']$, then by the Substitution Lemma 4.2.3 this gives

$$|P|^{\mathcal{M}^{\boldsymbol{b}'}}\boldsymbol{a} = |\psi_P|^{\mathcal{M}} g[\boldsymbol{\pi}'_P g/\boldsymbol{\pi}_P].$$

Now the assignment $g[\boldsymbol{\pi}'_P g/\boldsymbol{\pi}_P]$, i.e. $f[\boldsymbol{a}/\boldsymbol{x}_P, \boldsymbol{b}/\boldsymbol{\pi}'][\boldsymbol{\pi}'_P g/\boldsymbol{\pi}_P]$, agrees with $f[\boldsymbol{a}/\boldsymbol{x}_P, \boldsymbol{b}/\boldsymbol{\pi}]$ on the free variables of ψ_P, because $\boldsymbol{\pi}'$ is disjoint from ψ_P; each free variable of ψ_P belongs to one of the disjoint lists $\boldsymbol{x}_P$ and $\boldsymbol{\pi}_P$; and the update $[\boldsymbol{\pi}'_P g/\boldsymbol{\pi}_P]$ agrees with $[\boldsymbol{b}/\boldsymbol{\pi}]$ on $\boldsymbol{\pi}_P$, since $\boldsymbol{\pi}' g = \boldsymbol{b}$.

Consequently, by the Free Assignment Lemma 1.6.1, we get

$$|P|^{\mathcal{M}^{\boldsymbol{b}'}}\boldsymbol{a} = |\psi_P|^{\mathcal{M}} f[\boldsymbol{a}/\boldsymbol{x}_P, \boldsymbol{b}/\boldsymbol{\pi}] = |P|^{\mathcal{M}^{\boldsymbol{b}}}\boldsymbol{a}$$

by (4.4.2). That completes the proof that $\mathcal{M}^{\boldsymbol{b}'} = \mathcal{M}^{\boldsymbol{b}}$, and completes this case.

Finally, we have the case that $(\psi_P/P\boldsymbol{x}_P : P \in \Pi)$ is not free at θ. Then we reduce this to the second case just completed by modifying the substitution operator to one that is strongly free at θ. First we reletter the bound variables of the ψ_P's, choosing, for each $P \in \Pi$, a ψ^*_P congruent to ψ_P such that no bound variable of ψ^*_P occurs in θ. This ensures that the substitutions $(\psi^*_P/P\boldsymbol{x}_P)$ all satisfy (S1) at θ, i.e. each P-instance $P\boldsymbol{\tau}$ occurring in θ has $\boldsymbol{\tau}$

free for $\boldsymbol{x}$ in ψ_P^*. Note that $(\psi_P^*/P\boldsymbol{x}_P)$ has the same list $\boldsymbol{\pi}_P$ of parameters as $(\psi_P/P\boldsymbol{x}_P)$.

Next we proceed as in the second case above to replace each list $\boldsymbol{\pi}_P$ by a fresh list $\boldsymbol{\pi}'_P$ of parameters that is disjoint from θ and from $\boldsymbol{\pi}_Q$ and $\boldsymbol{x}_Q$ and ψ_Q^* for every $Q \in \Pi$, with $\{\boldsymbol{\pi}'_P : P \in \Pi\}$ being pairwise disjoint. This time we define ψ'_P to be $\psi_P^*(\boldsymbol{\pi}'_P/\boldsymbol{\pi}_P)$, and again take $\boldsymbol{\pi}'$ to be a concatenation of the $\boldsymbol{\pi}'_P$'s.

Now we have a substitution operator $(\psi'_P/P\boldsymbol{x}_P : P \in \Pi)$ that is strongly free at θ, so we apply the argument of the second case to this situation. Since ψ_P^* is congruent to ψ_P and $\mathcal{M}$ is a model, we have $|\psi_P^*|^{\mathcal{M}} = |\psi_P|^{\mathcal{M}}$ by Lemma 4.3.1. So replacing ψ_P by ψ_P^* in (4.4.2) does not change our premodel $\mathcal{M}^{\boldsymbol{b}}$. Then replacing ψ_P^* by $\psi'_P = \psi_P^*(\boldsymbol{\pi}'_P/\boldsymbol{\pi}_P)$ as in (4.4.4) again does not change $\mathcal{M}^{\boldsymbol{b}}$ and we obtain by the method of Lemma 4.4.1 that

$$|\theta|^{\mathcal{M}^{\boldsymbol{b}}} f = |\theta(\psi'_P/P\boldsymbol{x}_P : P \in \Pi)|^{\mathcal{M}} f[\boldsymbol{b}/\boldsymbol{\pi}']$$

in general. So putting $\theta^* = \theta(\psi_P^*(\boldsymbol{\pi}'_P/\boldsymbol{\pi}_P)/P\boldsymbol{x}_P : P \in \Pi)$ and $\boldsymbol{\pi}^* = \boldsymbol{\pi}'$ gives the desired result (4.4.3) in this final case. ⊣

COROLLARY 4.4.3. *Every truth set of $\mathcal{M}^{\boldsymbol{b}}$ is admissible in $\mathcal{S}$.*

PROOF. By (4.4.3), each truth set $|\theta|^{\mathcal{M}^{\boldsymbol{b}}} f$ is equal to $|\theta^*|^{\mathcal{M}} f[\boldsymbol{b}/\boldsymbol{\pi}^*]$, which belongs to *Prop* as $\mathcal{M}$ is a model on $\mathcal{S}$. ⊣

We can now complete the proof of equation (4.4.1), and hence the proof that the logic $\mathrm{L}_{\mathcal{S}}$ is closed under strongly free substitution. To recap: we started with a φ that is valid in $\mathcal{S}$, with the operator $(\psi_P/P\boldsymbol{x}_P : P \in \Pi)$ strongly free at φ, and an arbitrary $\boldsymbol{b} \in U^n$. Given a model $\mathcal{M}$ on $\mathcal{S}$, we defined the premodel $\mathcal{M}^{\boldsymbol{b}}$ which we now see from Corollary 4.4.3 is a model on $\mathcal{S}$. Thus as $\mathcal{S}$ validates φ we have φ valid in $\mathcal{M}^{\boldsymbol{b}}$, so for any $f \in U^{\omega}$, by Lemma 4.4.1,

$$|\varphi^{\Pi}|^{\mathcal{M}} f[\boldsymbol{b}/\boldsymbol{\pi}] = |\varphi|^{\mathcal{M}^{\boldsymbol{b}}} f = W,$$

which is the required (4.4.1).

That completes the proof that substitution for predicate letters is sound for validity in any model structure.

THEOREM 4.4.4. *The rule of substitution for predicate letters is sound for validity in any functional model structure $\mathfrak{S}$, i.e. the logic $\mathrm{L}_{\mathfrak{S}} = \{\varphi : \mathfrak{S} \models \varphi\}$ characterised by $\mathfrak{S}$ is closed under this rule.*

PROOF. The above soundness proof for validity of the rule in model structures goes through with one adjustment. Taking a φ that is valid in all models on $\mathfrak{S}$, we want to show that a substitution instance φ^{Π} is so valid as well. If $\mathcal{M}$ is any model on $\mathfrak{S}$, then $\mathcal{M}$ is a model on the structure $\mathcal{S}$ underlying $\mathfrak{S}$, and we can form the premodels $\mathcal{M}^{\boldsymbol{b}}$ on $\mathcal{S}$ by the definition (4.4.2). The one adjustment is that we now have to show that each $\mathcal{M}^{\boldsymbol{b}}$ is also a model on $\mathfrak{S}$, in order to conclude that φ is valid in $\mathcal{M}^b$, so that we can then use this fact to prove (4.4.1) and hence $\mathcal{M} \models \varphi^\Pi$.

For $\mathcal{M}^b$ to be a model on $\mathfrak{S}$ we require the function $|\theta|^{\mathcal{M}^b}$ to belong to the algebra *PropFun* of admissible propositional functions of $\mathfrak{S}$ whenever θ is atomic. In fact Lemma 4.4.2 directly shows that $|\theta|^{\mathcal{M}^b}$ is admissible for *any* formula θ, since $|\theta|^{\mathcal{M}^b}$ is equal to the simultaneous updating $|\theta^*|^{\mathcal{M}}[\boldsymbol{b}/\boldsymbol{\pi}^*]$ of the function $|\theta^*|^{\mathcal{M}}$. We have $|\theta^*|^{\mathcal{M}} \in \mathit{PropFun}$ as $\mathcal{M}$ is a model on $\mathfrak{S}$, and *PropFun* is closed under simultaneous updating by Lemma 4.2.4, so $|\theta|^{\mathcal{M}^b} \in \mathit{PropFun}$ as claimed. ⊣

Note that we could not treat the soundness proof for model structures as a special case of that for *functional* model structures, since we cannot guarantee that a given model structure $\mathcal{S}$ can be identified with a functional one (see Example 4.1.3).

4.5. Functional Canonicity

We will now see that the question of whether a logic L is closed under predicate substitution is equivalent to the question of whether it is characterised by its canonical functional model structure $\mathfrak{S}_{\mathrm{L}}$.

Recall that any logic L is complete for validity in $\mathfrak{S}_{\mathrm{L}}$: if $\mathfrak{S}_{\mathrm{L}} \models \varphi$ then $\mathfrak{M}_{\mathrm{L}} \models \varphi$, and so $\vdash_{\mathrm{L}} \varphi$ (Theorem 4.1.6). L will be called *functionally canonical* if, conversely, it is sound for validity in $\mathfrak{S}_{\mathrm{L}}$, and hence is characterised by $\mathfrak{S}_{\mathrm{L}}$: $\vdash_{\mathrm{L}} \varphi$ iff $\mathfrak{S}_{\mathrm{L}} \models \varphi$.

THEOREM 4.5.1. L *is functionally canonical iff it is closed under substitution for predicate letters.*

PROOF. If L is functionally canonical, then $\mathrm{L} = \{\varphi : \mathfrak{S}_{\mathrm{L}} \models \varphi\}$, which is closed under predicate substitution by Theorem 4.4.4.

For the converse, suppose L is closed under predicate substitution. Let $\vdash_{\mathrm{L}} \varphi$. We have to show that $\mathfrak{S}_{\mathrm{L}} \models \varphi$. Take any model $\mathcal{M}$ on $\mathfrak{S}_{\mathrm{L}}$, in order to show $\mathcal{M} \models \varphi$.

Let Π be the (finite) set of predicate letters occurring in φ. For each $P \in \Pi$ choose a δP-tuple $\boldsymbol{x}_P$ of distinct new variables. Now $|P\boldsymbol{x}_P|^{\mathcal{M}} \in \mathit{PropFun}_{\mathrm{L}}$, since $\mathcal{M}$ is a model on $\mathfrak{S}$, so there is a formula ψ_P with

$$|P\boldsymbol{x}_P|^{\mathcal{M}} = \alpha_{\psi_P} = |\psi_P|^{\mathfrak{M}_{\mathrm{L}}}.$$

We can assume that no bound variable of any ψ_P occurs in φ. This is justified by the fact that if, by rewriting bound variables, we change ψ_P to any formula congruent to it, this does not change the function $|\psi_P|^{\mathfrak{M}_{\mathrm{L}}}$.

We can also assume that no parameter of any $(\psi_P/P\boldsymbol{x}_P)$ occurs in φ, and even that there are no parameters for these operators. This is because the equation $|P\boldsymbol{x}_P|^{\mathcal{M}} = |\psi_P|^{\mathfrak{M}_{\mathrm{L}}}$ ensures that any value $|\psi_P|^{\mathfrak{M}_{\mathrm{L}}} f$ of $|\psi_P|^{\mathfrak{M}_{\mathrm{L}}}$

depends only on the values that f assigns to the members of $\boldsymbol{x}_P$. Indeed, if f and g agree on $\boldsymbol{x}_P$, then $|P\boldsymbol{x}_P|^{\mathcal{M}} f = |P\boldsymbol{x}_P|^{\mathcal{M}} g$, and hence $|\psi_P|^{\mathfrak{M}_\mathrm{L}} f = |\psi_P|^{\mathfrak{M}_\mathrm{L}} g$. Thus, as the parameters of $(\psi_P/P\boldsymbol{x}_P)$ are disjoint from $\boldsymbol{x}_P$, any value $|\psi_P|^{\mathfrak{M}_\mathrm{L}} f$ is unchanged if we change ψ_P by replacing parameters of $(\psi_P/P\boldsymbol{x}_P)$ by other terms, such as new variables disjoint from φ, or constants.

Overall then, we can assume that the operator $(\psi_P/P\boldsymbol{x}_P : P \in \Pi)$ is strongly free at φ, and even that it is parameterless, while maintaining $|P\boldsymbol{x}_P|^{\mathcal{M}} = |\psi_P|^{\mathfrak{M}_\mathrm{L}}$ for all $P \in \Pi$. In fact we only need this operator to be free at φ.

Now the universe $U_{\mathcal{L}}$ of the structure $\mathfrak{S}_\mathrm{L}$ is the set of closed terms of $\mathcal{L}$, so for each constant c, since $|\mathsf{c}|^{\mathcal{M}} \in U_{\mathcal{L}}$ there is some closed term τ_c such that $|\mathsf{c}|^{\mathcal{M}} = \tau_\mathsf{c}$. But $|\tau_\mathsf{c}|^{\mathfrak{M}_\mathrm{L}} = \tau_\mathsf{c}$, as τ_c is closed (see Lemma 1.9.3(4)), so we have $|\mathsf{c}|^{\mathcal{M}} = |\tau_\mathsf{c}|^{\mathfrak{M}_\mathrm{L}}$. The function $\mathsf{c} \mapsto \tau_\mathsf{c}$ induces a substitution on formulas: for each formula θ, let θ' be the result of simultaneously substituting τ_c for c in θ, for all constants c. Likewise, $\boldsymbol{\tau}'$ is the result of applying this substitution to a tuple $\boldsymbol{\tau}$ of terms. Then for any $\boldsymbol{\tau}$ we have $|\boldsymbol{\tau}|^{\mathcal{M}} f = |\boldsymbol{\tau}'|^{\mathfrak{M}_\mathrm{L}} f$ for all f, i.e. the term functions $|\boldsymbol{\tau}|^{\mathcal{M}}$ and $|\boldsymbol{\tau}'|^{\mathfrak{M}_\mathrm{L}}$ are identical.

Note that if $(\psi_P/P\boldsymbol{x}_P : P \in \Pi)$ is free at θ, it is also free at θ', since θ' differs from θ only by a replacement of constants by closed terms.

LEMMA 4.5.2. *Let θ be any formula whose predicate letters belong to Π. If $(\psi_P/P\boldsymbol{x}_P : P \in \Pi)$ is free at θ, then*

$$|\theta|^{\mathcal{M}} = |\theta'(\psi_P/P\boldsymbol{x}_P : P \in \Pi)|^{\mathfrak{M}_\mathrm{L}}.$$

PROOF. Put $\theta^{\Pi} = \theta(\psi_P/P\boldsymbol{x}_P : P \in \Pi)$ for each θ. Then we have to show that in general $|\theta|^{\mathcal{M}} = |\theta'^{\,\Pi}|^{\mathfrak{M}_\mathrm{L}}$. We proceed by induction on the formation of θ.

- Let θ be $P\boldsymbol{\tau}$ with $P \in \Pi$. Then $\boldsymbol{\tau}$ is free for $\boldsymbol{x}_P$ in θ, hence so is $\boldsymbol{\tau}'$, and we have $\theta' = P\boldsymbol{\tau}'$ and $\theta'^{\,\Pi} = \psi_P(\boldsymbol{\tau}'/\boldsymbol{x}_P)$. But for any $f \in {U_{\mathcal{L}}}^{\omega}$,

$$\begin{aligned} |P\boldsymbol{\tau}|^{\mathcal{M}} f &= |P\boldsymbol{x}_P(\boldsymbol{\tau}/\boldsymbol{x}_P)|^{\mathcal{M}} f \\ &= |P\boldsymbol{x}_P|^{\mathcal{M}} f[|\boldsymbol{\tau}|^{\mathcal{M}} f/\boldsymbol{x}_P], && \text{by Substitution Lemma 4.2.3,} \\ &= |\psi_P|^{\mathfrak{M}_\mathrm{L}} f[|\boldsymbol{\tau}|^{\mathcal{M}} f/\boldsymbol{x}_P], && \text{by definition of } \psi_P, \\ &= |\psi_P|^{\mathfrak{M}_\mathrm{L}} f[|\boldsymbol{\tau}'|^{\mathfrak{M}_\mathrm{L}} f/\boldsymbol{x}_P], && \text{as } |\boldsymbol{\tau}|^{\mathcal{M}} = |\boldsymbol{\tau}'|^{\mathfrak{M}_\mathrm{L}}, \\ &= |\psi_P(\boldsymbol{\tau}'/\boldsymbol{x}_P)|^{\mathfrak{M}_\mathrm{L}} f && \text{by Lemma 4.2.3.} \end{aligned}$$

  Hence the result $|\theta|^{\mathcal{M}} = |\theta'^{\,\Pi}|^{\mathfrak{M}_\mathrm{L}}$ holds in this case.
- The other base case is $\theta = \mathcal{E}\boldsymbol{\tau}$. Then $\theta'^{\,\Pi} = \mathcal{E}\boldsymbol{\tau}'$, and

$$|\mathcal{E}\boldsymbol{\tau}|^{\mathcal{M}} f = E(|\boldsymbol{\tau}|^{\mathcal{M}} f) = E(|\boldsymbol{\tau}'|^{\mathfrak{M}_\mathrm{L}} f) = |\mathcal{E}\boldsymbol{\tau}'|^{\mathfrak{M}_\mathrm{L}} f,$$

  so again $|\theta|^{\mathcal{M}} = |\theta'^{\,\Pi}|^{\mathfrak{M}_\mathrm{L}}$ holds.

- If the result holds for θ_1 and θ_2, and $(\psi_P/P\boldsymbol{x}_P : P \in \Pi)$ is free at $\theta_1 \wedge \theta_2$, then it is free at θ_1 and at θ_2, and
$$\begin{aligned}|\theta_1 \wedge \theta_2|^{\mathcal{M}} &= |\theta_1|^{\mathcal{M}} \cap |\theta_2|^{\mathcal{M}} \\ &= |\theta_1'^{\,\Pi}|^{\mathfrak{M}_{\mathrm{L}}} \cap |\theta_2'^{\,\Pi}|^{\mathfrak{M}_{\mathrm{L}}}, \quad \text{by induction hypothesis,} \\ &= |\theta_1'^{\,\Pi} \wedge \theta_2'^{\,\Pi}|^{\mathfrak{M}_{\mathrm{L}}},\end{aligned}$$
so the result holds for $\theta_1 \wedge \theta_2$ as $\theta_1'^{\,\Pi} \wedge \theta_2'^{\,\Pi} = (\theta_1 \wedge \theta_2)'^{\,\Pi}$.
- The inductive cases of $\neg$, and $\Box$ are similar to that of $\wedge$.
- If the result holds for θ, and $(\psi_P/P\boldsymbol{x}_P : P \in \Pi)$ is free at $\forall y\theta$, then it is free at θ , and
$$|\forall y\theta|^{\mathcal{M}} = \forall_y|\theta|^{\mathcal{M}} = \forall_y|\theta'^{\,\Pi}|^{\mathfrak{M}_{\mathrm{L}}} = |\forall y(\theta'^{\,\Pi})|^{\mathfrak{M}_{\mathrm{L}}} = |(\forall y\theta)'^{\,\Pi})|^{\mathfrak{M}_{\mathrm{L}}},$$
so the result holds for $\forall y\theta$.

That completes the proof of the Lemma. ⊣

Now to finish the proof of the Theorem. If φ has constants $\mathsf{c}_1, \dots, \mathsf{c}_n$, then
$$\varphi' = \varphi(\tau_{\mathsf{c}_1}/\mathsf{c}_1, \dots, \tau_{\mathsf{c}_n}/\mathsf{c}_n).$$
Since $\vdash_{\mathrm{L}} \varphi$, we have $\vdash_{\mathrm{L}} \varphi'$ by the rule Sub* of Lemma 1.2.3 (if φ has no constants, then $\varphi' = \varphi$, so $\vdash_{\mathrm{L}} \varphi'$ in any case). Hence $\vdash_{\mathrm{L}} \varphi'^{\,\Pi}$ because L is closed under predicate substitution. Therefore $\mathfrak{M}_{\mathrm{L}} \models \varphi'^{\,\Pi}$, so by the Lemma, for any $f \in U_{\mathcal{L}}{}^{\omega}$,
$$|\varphi|^{\mathcal{M}} f = |\varphi'^{\,\Pi}|^{\mathfrak{M}_{\mathrm{L}}} f = W_{\mathrm{L}}.$$
Thus $\mathcal{M} \models \varphi$ as required. ⊣

A large class of functionally canonical logics can be specified syntactically as follows. Define a *shape* to be any expression constructible from propositional variables $p, q, \dots$ and/or the constant formula F by means of the connectives $\wedge, \neg, \Box$ and quantifiers $\forall x$. Thus each propositional modal formula is itself a shape, and indeed the propositional formulas are precisely the $\forall$-free shapes. Notable examples of shapes involving quantifiers are the "Barcan shape" $\forall x\Box p \to \Box\forall xp$ and the commuting quantifiers shape $\forall x\forall yp \to \forall y\forall xp$.

If η is a shape in the propositional variables $p_1, \dots, p_n$, then $\eta(\mathcal{L})$ is the set of all $\mathcal{L}$-formulas of the form
$$\eta(\psi_1/p_1, \dots, \psi_n/p_n)$$
obtained by uniform substitution of $\mathcal{L}$-formulas ψ_i for p_i in η. If S is a set of shapes, then QES is the smallest $\mathcal{E}$-logic that includes the sets $\eta(\mathcal{L})$ for all $\eta \in$ S. Note that a propositional modal logic is a set of propositional modal formulas, hence a set of shapes, so the notation QES is consistent with the earlier usage.

When S contains only propositional modal formulas, the work we have already done shows that QES is functionally canonical. We observed in

Section 3.1 that QES is characterised by its canonical model structure $\mathcal{S}_{\text{QES}}$. But if a logic L is characterised by $\mathcal{S}_{\text{L}}$, then it is also characterised by $\mathfrak{S}_{\text{L}}$ (Theorem 4.1.6), so in this case QES is characterised by $\mathfrak{S}_{\text{QES}}$. We will now extend this conclusion to logics axiomatised by shapes that include $\forall$.

THEOREM 4.5.3. *For any set* S *of shapes, the logic* QES *is functionally canonical and closed under substitution for predicate letters.*

PROOF. The set $\{\varphi : \mathfrak{S}_{\text{QES}} \models \varphi\}$ is a logic, so it suffices to show that this set includes $\eta(\mathcal{L})$ for all $\eta \in$ S to conclude that it includes QES, making QES functionally canonical as required. The rest follows by Theorem 4.5.1.

So let $\varphi = \eta(\psi_1/p_1, \ldots, \psi_n/p_n)$ with $\eta \in$ S. Take any $\mathcal{L}$-model $\mathcal{M}$ on $\mathfrak{S}_{\text{QES}}$ in order to show $\mathcal{M} \models \varphi$. Then for each $i \leq n$ we have $|\psi_i|^{\mathcal{M}} \in \mathit{PropFun}_{\text{QES}}$, so $|\psi_i|^{\mathcal{M}} = |\theta_i|^{\mathfrak{M}_{\text{QES}}}$ for some $\mathcal{L}$-formula θ_i.

An induction on the formation of shapes ζ in $p_1, \ldots, p_n$ then shows that in general

$$|\zeta(\psi_1/p_1, \ldots, \psi_n/p_n)|^{\mathcal{M}} = |\zeta(\theta_1/p_1, \ldots, \theta_n/p_n)|^{\mathfrak{M}_{\text{QES}}}.$$

This uses the facts, in the algebra of $\mathit{PropFun}_{\text{QES}}$, that

- $|\mathsf{F}|^{\mathcal{M}} = \alpha_{\emptyset}$,
- $|\neg\psi|^{\mathcal{M}} = -|\psi|^{\mathcal{M}}$,
- $|\psi_1 \wedge \psi_2|^{\mathcal{M}} = |\psi_1|^{\mathcal{M}} \cap |\psi_2|^{\mathcal{M}}$,
- $|\Box\psi|^{\mathcal{M}} = [R]|\psi|^{\mathcal{M}}$,
- $|\forall x\psi|^{\mathcal{M}} = \forall_x|\psi|^{\mathcal{M}}$,

and likewise with $\mathfrak{M}_{\text{QES}}$ in place of $\mathcal{M}$, as well as the fact that substitution for propositional variables in a shape commutes with the quantifiers, i.e.

$$(\forall x\zeta)(\psi_1/p_1, \ldots, \psi_n/p_n) = \forall x(\zeta(\psi_1/p_1, \ldots, \psi_n/p_n)),$$

and similarly commutes with the propositional connectives.

In particular, when $\zeta = \eta$, we get

$$|\varphi|^{\mathcal{M}} = |\eta(\theta_1/p_1, \ldots, \theta_n/p_n)|^{\mathfrak{M}_{\text{QES}}}.$$

But $\eta(\theta_1/p_1, \ldots, \theta_n/p_n)$ belongs to $\eta(\mathcal{L})$, so is a QES-theorem and hence is valid in $\mathfrak{M}_{\text{QES}}$. Therefore φ is valid in $\mathcal{M}$ as required. ⊣

Here is an application of this result to the Barcan Formula.

THEOREM 4.5.4. (1) *For any set* S *of shapes, the logic* QES + BF *is functionally canonical and closed under substitution for predicate letters.*

(2) *If* S *is a set of modal formulas, then* QES + BF *is characterised by the class of all E-admissible functional model structures whose underlying general frame validates* S *and in which the relation*

$$\forall_x[R]\alpha \subseteq [R]\forall_x\alpha \tag{4.5.1}$$

holds for all $\alpha \in \mathit{PropFun}$.

PROOF. Let S′ be S together with the Barcan shape. Then QES + BF is just QES′, and so result (1) is given by the previous Theorem.

For (2), we first show that any functional structure $\mathfrak{S}$ satisfying (4.5.1) must validate BF. For if $\mathcal{M}$ is any model on $\mathfrak{S}$, and we take α in (4.5.1) to be $|\varphi|^{\mathcal{M}}$, then for all $f \in U^{\omega}$ we have

$$|\forall x\Box\varphi|^{\mathcal{M}}f = \forall_x[R]|\varphi|^{\mathcal{M}}f \subseteq [R]\forall_x|\varphi|^{\mathcal{M}}f = |\Box\forall x\varphi|^{\mathcal{M}}f,$$

and so $\mathcal{M} \models \forall x\Box\varphi \to \Box\forall x\varphi$.

Thus if L = QES + BF, then, combining with earlier results about QES (Lemma 1.10.1, Theorem 3.1.5), we get that L is sound for the class of E-admissible functional structures satisfying (4.5.1) whose underlying general frame validates S. To prove that it is also complete for this class, it suffices to show that $\mathfrak{S}_{\mathrm{L}}$ satisfies (4.5.1) whenever $\vdash_{\mathrm{L}}$ BF (since we already know that the general frame of $\mathfrak{S}_{\mathrm{L}}$ validates S when L includes QES). Now an arbitrary member of $\mathit{PropFun}_{\mathrm{L}}$ is of the form α_φ for some φ, and from Lemma 4.1.4 we have

$$\forall_x[R]\alpha_\varphi = \alpha_{\forall x\Box\varphi} \quad \text{and} \quad [R]\forall_x\alpha_\varphi = \alpha_{\Box\forall x\varphi}.$$

But by BF, $\vdash_{\mathrm{L}} \forall x\Box\varphi \to \Box\forall x\varphi$, so $\alpha_{\forall x\Box\varphi} \subseteq \alpha_{\Box\forall x\varphi}$ by Corollary 4.1.5(5).

This shows that $\forall_x[R]\alpha_\varphi \subseteq [R]\forall_x\alpha_\varphi$ for all formulas φ,which suffices to show that $\mathfrak{S}_{\mathrm{L}}$ satisfies (4.5.1). ⊣

Now if S is a propositional logic that is canonical, then the Kripke frame underlying $\mathcal{S}_{\mathrm{QES}}$ validates S (Theorem 1.10.6). But this Kripke frame is also the one that underlies the canonical functional structure $\mathfrak{S}_{\mathrm{QES}}$ characterising QES. This gives the following stronger result:

THEOREM 4.5.5. *If* S *is a canonical propositional logic, then* QES + BF *is characterised by the class of all E-admissible functional model structures that satisfy* (4.5.1) *and whose underlying Kripke frame validates* S. ⊣

For languages without the existence predicate $\mathcal{E}$, we can define QS to be the smallest set of $\mathcal{E}$-free $\mathcal{L}$-formulas that forms a logic including the sets $\eta(\mathcal{L})$ for all shapes η in the set S. Then the theorems of this section hold with QS in place of QES, and with the reference to E-admissibility deleted. In particular,

- If S is any canonical propositional logic, then QS + BF is characterised by the class of all functional model structures satisfying (4.5.1) whose underlying Kripke frame validates S.

The Commuting Quantifiers axiom CQ can also be handled in this fashion, characterising it by validity in functional model structures in which *PropFun* satisfies

$$\forall_x\forall_y\alpha \subseteq \forall_y\forall_x\alpha. \tag{4.5.2}$$

This can be combined with all the results given here about BF. For instance:

- If S is any canonical propositional logic, then QS + CQ + BF is characterised by the class of all functional model structures satisfying (4.5.2) and(4.5.1) whose underlying Kripke frame validates S.

We can also combine these characterisations with others we have obtained. For instance, if S is a set of modal formulas,

- QES+UI+BF is characterised by the class of all E-admissible functional model structures that satisfy (4.5.1), are based on a general S-frame, and have one universal domain.

Again, deleting E-admissibility gives a characterisation of QS + UI + BF.

It might be thought that characterisations using the condition (4.5.1) do little to advance our understanding of the Barcan Formula, since this is essentially a translation of BF into structural/algebraic form. On the other hand, for QS + BF and QES + BF, and their extensions by CQ, the results given here would appear to be the first characterisations of any kind that are based on possible-worlds style relational structures.

Chapter 5

IDENTITY

We now extend our language for quantified modal logic by adding a predicate symbol $\approx$ for an *identity* relation, allowing us to express assertions about the identity of individuals. A two-sorted language is developed, with one sort of term standing for individual concepts, represented as partial functions from worlds to individuals, and the other sort specialising to concepts that are rigid, i.e. have the same value in accessible worlds. The identity predicate allows the existence predicate $\mathcal{E}$ to be defined, taking $\mathcal{E}\tau$ to be the self-identity formula $\tau \approx \tau$, whose corresponding proposition/truth set is the domain of the partial function interpreting τ. The use of admissibility is extended from propositions to individuals by requiring each model to have a designated set of admissible individual concepts, within which there is a designated set of admissible rigid ones.

We axiomatise the set of formulas that are valid in these models, using a new inference rule that allows deduction of assertions of non-existence ($\neg\mathcal{E}\tau$). The logic characterised by Kripkean models is then treated separately. The final section of the chapter gives a semantic analysis of Russelian definite description terms $\iota x.\varphi$ in this context, and shows how to construct canonical models and axiomatisations for logics in languages that have these description terms as well as the identity predicate.

5.1. Intension versus Extension

The theory of identity in the modal context has been much discussed. It lies at the heart of our understanding of the nature of the entities we reason about. An important issue is whether true identities should be *necessarily* true. To illustrate the point, let τ be the term "the FIFA World Cup holder", and let τ' be "Spain". Then at the time of writing, the identity statement $\tau \approx \tau'$ is true. But this is a contingent truth rather than a necessary one, since some other country might well have won the 2010 tournament. Thus $\Box(\tau \approx \tau')$ is false, and the example invalidates the scheme

NI: $\quad \tau \approx \tau' \to \Box(\tau \approx \tau')$

of *Necessity of Identity*. Moreover, accepting that every object is necessarily identical with itself, i.e. $\Box(\tau \approx \tau)$ is true, the example falsifies

$$\tau \approx \tau' \to (\Box(\tau \approx \tau) \to \Box(\tau \approx \tau')).$$

This is an instance of the scheme

$$\text{SI}: \quad \tau \approx \tau' \to (\varphi \to \varphi(\tau'//\tau)),$$

where $\varphi(\tau'//\tau)$ denotes any formula obtained from φ by substitution of τ' for some free occurrences of τ. SI expresses the principle of *Substitutivity of Identicals*, a fundamental law of non-modal quantificational logic. This law can fail when the substitution takes place within the scope of the modal connective $\Box$.

To clarify this situation we invoke the well-known distinction between the *intension* and the *extension* of an expression. The extension is generally taken to be the particular entity that the expression denotes, and in the modal setting this is context-dependent. An expression has an extension in each possible world—at least in each world in which it makes sense—and this extension may vary from world to world. The intension is variously described as the meaning, the sense, or the content of an expression, or the concept that it evokes. Carnap [1947] proposed that this intension be identified with a function from possible worlds or states of affairs to denotations, namely the function that assigns to each world the extension of the expression in that world.

For example, a sentence has a truth-value as its extension: each sentence is either true or false in each world. The intension of a sentence is the function itself from worlds to truth-values that assigns to each world the truth-value that the sentence has in that world. But we can identify this function with the set of worlds to which it assigns the value *true*. In this way, the intension of a sentence becomes identified with a set of worlds, i.e. a proposition, namely the set of worlds in which the sentence is true. This proposition is the meaning of the sentence, according to this analysis.

Now for terms denoting individuals, there are some, like "8", or "Spain", that denote the same individual in all worlds. We might say that such terms are *extensional*. They are also called *rigid designators*. We will use a slightly weaker notion of extensionality here that requires rigidity only between worlds that are accessible, i.e. if wRu then a rigid designator must denote the same individual in both w and u. This is enough to ensure that a true identity between extensional terms remains true in all accessible worlds, hence is *necessarily* true.

The expression "the FIFA World Cup holder" is, by contrast, *intensional*: the particular country it denotes is genuinely context-dependent and varies according to the situation, or possible world, in which the expression is being used. It may take different values in two worlds that are accessible to each other. Other well-known examples of terms that are intensional in this sense include Russell's "the present King of France", which presently has no denotation, and Frege's "the Morning Star", which happens to denote the planet

Venus, but could have denoted some other astronomical body. Thus the intension of an intensional term is a function from worlds to individuals that is not rigid, i.e. assigns different values to some accessible worlds. We will refer to a function from worlds to individuals as an *individual concept*. The case of the present King of France indicates that we should allow an individual concept to be a *partial* function on the set of all worlds. We take an "object" to be a such a function that is rigid.

On this account, the scheme NI is valid when τ and τ' are rigid designators. On the other hand, an identity between intensional terms, or between an intensional and an extensional term, is a matter of contingency and may violate NI and the principle of substitutivity of identicals.

In this chapter we formalise these ideas into a *two-sorted* language for quantified modal logic, and provide an admissible semantics for it. The set of quantifiable variables will range over individual concepts, and each model structure will have a nominated set $\mathbb{C}$ of such partial functions. There will also be a second sort of variable, called *object variables*. These range over a nominated subset $\mathbb{O}$ of $\mathbb{C}$ that consists of rigid functions. In addition, we allow constants of both sorts, denoting members of $\mathbb{C}$ and of $\mathbb{O}$. An identity $\tau \approx \tau'$ is true at a world w when the terms τ and τ' have the same value at w, which means that the partial functions denoted by τ and τ' are both defined at w and take the same value there. We can then distinguish cases of the scheme NI according to the sort of the terms involved. NI is valid under this semantics if τ and τ' are rigid concept terms, but not if they are general concept terms. Likewise, the principle SI of substitutivity of identicals is valid when the identity $\tau \approx \tau'$ involves rigid concept terms, but is only valid for general concept terms when the formula φ is $\Box$-free. Thus the validity of this principle depends on whether terms are interpreted extensionally or intensionally.

In the following sections we set out the details of this theory and, building on the work of earlier chapters, axiomatise the resulting logic by a suitable canonical model construction.

There is an extensive philosophical and technical literature that deals with identity in quantified modal logic, including (but not confined to) Barcan [1947], Hughes and Cresswell [1968], [1996], Hintikka [1969], Scott [1970], Thomason [1970], Kripke [1971], [1980], [1992], Linsky [1971], Bressan [1972], Forbes [1985], Marcus [1993], Fitting and Mendelsohn [1998], Garson [2001], Cocchiarella [2001], Stalnaker [2003], Føllesdal [2004], Corsi [2006], Priest [2008] and Gabbay, Shehtman, and Skvortsov [2009].

5.2. Syntax

Our language is based as usual on a denumerably infinite set InVar of individual variables, but these are now to be thought of as ranging over individual *concepts*. In addition, InVar has a distinguished infinite subset ObVar of *object variables*, with its complement InVar − ObVar also being infinite. Thus an object variable is a special kind of concept variable.

A signature $\mathcal{L}$ consists of the following:

- A set Con of *individual concept constants*.
- A set ObCon *object constants*, with ObCon $\subseteq$ Con.
- A set of finitary predicate symbols P.

To focus on the primary features of our approach, we have not included function symbols in a signature. Hence a *term* is either a variable or a constant: the $\mathcal{L}$-terms are just the members of InVar $\cup$ Con. An *object term* is defined to be either an object variable or an object constant, so the object terms for $\mathcal{L}$ are just the members of ObVar $\cup$ ObCon. Thus all terms are individual concept terms, while an object term is a special kind of concept term.

We continue to use Roman letters $x, y, \ldots$ for members of InVar, and τ for a typical term. For constants we will use c, d and also k. We sometimes write τ : Ob, and say that τ *is of sort* Ob, when τ is an object term.

Atomic formulas for $\mathcal{L}$ are now of two kinds:

- Any expression $P\tau_1 \cdots \tau_n$ where P is an n-ary predicate symbol from $\mathcal{L}$, and $\tau_1, \ldots, \tau_n$ are $\mathcal{L}$-terms.
- Any expression $\tau \approx \tau'$, where τ and τ' are $\mathcal{L}$-terms. Such an expression is *an identity*, or an *identity formula*.

The set of *$\mathcal{L}$-formulas* is generated from these atomic ones and the constant formula F by the propositional connectives $\wedge$, $\neg$, and $\Box$ and the quantifiers $\forall x$ for each $x \in$ InVar. We usually write $\tau \not\approx \tau'$ as an abbreviation for $\neg(\tau \approx \tau')$.

A *self-identity* formula $\tau \approx \tau$ will often be written $\mathcal{E}\tau$, and can be read "τ exists". Thus in the present language with identity, the existence predicate $\mathcal{E}$ becomes definable. The definition will be shown to be appropriate by our interpretation of $\approx$ (see (5.4.2)).

This has been a somewhat one-sided description of a two-sorted language. Presentations of many-sorted logic usually require a strict syntactic separation of sorts, with distinct sorts having disjoint sets of constants as well as disjoint sets of variables, and an identity $\tau \approx \tau'$ only being well-formed when the terms τ and τ' are of the same sort.

Here we have taken a different approach. We do indeed have two sorts of terms, namely those that are of sort Ob and those that are not, but we have not given a distinctive name to the non-object terms. This is because we are most interested in singling out the object terms as a special subset of the set of *all* terms. This reflects our treatment of extensional objects as a special subset of the set of all individual concepts, namely as the rigid concepts. So we definitely want to allow identities between terms of different sort, as in "the FIFA World Cup holder $\approx$ Spain".

The way this all works should become clearer when we introduce the formal semantics for this new language in Section 5.4. That semantics will allow us to say when an individual concept is identical to a rigid one at a particular world.

5.3. Identity and Rigid Functions

We write $E\alpha$ for the *domain* of a function α. This is the set of points w for which the function value $\alpha(w)$ is defined. α is a *partial function from set W to set X* if $E\alpha$ is a subset of W and the value $\alpha(w)$ belongs to X for all $w \in E\alpha$.

In set theory the domain of α is typically referred to by a notation like "dom α", but here we wish to emphasis that this domain is the set on which α *exists* and that, when W is a set of possible worlds, $E\alpha$ may be viewed as a proposition asserting this existence.

A basic example of a partial function from W to X is the one that is *undefined* at every member of W. This is the *empty function*, with empty domain. It will be denoted $\emptyset_X^W$ to emphasise that it is being viewed as a partial function from W to X.

A binary relation $\simeq$ of identity between values of partial functions is introduced by defining the expression

$$\alpha(w) \simeq \beta(u)$$

to mean that $\alpha(w)$ and $\beta(u)$ are *both* defined and are equal, i.e. $w \in E\alpha$ and $u \in E\beta$ and $\alpha(w) = \beta(u)$. When this holds we may say that α and β are *identical at* w. We also write

$$\alpha(w) \simeq a$$

to mean that $\alpha(w)$ is defined and is equal to a.

The relation $\alpha(w) \simeq \beta(u)$ can fail in several ways: if $\alpha(w)$ and $\beta(u)$ are both undefined, or if one is defined and the other is not, or if they are both defined but not equal. $\alpha(w) \simeq a$ fails in two ways: either $w \notin E\alpha$, or $w \in E\alpha$ but $\alpha(w) \neq a$.

For each pair α, β of partial functions on a set W we define

$$[\![\alpha \simeq \beta]\!] = \{w \in W : \alpha(w) \simeq \beta(w)\}.$$

$[\![\alpha \simeq \beta]\!]$ is the *identity set* of α and β, the subset of W on which the two functions are both defined and agree. If W is a set of worlds, then $[\![\alpha \simeq \beta]\!]$ may be thought of as a proposition asserting the identity of α and β. Identity sets will be used to interpret identity formulas $\tau \approx \tau'$.

Connected with $\simeq$ is a weaker relation $\equiv$, introduced by defining

$$\alpha(w) \equiv \beta(u)$$

to mean that either $\alpha(w) \simeq \alpha(u)$, or else $\alpha(w)$ and $\beta(u)$ are *both* undefined. When this holds we may say that α and β are *indistinguishable at* w, or that $\alpha(w)$ and $\beta(u)$ are indistinguishable (which may mean that they do not exist). Define

$$[\![\alpha \equiv \beta]\!] = \{w \in W : \alpha(w) \equiv \beta(w)\}$$

to be the *indistinguishability set* of α and β. Then we have

$$\begin{aligned}[\![\alpha \equiv \beta]\!] &= [\![\alpha \simeq \beta]\!] \cup (-E\alpha \cap -E\beta)\\ &= (E\alpha \cup E\beta) \Rightarrow [\![\alpha \simeq \beta]\!].\end{aligned}$$

In our formal language, indistinguishability is expressible by the formula $\tau \approx \tau' \vee (\neg\mathcal{E}\tau \wedge \neg\mathcal{E}\tau')$.

Note that any pair α, β of partial functions induces a partition of W into the five disjoint sets

$$\begin{aligned}&[\![\alpha \simeq \beta]\!]\\ &-E\alpha \cap -E\beta\\ &E\alpha - E\beta\\ &E\beta - E\alpha\\ &E\alpha \cap E\beta - [\![\alpha \simeq \beta]\!]\end{aligned} \tag{5.3.1}$$

(it may be helpful to draw a Venn diagram of them). The first two together make up $[\![\alpha \equiv \beta]\!]$, while the others represent the three ways that α and β may be distinguishable at a point.

The indistinguishability relation can be used to state when two partial functions themselves are identical, since we have $\alpha = \beta$ (i.e. α and β are the same function) iff $\alpha(w) \equiv \beta(w)$ for all $w \in W$.

Now if (W, R) is a Kripke frame, then a partial function α on W will be called *rigid* if, for all $w, u \in W$,

$$\text{if } wRu, \text{ then } \alpha(w) \equiv \alpha(u).$$

In other words, if wRu, then either $\alpha(w)$ and $\alpha(u)$ are *both* defined and are equal, or else $\alpha(w)$ and $\alpha(u)$ are both undefined. Some immediate examples of rigid functions are any *empty* partial function on W, and any *constant* partial function, i.e. one taking a single value on its domain.

We can better understand the behaviour of a rigid function α by taking the partition of a frame into its *R-components*, as described at the beginning of Section 2.4, and considering how α acts on each component. Recall that the components of (W, R) are the equivalence classes under R^{er}, the smallest equivalence relation that includes R. Here $wR^{\text{er}}u$ iff there is a finite sequence $w = u_0, \ldots, u_n = u$ such that for each $i < n$, either $u_i R u_{i+1}$ or $u_{i+1} R u_i$. Such a sequence *R-connects* w and u : we can pass from one to the other in finitely many steps by going back and/or forth along R.

Now for such a sequence it is readily seen, by induction along the sequence, that if α is rigid then for each $i \leq n$, either $\alpha(u_0) \simeq \alpha(u_i)$ or else α is undefined at both u_0 and u_i. Hence for each w, if u is any member of the R-connected component of w, then $\alpha(w)$ and $\alpha(u)$ are indistinguishable. It follows that the restriction of α to a particular component is either

(i) totally defined and constant, i.e. defined at all members of the component and takes the same value at all members; or
(ii) totally undefined, i.e. is the empty partial function on the component.

Thus a rigid partial function can be completely described by saying that its domain is a union of disjoint components, and it takes a constant value on each component, with distinct components possibly being assigned different constant values. This gives a recipe for constructing all rigid partial functions α on a frame: choose a set of components and take its union to be the domain of α, then choose a constant value for α on each chosen component.

5.4. Model Structures and Models

A model structure for our two-sorted language with identity is a system

$$\mathcal{S} = (W, R, Prop, U, D, \mathbb{C}, \mathbb{O}),$$

meeting the following description:

- $(W, R, Prop, U, D)$ is a model structure for the one-sorted language without identity, as defined at the start of Section 1.5. So $(W, R, Prop)$ is a general frame, U is a set, and D assigns to each $w \in W$ a domain of individuals $Dw \subseteq U$.
- $\mathbb{C}$ is a set of *partial* functions from W to U, such that each $\alpha \in \mathbb{C}$ has $\alpha(w) \in Dw$ for all $w \in E\alpha$.
- $\mathbb{O} \subseteq \mathbb{C}$, and each $\alpha \in \mathbb{O}$ is a *rigid* partial function.
- For all $w \in W$ and all $a \in Dw$, there exists an $\alpha \in \mathbb{O}$ with $\alpha(w) \simeq a$.

$\mathbb{C}$ is the set of *admissible individual concepts* of $\mathcal{S}$, and $\mathbb{O}$ the set of *admissible objects*. The last condition in the description of $\mathcal{S}$ may be expressed by saying that $\mathcal{S}$ is *full of objects*. It states that each individual in the domain of w is the value at w of some admissible object. This requirement forces the domain function D to be constant on R-connected components, since it ensures that

$$wRu \quad \text{implies} \quad Dw = Du. \tag{5.4.1}$$

To see this, note that if $a \in Dw$ is equal to $\alpha(w)$ with $\alpha \in \mathbb{O}$, then since α is rigid and wRu we have $\alpha(w) \equiv \alpha(u)$, and so $a = \alpha(w) \simeq a(u) \in Du$. Conversely, if $a = \alpha(u) \in Du$ with $\alpha \in \mathbb{O}$, then $a = \alpha(w) \in Dw$.

A *premodel* $\mathcal{M} = (\mathcal{S}, |-|^{\mathcal{M}})$ for $\mathcal{L}$, based on this new kind of model structure, is given by an interpretation function $|-|^{\mathcal{M}}$ on $\mathcal{L}$ that assigns:

- to each constant $\mathsf{c} \in \mathsf{Con}$ of $\mathcal{L}$ a member $|\mathsf{c}|^{\mathcal{M}}$ of the set $\mathbb{C}$ of admissible concepts of $\mathcal{S}$, provided that if $\mathsf{c} \in \mathsf{ObCon}$, then $|\mathsf{c}|^{\mathcal{M}} \in \mathbb{O}$; and
- to each n-ary predicate symbol P of $\mathcal{L}$ a function $|P|^{\mathcal{M}} : \mathbb{C}^n \to \wp W$, satisfying the following requirements.

  Extensionality:
  If $\alpha_i(w) \simeq \beta_i(w)$ for all $i \leq n$, then

$$w \in |P|^{\mathcal{M}}(\alpha_1, \ldots, \alpha_n) \quad \text{iff} \quad w \in |P|^{\mathcal{M}}(\beta_1, \ldots, \beta_n).$$

  Existence:
  If $w \in |P|^{\mathcal{M}}(\alpha_1, \ldots, \alpha_n)$, then for all $i \leq n$, $w \in E\alpha_i$.

These Extensionality and Existence requirements will ensure the validity of the scheme

$$\tau \approx \tau' \to (\varphi \to \varphi(\tau'//\tau))$$

of Substitutivity of Identicals in the case that φ has no occurrence of the modality $\Box$.

To assign values to variables, we define a *valuation* to be any function $f : \mathsf{InVar} \to \mathbb{C}$, from the set of variables to the set of admissible concepts of $\mathcal{S}$, satisfying the proviso that whenever $x \in \mathsf{ObVar}$, then $fx \in \mathbb{O}$. The set of all such valuations will be denoted $Val_{\mathcal{S}}$.

Note that if $f \in Val_{\mathcal{S}}$ and $x \in \mathsf{ObVar}$, then an updating $f[\alpha/x]$ of f will belong to $Val_{\mathcal{S}}$ iff $\alpha \in \mathbb{O}$. That will be an important consideration in our interpretation of the quantifier $\forall x$.

In a premodel $\mathcal{M}$, each valuation f assigns to each $\mathcal{L}$-term τ a value $|\tau|^{\mathcal{M}} f \in \mathbb{C}$, so overall $\mathcal{M}$ interprets τ as a function $|\tau|^{\mathcal{M}} : Val_{\mathcal{S}} \to \mathbb{C}$. These term functions are defined by:

$$|x|^{\mathcal{M}} f = f(x) \quad \text{for } x \in \mathsf{InVar},$$
$$|\mathsf{c}|^{\mathcal{M}} f = |\mathsf{c}|^{\mathcal{M}} \quad \text{for } \mathsf{c} \in \mathsf{Con}.$$

This makes $|\tau|^{\mathcal{M}} f$ a member of $\mathbb{C}$ for all terms τ and all $f \in Val_{\mathcal{S}}$. Moreover, by the provisos in the definitions of "premodel" and "valuation", we have $|\tau|^{\mathcal{M}} f \in \mathbb{O}$ whenever τ is an object term, i.e. a member of ObVar or ObCon.

To work with the term functions $|\tau|^{\mathcal{M}}$ we introduce a new notation for expressing functional values. We have been using both the notations "$\alpha(w)$" and "αw" for the value of a function α at a point w. Now we adopt a third option "$\alpha.w$" to mean the same thing. Since the value $|\tau|^{\mathcal{M}} f$ of a term function is itself a function (as a member of $\mathbb{C}$), we can write $|\tau|^{\mathcal{M}} f.w$ for the value of this function at w, rather than the more cumbersome $(|\tau|^{\mathcal{M}}(f))(w)$.

A premodel now assigns a propositional function of the form

$$|\varphi|^{\mathcal{M}} : Val_{\mathcal{S}} \to \wp W$$

to each $\mathcal{L}$-formula, with the associated satisfaction relation $\mathcal{M}, w, f \models \varphi$ meaning that $w \in |\varphi|^{\mathcal{M}} f$. For standard atomic formulas we define

$$|P\tau_1 \cdots \tau_n|^{\mathcal{M}} f = |P|^{\mathcal{M}}(|\tau_1|^{\mathcal{M}} f, \ldots, |\tau_n|^{\mathcal{M}} f),$$

as usual. For identity formulas we define

$$|\tau \approx \tau'|^{\mathcal{M}} f = [\![\, |\tau|^{\mathcal{M}} f \simeq |\tau'|^{\mathcal{M}} f \,]\!],$$

where $[\![\, |\tau|^{\mathcal{M}} f \simeq |\tau'|^{\mathcal{M}} f \,]\!]$ is the identity set

$$\{w \in W : |\tau|^{\mathcal{M}} f.w \simeq |\tau'|^{\mathcal{M}} f.w\}$$

of the admissible concepts $|\tau|^{\mathcal{M}} f$ and $|\tau'|^{\mathcal{M}} f$, as defined in the previous section. In terms of satisfaction, the definition of $|\tau \approx \tau'|^{\mathcal{M}} f$ says precisely that

$$\mathcal{M}, w, f \models \tau \approx \tau' \quad \text{iff} \quad |\tau|^{\mathcal{M}} f.w \simeq |\tau'|^{\mathcal{M}} f.w\,.$$

Thus an identity formula $\tau \approx \tau'$ is satisfied at w under valuation f when the individual concepts interpreting τ and τ' under f are both defined and take identical values at w.

The inductive definitions of the truth sets $|\varphi \wedge \psi|^{\mathcal{M}} f$, $|\neg\varphi|^{\mathcal{M}} f$ and $|\Box\varphi|^{\mathcal{M}} f$ are as previously. For the quantifiers, if x is of sort Ob we define

$$|\forall x\varphi|^{\mathcal{M}} f = \bigsqcap_{\alpha \in \mathbb{O}} \big(E\alpha \Rightarrow |\varphi|^{\mathcal{M}} f[\alpha/x]\big),$$

while for $x \notin \mathsf{ObVar}$ we put

$$|\forall x\varphi|^{\mathcal{M}} f = \bigsqcap_{\alpha \in \mathbb{C}} \big(E\alpha \Rightarrow |\varphi|^{\mathcal{M}} f[\alpha/x]\big).$$

Here as usual, $\bigsqcap$ is the "admissible conjunction" operation determined by *Prop*. Thus $\forall x$ expresses "for all existing admissible objects" when $x : \mathsf{Ob}$, and otherwise expresses "for all existing admissible concepts". So this is an *actualist* interpretation of the quantifiers. Note that in the case that $x : \mathsf{Ob}$, the restriction to α in $\mathbb{O}$ ensures that the updatings $f[\alpha/x]$ are in $Val_{\mathcal{S}}$.

In terms of satisfaction, for $x : \mathsf{Ob}$ we have

$$\begin{aligned} \mathcal{M}, w, f \models \forall x\varphi \quad \text{iff} \quad & \text{there is an } X \in \textit{Prop} \text{ such that } w \in X \text{ and} \\ & X \subseteq \textstyle\bigcap_{\alpha \in \mathbb{O}} \big(E\alpha \Rightarrow |\varphi|^{\mathcal{M}} f[\alpha/x]\big); \end{aligned}$$

and likewise with $\mathbb{O}$ replaced by $\mathbb{C}$ when $x \notin \mathsf{ObVar}$.

A formula φ is *valid in* the premodel $\mathcal{M}$, written $\mathcal{M} \models \varphi$, if $|\varphi|^{\mathcal{M}} f = W$ for all $f \in Val_{\mathcal{S}}$.

Our semantics for identity has the special case

$$\mathcal{M}, w, f \models \tau \approx \tau \quad \text{iff} \quad |\tau|^{\mathcal{M}} f.w \simeq |\tau|^{\mathcal{M}} f.w\,.$$

But in general, the relation $\alpha.w \simeq \alpha.w$ holds just when $w \in E\alpha$, so

$$\mathcal{M}, w, f \models \tau \approx \tau \quad \text{iff} \quad w \in E|\tau|^{\mathcal{M}} f. \tag{5.4.2}$$

Thus, with our definition of $\mathcal{E}\tau$ as the formula $\tau \approx \tau$, we get $\mathcal{E}\tau$ asserting that τ "exists" at w, in the sense that the individual concept interpreting τ is defined there and so interprets τ as a member of the domain Dw. Moreover, we have

$$|\mathcal{E}\tau|^{\mathcal{M}} f = E|\tau|^{\mathcal{M}} f, \tag{5.4.3}$$

the same equation as in the theory without identity of Chapter 3.

Note also that the formula $\tau \not\approx \tau$ is just $\neg\mathcal{E}\tau$, and thus asserts that τ is undefined. So this formula may well be satisfiable, unlike in classical non-modal theories of identity, where it is universally false. We will make significant use of such formulas in a completeness theorem later in this chapter.

In premodels we can show that satisfaction of a formula depends only on the values assigned to *free* variables:

LEMMA 5.4.1 (Free Assignment). *In any premodel $\mathcal{M}$, for any formula φ, if valuations $f, g \in Val_{\mathcal{S}}$ agree on all free variables of φ, then $|\varphi|^{\mathcal{M}} f = |\varphi|^{\mathcal{M}} g$.*

PROOF. This is proven by induction on φ as in Lemma 1.6.1. In the case that φ is $\forall x\psi$, with the Lemma assumed to hold for ψ, we consider first the case that x is of sort Ob. If valuations f and g agree on all free variables of φ, then they agree on all free variables of ψ except perhaps for x. Hence for each α from $\mathbb{O}$, $f[\alpha/x]$ and $g[\alpha/x]$ agree on all free variables of ψ, *and moreover $f[\alpha/x]$ and $g[\alpha/x]$ belong to $Val_{\mathcal{S}}$ as required*, since $\alpha \in \mathbb{O}$, so $|\psi|^{\mathcal{M}} f[\alpha/x] = |\psi|^{\mathcal{M}} g[\alpha/x]$ by induction hypothesis. Thus

$$|\varphi|^{\mathcal{M}} f = \prod_{\alpha \in \mathbb{O}} \left(E\alpha \Rightarrow |\psi|^{\mathcal{M}} f[\alpha/x]\right) = \prod_{\alpha \in \mathbb{O}} \left(E\alpha \Rightarrow |\psi|^{\mathcal{M}} g[\alpha/x]\right) = |\varphi|^{\mathcal{M}} g,$$

so the Lemma holds for φ. In the case that x is not of sort Ob, the argument goes through similarly, using $\mathbb{C}$ in place of $\mathbb{O}$. ⊣

This result is needed in proving the appropriate version of the Substitution Lemma:

LEMMA 5.4.2 (Substitution). *Let τ be free for x in φ, with τ : Ob if x : Ob. Then in any premodel $\mathcal{M}$, for any $f \in Val_{\mathcal{S}}$,*

$$|\varphi(\tau/x)|^{\mathcal{M}} f = |\varphi|^{\mathcal{M}} f[\,|\tau|^{\mathcal{M}} f/x].$$

PROOF. By the inductive method of Lemma 1.6.2, with attention in the inductive case that φ is of the form $\forall y\psi$ to whether y is an object variable or not. The requirement that τ : Ob when x : Ob ensures that $|\tau|^{\mathcal{M}} f \in \mathbb{O}$ when x : Ob , and hence that $f[\,|\tau|^{\mathcal{M}} f/x]$ is a well-defined valuation in that case. The details are left to the reader. ⊣

We will say that a formula φ is *admissible in* $\mathcal{M}$ if the function $|\varphi|^{\mathcal{M}}$ has the form $Val_{\mathcal{S}} \to Prop$, i.e. the truth set $|\varphi|^{\mathcal{M}} f$ belongs to *Prop* for all $f \in Val_{\mathcal{S}}$. Then $\mathcal{M}$ is a *model* if every formula is admissible in $\mathcal{M}$. φ will be called *valid in* a model structure $\mathcal{S}$, written $\mathcal{S} \models \varphi$, if φ is valid in all ***models*** on $\mathcal{S}$.

For our present language, the appropriate version of Existing Instantiation is the following scheme:

$$\mathrm{EI}^+\colon \quad \forall x\varphi \to (\mathcal{E}\tau \to \varphi(\tau/x)), \qquad \text{with } \tau \text{ free for } x \text{ in } \varphi, \text{ and } \tau : \mathsf{Ob} \text{ if } x : \mathsf{Ob}.$$

LEMMA 5.4.3. *In any premodel $\mathcal{M}$, the scheme* EI$^+$*is valid. If $\mathcal{M}$ is a model, then validity in $\mathcal{M}$ is preserved by the rule*

$$\mathcal{E}\forall\text{-Intro:} \quad \frac{\varphi \to (\mathcal{E}x \to \psi)}{\varphi \to \forall x\psi}, \qquad \textit{if } x \textit{ is not free in } \varphi.$$

Hence this rule preserves validity in any model structure.

PROOF. The arguments used in Theorem 3.1.4, that showed validity for EI and $\mathcal{E}\forall$-Intro there, continue to work here, given the equation (5.4.3) and the Substitution Lemma 5.4.2, and with appropriate attention to the sort of the variable x. ⊣

Using atomic formulas $Px_1 \cdots x_n$ with $x_i \notin \mathsf{ObVar}$ we can show that in any model the functions $|P|^{\mathcal{M}}$ all take all their values in *Prop* (see the argument for (1.6.4)). In addition we have the following results, which will be used in the next section.

LEMMA 5.4.4. *In any model $\mathcal{M}$, if α and β are any members of $\mathbb{C}$, then Prop contains*

(1) *the existence sets $E\alpha$ and $E\beta$;*
(2) *the identity set $[\![\alpha \simeq \beta]\!]$; and*
(3) *the indistinguishability set $[\![\alpha \equiv \beta]\!]$.*

PROOF. Take distinct variables x, y not in ObVar. Given any $\alpha, \beta \in \mathbb{C}$, pick any valuation f that has $fx = \alpha$ and $fy = \beta$. Then $|x|^{\mathcal{M}}f = \alpha$ and $|y|^{\mathcal{M}}f = \beta$.

(1) $E\alpha = E|x|^{\mathcal{M}}f = |\mathcal{E}x|^{\mathcal{M}}f$ by (5.4.3). But $|\mathcal{E}x|^{\mathcal{M}}f \in \mathit{Prop}$ as $\mathcal{M}$ is a model, so $E\alpha \in \mathit{Prop}$. Likewise for $E\beta$.
(2) $[\![\alpha \simeq \beta]\!] = [\![\, |x|^{\mathcal{M}}f \simeq |y|^{\mathcal{M}}f \,]\!] = |x \approx y|^{\mathcal{M}}f \in \mathit{Prop}$.
(3) $[\![\alpha \equiv \beta]\!] = [\![\alpha \simeq \beta]\!] \cup (-E\alpha \cap -E\beta)$, which belongs to *Prop* because $E\alpha$, $E\beta$ and $[\![\alpha \simeq \beta]\!]$ do by (1) and (2), and *Prop* is closed under the Boolean set operations. In fact by this reasoning, *Prop* contains all members of the partition listed in (5.3.1). ⊣

5.5. Validity of Substitutivity of Identicals

We now work in an arbitrary model $\mathcal{M}$, and investigate what instances of Substitutivity of Identicals it validates. $\mathcal{M}$ will remain fixed until further notice.

LEMMA 5.5.1. *Take any* $\alpha, \beta \in \mathbb{C}$, *any* $w \in W$, *any* $f \in Val_{\mathcal{S}}$, *any term* τ, *and any variable* x, *provided that* $x : \mathsf{Ob}$ *if* $\alpha, \beta \in \mathbb{O}$. *Then if* α *and* β *are indistinguishable at* w, *so are* $|\tau|^{\mathcal{M}} f[\alpha/x]$ *and* $|\tau|^{\mathcal{M}} f[\beta/x]$. *In other words*

$$\alpha(w) \equiv \beta(w) \quad \textit{implies} \quad |\tau|^{\mathcal{M}} f[\alpha/x].w \equiv |\tau|^{\mathcal{M}} f[\beta/x].w.$$

PROOF. Let $\alpha(w) \equiv \beta(w)$. If τ is x, then the desired conclusion just restates the assumption $\alpha(w) \equiv \beta(w)$.

If $\tau \neq x$, then $|\tau|^{\mathcal{M}} f[\alpha/x] = |\tau|^{\mathcal{M}} f = |\tau|^{\mathcal{M}} f[\beta/x]$, so $|\tau|^{\mathcal{M}} f[\alpha/x]$ and $|\tau|^{\mathcal{M}} f[\beta/x]$ are one and the same function, hence are indistinguishable at every point. ⊣

We are now ready to prove a key result, stating that if α and β are rigid and indistinguishable at w as partial functions, then they are *semantically* indistinguishable at w in terms of satisfaction of formulas.

LEMMA 5.5.2 (Indistinguishability). *Let* φ *be any formula. Then*:

(1) *For all* $\alpha, \beta \in \mathbb{O}$, *all* $w \in W$, *and all* $f \in Val_{\mathcal{S}}$, *if* $\alpha(w) \equiv \beta(w)$, *then for any* $x : \mathsf{Ob}$,

$$\mathcal{M}, w, f[\alpha/x] \models \varphi \quad \textit{iff} \quad \mathcal{M}, w, f[\beta/x] \models \varphi.$$

(2) *If* φ *is* $\Box$*-free, then* (1) *holds also for all* $\alpha, \beta \in \mathbb{C}$ *and any variable* x, *provided that if* $\alpha \notin \mathbb{O}$ *or* $\beta \notin \mathbb{O}$, *then* $x \notin \mathsf{ObVar}$.

PROOF. We prove (1) by induction on the formation of φ, using the assumption on $\alpha, \beta \in \mathbb{O}$ throughout that $\alpha(w) \equiv \beta(w)$. Since $\equiv$ is symmetric, it is enough to show in each case that the satisfaction relations in (1) hold from left to right, i.e. that $\mathcal{M}, w, f[\alpha/x] \models \varphi$ implies $\mathcal{M}, w, f[\beta/x] \models \varphi$, since the converse then holds by interchange of α and β.

Note that another way to formulate the result for a given φ is that

$$[\![\alpha \equiv \beta]\!] \cap |\varphi|^{\mathcal{M}} f[\alpha/x] \subseteq |\varphi|^{\mathcal{M}} f[\beta/x],$$

and likewise with α and β interchanged. We will use this perspective in the inductive case for the quantifier.

If φ is an identity $\tau \approx \tau'$, then if $\mathcal{M}, w, f[\alpha/x] \models \tau \approx \tau'$, we have

$$|\tau|^{\mathcal{M}} f[\alpha/x].w \simeq |\tau'|^{\mathcal{M}} f[\alpha/x].w.$$

In particular, $|\tau|^{\mathcal{M}} f[\alpha/x].w$ is defined. But $|\tau|^{\mathcal{M}} f[\alpha/x].w$ is indistinguishable from $|\tau|^{\mathcal{M}} f[\beta/x].w$ by the previous Lemma, because $\alpha(w) \equiv \beta(w)$. Hence $|\tau|^{\mathcal{M}} f[\beta/x].w$ is defined and

$$|\tau|^{\mathcal{M}} f[\alpha/x].w \simeq |\tau|^{\mathcal{M}} f[\beta/x].w.$$

Similarly,

$$|\tau'|^{\mathcal{M}} f[\alpha/x].w \simeq |\tau'|^{\mathcal{M}} f[\beta/x].w.$$

Altogether these relations imply

$$|\tau|^{\mathcal{M}} f[\beta/x].w \simeq |\tau'|^{\mathcal{M}} f[\beta/x].w,$$

hence $\mathcal{M}, w, f[\beta/x] \models \tau \approx \tau'$ as required.

The other base case is when φ is of the form $P\tau_1 \cdots \tau_n$. Then if we have $\mathcal{M}, w, f[\alpha/x] \models P\tau_1 \cdots \tau_n$, we get

$$w \in |P|^{\mathcal{M}}(|\tau_1|^{\mathcal{M}} f[\alpha/x], \dots, |\tau_n|^{\mathcal{M}} f[\alpha/x]).$$

Hence for all $i \leq n$, we infer that $|\tau_i|^{\mathcal{M}} f[\alpha/x].w$ is defined by the Existence condition on $|P|^{\mathcal{M}}$ in the definition of premodel. But $|\tau_i|^{\mathcal{M}} f[\alpha/x].w \equiv |\tau_i|^{\mathcal{M}} f[\beta/x].w$ by the previous Lemma, because $\alpha(w) \equiv \beta(w)$, so this implies that

$$|\tau_i|^{\mathcal{M}} f[\alpha/x].w \simeq |\tau_i|^{\mathcal{M}} f[\beta/x].w.$$

Since this holds for all $i \leq n$, by the Extensionality condition on $|P|^{\mathcal{M}}$ in the definition of premodel we get

$$w \in |P|^{\mathcal{M}}(|\tau_1|^{\mathcal{M}} f[\beta/x], \dots, |\tau_n|^{\mathcal{M}} f[\beta/x]),$$

which means that $\mathcal{M}, w, f[\beta/x] \models P\tau_1 \cdots \tau_n$, as required.

The case that φ is F, and the inductive cases for φ of the form $\psi_1 \wedge \psi_2$ and $\neg\psi$, are straightforward.

Now suppose that φ is $\forall y\psi$, and assume the result for ψ. First, if $y = x$, then x is not free in $\forall y\psi$, so $f[\alpha/x]$ and $f[\beta/x]$ agree on all free variables of $\forall y\psi$, so the Free Assignment Lemma 5.4.1 gives

$$\mathcal{M}, w, f[\alpha/x] \models \forall y\psi \quad \text{iff} \quad \mathcal{M}, w, f[\beta/x] \models \forall y\psi.$$

So we can assume that $y \neq x$. If $\mathcal{M}, w, f[\alpha/x] \models \forall y\psi$, then by the semantics of $\forall$ there is some $X \in \mathit{Prop}$ with

$$w \in X \subseteq \bigcap_{\gamma \in \mathbb{X}} (E\gamma \Rightarrow |\psi|^{\mathcal{M}} f[\alpha/x][\gamma/y]), \tag{5.5.1}$$

where $\mathbb{X}$ is $\mathbb{O}$ or $\mathbb{C}$ according as $x : \mathsf{Ob}$ or not. Now the induction hypothesis on ψ implies that for any $g \in \mathit{Val}_{\mathcal{S}}$,

$$[\![\alpha \equiv \beta]\!] \cap |\psi|^{\mathcal{M}} g[\alpha/x] \subseteq |\psi|^{\mathcal{M}} g[\beta/x]. \tag{5.5.2}$$

Also, as $x \neq y$ we always have

$$f[\alpha/x][\gamma/y] = f[\gamma/y][\alpha/x]. \tag{5.5.3}$$

From all this we can show that

$$[\![\alpha \equiv \beta]\!] \cap X \subseteq \bigcap_{\gamma \in \mathbb{X}} \big(E\gamma \Rightarrow |\psi|^{\mathcal{M}} f[\beta/x][\gamma/y]\big). \tag{5.5.4}$$

To see why, suppose that $u \in [\![\alpha \equiv \beta]\!] \cap X$. Then for any $\gamma \in \mathbb{X}$, let $u \in E\gamma$. As $u \in X$, by (5.5.1) we have $u \in \big(E\gamma \Rightarrow |\psi|^{\mathcal{M}} f[\alpha/x][\gamma/y]\big)$. It follows that $u \in |\psi|^{\mathcal{M}} f[\alpha/x][\gamma/y]$, so $u \in |\psi|^{\mathcal{M}} f[\gamma/y][\alpha/x]$ by (5.5.3). Since also $u \in [\![\alpha \equiv \beta]\!]$, putting $g = f[\gamma/y]$ in (5.5.2) then yields $u \in |\psi|^{\mathcal{M}} f[\gamma/y][\beta/x]$, hence $u \in |\psi|^{\mathcal{M}} f[\beta/x][\gamma/y]$ by (5.5.3) again. Altogether we have shown that

$$u \in \big(E\gamma \Rightarrow |\psi|^{\mathcal{M}} f[\beta/x][\gamma/y]\big)$$

for any $\gamma \in \mathbb{X}$, which proves (5.5.4).

Now $[\![\alpha \equiv \beta]\!] \in Prop$ by Lemma 5.4.4, so from our assumption that $\alpha(w) \equiv \beta(w)$ we have

$$w \in [\![\alpha \equiv \beta]\!] \cap X \in Prop.$$

By (5.5.4) and the semantics of $\forall$, this implies that $\mathcal{M}, w, f[\beta/x] \models \forall y\psi$, as required.

The final case is where φ is $\Box\psi$, and here is exactly where we need to know that α and β belong to $\mathbb{O}$, so are *rigid*. Suppose that $\mathcal{M}, w, f[\alpha/x] \models \Box\psi$. Take any $u \in W$ with wRu. Then $\mathcal{M}, u, f[\alpha/x] \models \psi$. Now by rigidity of α and β we have $\alpha(w) \equiv \alpha(u)$ and $\beta(w) \equiv \beta(u)$. But $\alpha(w) \equiv \beta(w)$ by assumption. Hence $\alpha(u) \equiv \beta(u)$, so the induction hypothesis on ψ allows us to conclude that $\mathcal{M}, u, f[\beta/x] \models \psi$. This shows that $\mathcal{M}, w, f[\beta/x] \models \Box\psi$.

That completes the proof of (1). For (2), we noted in the last paragraph that the condition that $\alpha, \beta \in \mathbb{O}$ was used only in the inductive case for $\Box$. So if φ is $\Box$-free, then the proof of (1) for φ works also if $\alpha \notin \mathbb{O}$ or $\beta \notin \mathbb{O}$, provided in that case that $x \notin \mathsf{ObVar}$ so that $f[\alpha/x]$ and $f[\beta/x]$ are both legitimate valuations. ⊣

Here is a first general result on substitution of identicals:

THEOREM 5.5.3. *Any formula of the form*

$$\tau \approx \tau' \to (\varphi(\tau/x) \leftrightarrow \varphi(\tau'/x))$$

is valid if τ and τ' are object terms and $x : \mathsf{Ob}$.

If φ is $\Box$-free, then any such formula is valid for all terms τ and τ', provided that if either of τ and τ' is not an object term, then x is not an object variable.

PROOF. Suppose $\mathcal{M}, w, f \models \tau \approx \tau'$, with $\mathcal{M}$ a model and τ and τ' object terms. Put $\alpha = |\tau|^{\mathcal{M}} f$ and $\beta = |\tau'|^{\mathcal{M}} f$. Then $\alpha, \beta \in \mathbb{O}$ and $\alpha(w) \simeq \beta(w)$, so $\alpha(w) \equiv \beta(w)$. Hence

$\mathcal{M}, w, f \models \varphi(\tau/x)$

iff $\mathcal{M}, w, f[\alpha/x] \models \varphi$	by the Substitution Lemma 5.4.2
iff $\mathcal{M}, w, f[\beta/x] \models \varphi$	by the Indistinguishability Lemma 5.5.2(1)
iff $\mathcal{M}, w, f \models \varphi(\tau'/x)$	by the Substitution Lemma.

Therefore $\mathcal{M}, w, f \models \varphi(\tau/x) \leftrightarrow \varphi(\tau'/x)$ as required.

If φ is $\Box$-free, then we can use Indistinguishability Lemma 5.5.2(2) here and allow α and β here to be any members of $\mathbb{C}$, hence τ and τ' any terms (with the given proviso). ⊣

More generally than uniform substitution, we are interested in arbitrary replacement of some term by another that is asserted to have an identical interpretation. We already introduced the notation $\varphi(\tau'//\tau)$ to stand for any formula obtained from φ by replacing any number, zero or more, of free occurrences of τ by τ'. We can also apply this construction to a term σ, to form terms $\sigma(\tau'//\tau)$. Since σ is either a variable or a constant, with no subterms, in practice $\sigma(\tau'//\tau)$ is either just σ itself or τ'.

LEMMA 5.5.4. *For any terms τ, τ', and σ, if $|\tau|^{\mathcal{M}}f.w \simeq |\tau'|^{\mathcal{M}}f.w$ and $|\sigma|^{\mathcal{M}}f.w$ is defined, then $|\sigma|^{\mathcal{M}}f.w \simeq |\sigma(\tau'//\tau)|^{\mathcal{M}}f.w$.*

PROOF. If $\sigma(\tau'//\tau) = \sigma$, then the conclusion is immediate if $|\sigma|^{\mathcal{M}}f.w$ is defined. But if $\sigma(\tau'//\tau) \neq \sigma$, we must have $\sigma = \tau$ and $\sigma(\tau'//\tau) = \tau'$, so the conclusion follows if $|\tau|^{\mathcal{M}}f.w \simeq |\tau'|^{\mathcal{M}}f.w$. ⊣

Now we can prove our main result on the validity of Substitutivity of Identicals:

THEOREM 5.5.5. *Let φ be any $\Box$-free formula. Then for any terms τ and τ', the formula*

$$\tau \approx \tau' \to (\varphi \leftrightarrow \varphi(\tau'//\tau))$$

is valid, provided τ' is free for τ in φ.

PROOF. As usual, by induction on φ. Let $\mathcal{M}$ be any model.

If φ is $P\sigma_1 \cdots \sigma_n$, then $\varphi(\tau'//\tau)$ is $P\sigma_1(\tau'//\tau)\cdots\sigma_n(\tau'//\tau)$. Suppose $|\tau|^{\mathcal{M}}f.w \simeq |\tau'|^{\mathcal{M}}f.w$. Now if $\mathcal{M}, w, f \models \varphi$, then

$$w \in |P|^{\mathcal{M}}(|\sigma_1|^{\mathcal{M}}f, \dots, |\sigma_n|^{\mathcal{M}}f).$$

Hence by the Existence condition on premodels, $|\sigma_i|^{\mathcal{M}}f.w$ is defined for all $i \leq n$, so by Lemma 5.5.4 just proved, $|\sigma_i|^{\mathcal{M}}f.w \simeq |\sigma_i(\tau'//\tau)|^{\mathcal{M}}f.w$. Then by the Extensionality condition on premodels,

$$w \in |P|^{\mathcal{M}}(|\sigma_1(\tau'//\tau)|^{\mathcal{M}}f, \dots, |\sigma_n(\tau'//\tau)|^{\mathcal{M}}f) = |\varphi(\tau'//\tau)|^{\mathcal{M}}f,$$

so $\mathcal{M}, w, f \models \varphi(\tau'//\tau)$.

But we can reverse this argument, for if $\mathcal{M}, w, f \models \varphi(\tau'//\tau)$, then as $|\tau'|^{\mathcal{M}}f.w \simeq |\tau|^{\mathcal{M}}f.w$, we can use the argument just given to show that $\mathcal{M}, w, f \models \psi$ where ψ is any formula obtained from $\varphi(\tau'//\tau)$ by replacing some occurrences of τ' by τ. In particular $\mathcal{M}, w, f \models \varphi$, since φ can be obtained in this way by putting τ back in all the places in $\varphi(\tau'//\tau)$ where τ' replaced τ in φ. Hence altogether $\mathcal{M}, w, f \models \varphi \leftrightarrow \varphi(\tau'//\tau)$.

Next we have the case that φ is an identity $\sigma \approx \sigma'$. Then $\varphi(\tau'//\tau)$ is $\sigma(\tau'//\tau) \approx \sigma'(\tau'//\tau)$. Let $|\tau|^{\mathcal{M}}f.w \simeq |\tau'|^{\mathcal{M}}f.w$. If $\mathcal{M}, w, f \models \varphi$, then $|\sigma|^{\mathcal{M}}f.w \simeq |\sigma'|^{\mathcal{M}}f.w$. Hence $|\sigma|^{\mathcal{M}}f.w$ is defined, so by Lemma 5.5.4 we get $|\sigma|^{\mathcal{M}}f.w \simeq |\sigma(\tau'//\tau)|^{\mathcal{M}}f.w$. Likewise $|\sigma'|^{\mathcal{M}}f.w \simeq |\sigma'(\tau'//\tau)|^{\mathcal{M}}f.w$. It follows that $|\sigma(\tau'//\tau)|^{\mathcal{M}}f.w \simeq |\sigma'(\tau'//\tau)|^{\mathcal{M}}f.w$, and so $\mathcal{M}, w, f \models \varphi(\tau'//\tau)$. As in the previous case, this argument can be reversed.

The case of F, and the inductive cases of conjunction and negation, are straightforward.

The final case is when φ is $\forall x\psi$, and the result is assumed for ψ. First, if τ is x, then $\varphi(\tau'//\tau)$ is φ, and the formula to be proved valid is a tautology. Otherwise, $\varphi(\tau'//\tau)$ is $\forall x(\psi(\tau'//\tau))$, with τ' free for τ in ψ, given the assumption that it is free for τ in φ.

Again let $|\tau|^{\mathcal{M}}f.w \simeq |\tau'|^{\mathcal{M}}f.w$. Put $\alpha = |\tau|^{\mathcal{M}}f$ and $\beta = |\tau'|^{\mathcal{M}}f$. If $\mathcal{M}, w, f \models \varphi$, then there is some $X \in Prop$ with

$$w \in X \subseteq \bigcap_{\gamma \in \mathbb{X}} \big(E\gamma \Rightarrow |\psi|^{\mathcal{M}} f[\gamma/x]\big), \tag{5.5.5}$$

where $\mathbb{X}$ is $\mathbb{O}$ or $\mathbb{C}$ according as x : Ob or not. Now the induction hypothesis on ψ gives

$$\mathcal{M} \models \tau \approx \tau' \to (\psi \leftrightarrow \psi(\tau'//\tau)).$$

This implies that for any $g \in Val_S$,

$$[\![\alpha \simeq \beta]\!] \cap |\psi|^{\mathcal{M}} g \subseteq |\psi(\tau'//\tau)|^{\mathcal{M}} g. \tag{5.5.6}$$

From this we can show that

$$[\![\alpha \simeq \beta]\!] \cap X \subseteq \bigcap_{\gamma \in \mathbb{X}} \big(E\gamma \Rightarrow |\psi(\tau'//\tau)|^{\mathcal{M}} f[\gamma/x]\big). \tag{5.5.7}$$

This holds because for each $\gamma \in \mathbb{X}$, by (5.5.5) $X \subseteq E\gamma \Rightarrow |\psi|^{\mathcal{M}} f[\gamma/x]$, hence $X \cap E\gamma \subseteq |\psi|^{\mathcal{M}} f[\gamma/x]$. But by (5.5.6), $[\![\alpha \simeq \beta]\!] \cap |\psi|^{\mathcal{M}} f[\gamma/x] \subseteq |\psi(\tau'//\tau)|^{\mathcal{M}} f[\gamma/x]$, so combining these,

$$[\![\alpha \simeq \beta]\!] \cap X \cap E\gamma \subseteq [\![\alpha \simeq \beta]\!] \cap |\psi|^{\mathcal{M}} f[\gamma/x] \subseteq |\psi(\tau'//\tau)|^{\mathcal{M}} f[\gamma/x].$$

Hence $[\![\alpha \simeq \beta]\!] \cap X \subseteq E\gamma \Rightarrow |\psi(\tau'//\tau)|^{\mathcal{M}} f[\gamma/x]$, for all $\gamma \in \mathbb{X}$, proving (5.5.7). Since $w \in [\![\alpha \simeq \beta]\!] \cap X \in Prop$, by the semantics of $\forall$ this shows that $\mathcal{M}, w, f \models \forall x(\psi(\tau'//\tau))$, i.e. $\mathcal{M}, w, f \models \varphi(\tau'//\tau)$ as required.

Once more the argument reverses: since $|\tau'|^{\mathcal{M}}f.w \simeq |\tau|^{\mathcal{M}}f.w$, if we have $\mathcal{M}, w, f \models \varphi(\tau'//\tau)$, then we can show that $\mathcal{M}, w, f \models \varphi$ by putting τ back in all the places in $\varphi(\tau'//\tau)$ where τ' replaced τ in φ. Those places contain free occurrences of τ', since τ' is free for τ in φ. That completes this inductive case, and hence the proof of the Theorem. ⊣

THEOREM 5.5.6. *For any object terms τ and τ', the following are valid.*

(1) $\tau \approx \tau' \to \Box(\tau \approx \tau')$.
(2) $\tau \not\approx \tau' \to \Box(\tau \not\approx \tau')$.

PROOF. Let $\mathcal{M}$ be any model.

(1) Suppose $\mathcal{M}, w, f \models \tau \approx \tau'$, i.e. $|\tau|^{\mathcal{M}}f.w \simeq |\tau'|^{\mathcal{M}}f.w$. Let wRu. Since τ is of sort Ob, $|\tau|^{\mathcal{M}}f$ is in $\mathbb{O}$, hence is rigid, so as $|\tau|^{\mathcal{M}}f.w$ is defined, we have $|\tau|^{\mathcal{M}}f.w \simeq |\tau|^{\mathcal{M}}f.u$. Likewise $|\tau'|^{\mathcal{M}}f.w \simeq |\tau'|^{\mathcal{M}}f.u$, as τ' : Ob. Hence $|\tau|^{\mathcal{M}}f.u \simeq |\tau'|^{\mathcal{M}}f.u$, and therefore $\mathcal{M}, u, f \models \tau \approx \tau'$. This shows that $\mathcal{M}, w, f \models \Box(\tau \approx \tau')$.
(2) Suppose $\mathcal{M}, w, f \not\models \Box(\tau \not\approx \tau')$. Then for some $u \in W$, wRu and $\mathcal{M}, u, f \not\models (\tau \not\approx \tau')$, hence $|\tau|^{\mathcal{M}}f.u \simeq |\tau'|^{\mathcal{M}}f.u$. Similarly to (1), we then use rigidity to show that $|\tau|^{\mathcal{M}}f.w \simeq |\tau'|^{\mathcal{M}}f.w$, and therefore that $\mathcal{M}, w, f \not\models (\tau \not\approx \tau')$. ⊣

We turn now to a discussion of the soundness of a certain rule of inference, involving assertions of non-identity and non-existence, that will be used in axiomatising the logic characterised by the present semantics. If c is an object constant that does not occur in a term τ, then this rule allows us to infer $\neg\mathcal{E}\tau$ from $\tau \not\approx \mathsf{c}$. The notion of validity that the rule preserves is not that of validity in a single model, but rather validity in a model structure $\mathcal{S}$, i.e. if $\mathcal{S} \models \tau \not\approx \mathsf{c}$, then $\mathcal{S} \models \neg\mathcal{E}\tau$. Informally, the reason why this holds is that if c does not occur in τ, then we are free to alter the interpretation of c without altering that of τ. So if $\tau \not\approx \mathsf{c}$ is valid no matter what object is assigned to c, then τ cannot be identified with any object, so must be totally undefined, and hence $\mathcal{E}\tau$ cannot be satisfied.

Formally, the correctness of this argument depends on $\mathcal{S}$ being *full of objects*. Putting it contrapositively, if $\neg\mathcal{E}\tau$ is not valid in $\mathcal{S}$, then $|\tau|^{\mathcal{M}} f.w$ is defined for some f and w in some model $\mathcal{M}$ on $\mathcal{S}$. Then since $\mathcal{S}$ is full of objects, there is some $\alpha \in \mathbb{O}$ such that $\alpha(w) \simeq |\tau|^{\mathcal{M}} f.w$. Interpreting c as α and leaving the interpretation of τ unchanged then gives a model $\mathcal{M}^*$ on $\mathcal{S}$ having $\mathcal{M}^*, w, f \models \tau \approx \mathsf{c}$, so $\tau \not\approx \mathsf{c}$ is not valid in $\mathcal{S}$.

In fact we are going to produce a stronger version of this rule, by invoking the notion of *template* introduced in Section 3.2. Recall that each member ρ of the set Tem of templates contains a single occurrence of the symbol #, such that if # is replaced by a formula ψ, the result is a formula which we refer to as $\rho(\psi)$. We will show that it is valid to infer $\rho(\neg\mathcal{E}\tau)$ from $\rho(\tau \not\approx \mathsf{c})$ in any model structure, if c does not occur in ρ or τ. This rule will enable us to built individual concepts out of the values of object terms in a canonical model.

The construction of a premodel $\mathcal{M}^*$ differing from another $\mathcal{M}$ only in its interpretation of one constant was used extensively in the soundness proofs of Section 1.7, particularly for the rule GC, where it was shown that if $\mathcal{M}$ is a model then so is $\mathcal{M}^*$ (see Lemma 1.7.5 and Corollary 1.7.6). Essentially the same arguments show that this applies also to our present notion of premodel for logic with identity, and we make use of that fact in what follows.

THEOREM 5.5.7. *Let τ be any term, and* c *any object constant that does not occur in τ. Then for any model structure $\mathcal{S}$ and any template ρ not containing* c, *if $\mathcal{S} \models \rho(\tau \not\approx \mathsf{c})$, then $\mathcal{S} \models \rho(\neg\mathcal{E}\tau)$.*

PROOF. Fix τ, c and $\mathcal{S}$. Let $\mathrm{C}(\rho)$ be the following condition on a template ρ:

> For any $w \in W$ and $f \in \mathit{Val}_{\mathcal{S}}$, if $\mathcal{M}, w, f \not\models \rho(\neg\mathcal{E}\tau)$ for some model $\mathcal{M}$ on $\mathcal{S}$, then there is a model $\mathcal{M}^*$ on $\mathcal{S}$, differing from $\mathcal{M}$ only in its interpretation of c, such that $\mathcal{M}^*, w, f \not\models \rho(\tau \not\approx \mathsf{c})$.

Every template ρ not containing c satisfies this condition $\mathrm{C}(\rho)$, as we will prove by induction on the formation of ρ. But that suffices to prove the Theorem, since it implies that if $\rho(\neg\mathcal{E}\tau)$ is falsifiable in some model on $\mathcal{S}$, then $\rho(\tau \not\approx \mathsf{c})$ is falsifiable in some other model on $\mathcal{S}$ (at the same w and f).

The base of the induction is when $\rho = \#$. Suppose that $\mathcal{M}, w, f \not\models \neg\mathcal{E}\tau$ in some model $\mathcal{M}$ on $\mathcal{S}$. Then $|\tau|^{\mathcal{M}} f.w$ is defined, and belongs to Dw. Since $\mathcal{S}$ is full of objects, there is some $\alpha \in \mathbb{O}$ with $\alpha(w) \simeq |\tau|^{\mathcal{M}} f.w$. Define $\mathcal{M}^*$ to be the premodel on $\mathcal{S}$ that is the same as $\mathcal{M}$ except that $|\mathsf{c}|^{\mathcal{M}^*} = \alpha$. Then as c does not occur in τ,

$$|\tau|^{\mathcal{M}^*} f.w \simeq |\tau|^{\mathcal{M}} f.w \simeq |\mathsf{c}|^{\mathcal{M}^*} f.w,$$

so $\mathcal{M}^*, w, f \models \tau \approx \mathsf{c}$, hence $\mathcal{M}^*, w, f \not\models \tau \not\approx \mathsf{c}$. But $\mathcal{M}^*$ can be shown to be a model, since $\mathcal{M}$ is, by the methods of Lemma 1.7.5 and Corollary 1.7.6. Hence $C(\#)$ holds.

Now consider a template $\varphi \to \rho$ not containing c, so ρ does not contain c, and suppose inductively that $\mathrm{C}(\rho)$ holds. Let $\mathcal{M}, w, f \not\models \varphi \to \rho(\neg\mathcal{E}\tau)$ for some model $\mathcal{M}$ on $\mathcal{S}$. Then $\mathcal{M}, w, f \models \varphi$ and $\mathcal{M}, w, f \not\models \rho(\neg\mathcal{E}\tau)$. By $\mathrm{C}(\rho)$ there is some model $\mathcal{M}^*$ on $\mathcal{S}$ differing from $\mathcal{M}$ only on c, with $\mathcal{M}^*, w, f \not\models \rho(\tau \not\approx \mathsf{c})$. Then $\mathcal{M}^*, w, f \models \varphi$ as $\mathcal{M}, w, f \models \varphi$ and c does not occur in φ, so $\mathcal{M}^*, w, f \not\models \varphi \to \rho(\tau \not\approx \mathsf{c})$. This proves $\mathrm{C}(\varphi \to \rho)$.

Finally, consider a template $\Box\rho$ not containing c, and suppose inductively that $\mathrm{C}(\rho)$ holds. Let $\mathcal{M}, w, f \not\models \Box\rho(\neg\mathcal{E}\tau)$ for some model $\mathcal{M}$ on $\mathcal{S}$. Then there is some u with wRu and $\mathcal{M}, u, f \not\models \rho(\neg\mathcal{E}\tau)$. By $\mathrm{C}(\rho)$ there is some model $\mathcal{M}^*$ on $\mathcal{S}$ differing from $\mathcal{M}$ only on c, with $\mathcal{M}^*, u, f \not\models \rho(\tau \not\approx \mathsf{c})$. Hence $\mathcal{M}^*, w, f \not\models \Box\rho(\tau \not\approx \mathsf{c})$. This proves $\mathrm{C}(\Box\rho)$, completing the inductive argument, and hence the proof of the Theorem. ⊣

5.6. Axiomatisation and Completeness

In this section we axiomatise the set of formulas that are valid in all model structures for identity. Here are the new axioms we need:

EI$^+$: $\forall x\varphi \to (\mathcal{E}\tau \to \varphi(\tau/x))$, with τ free for x in φ, and $\tau : \mathsf{Ob}$ if $x : \mathsf{Ob}$.

EX: $P\tau_1 \cdots \tau_n \to \mathcal{E}\tau_1 \wedge \cdots \wedge \mathcal{E}\tau_n$

WR: $\tau \approx \tau' \to \tau \approx \tau$

AtSI: $\tau \approx \tau' \to (\varphi \to \varphi(\tau'//\tau))$, where φ is atomic.

NI: $\tau \approx \tau' \to \Box(\tau \approx \tau')$, where $\tau, \tau' : \mathsf{Ob}$.

NNI: $\tau \not\approx \tau' \to \Box(\tau \not\approx \tau')$, where $\tau, \tau' : \mathsf{Ob}$.

EI$^+$ was shown to be valid in Lemma 5.4.3. EX is valid in all premodels by their Existence condition on $|P|^{\mathcal{M}}$. WR expresses *Weak Reflexivity* of identity: if τ is identical to something then it is identical to itself. It can also be written as $\tau \approx \tau' \to \mathcal{E}\tau$, stating that τ can only be identical to something if it exists. Validity of WR is an immediate consequence of the semantics of $\tau \approx \tau'$.

AtSI is *Substitutivity of Identicals* for atomic formulas only, and its validity follows from that of the scheme in Theorem 5.5.5. NI (Necessity of Identity) and NNI (Necessity of Non-Identity) are only for object terms, and were shown valid in Theorem 5.5.6.

A *logic with identity* is defined to be any set L of formulas that includes all instances of the above schemes, together with all Boolean tautologies and instances of the scheme K, and is closed under the rules of Modus Ponens (MP), Necessitation (N), Generalisation on Constants (GC), Term Instantiation (TI), $\mathcal{E}\forall$-Introduction, and *Template Non-Existence*, which is the rule

$$\text{TNE:} \quad \frac{\rho(\tau \not\approx \mathsf{c})}{\rho(\neg\mathcal{E}\tau)}, \quad \text{if } \rho \in \mathsf{Tem},\ \mathsf{c} : \mathsf{Ob} \text{ and } \mathsf{c} \text{ is not in } \rho \text{ or } \tau.$$

THEOREM 5.6.1. *A logic with identity is a logic as defined in Section* 1.2.

PROOF. It is required to show that a logic with identity includes the schemes AI, UD and VQ, and is closed under the rule UG. Now these properties were proved in Theorem 3.1.2 for any $\mathcal{E}$-logic in the language without identity, and moreover they were shown to follow using only PC, the axiom EI and the rule $\mathcal{E}\forall$-Intro. The same proofs hold here, using EI^+ and $\mathcal{E}\forall$-Intro. ⊣

A significant consequence of this Theorem is that any logic with identity has all the properties specified in Lemma 1.2.1, and is closed under the rules listed in Lemma 1.2.3, including TI^* and GC^*.

THEOREM 5.6.2. *If $\mathcal{S}$ is any model structure for the language with identity, then the set $\mathrm{L}_{\mathcal{S}} = \{\varphi : \mathcal{S} \models \varphi\}$ of formulas valid in all models on $\mathcal{S}$ is a logic with identity.*

PROOF. We noted above that the new axioms for a logic with identity are all valid in $\mathcal{S}$, hence included in $\mathrm{L}_{\mathcal{S}}$. So too are all tautologies and instances of the scheme K.

The rules $\mathcal{E}\forall$-Intro and TNE were shown to be sound for validity in all model structures in the previous section, so $\mathrm{L}_{\mathcal{S}}$ is closed under these rules. It is also closed under MP and N as usual; under GC by the methods used for Theorem 1.7.7; and under TI by the proof given in Theorem 1.7.1. ⊣

If S is any set of propositional modal formulas, we use the name $\mathrm{QS}^{\approx}$ for the *smallest* logic with identity that contains every $\mathcal{L}$-formula that is a substitution-instance of a member of S. The smallest of all logics with identity is $\mathrm{QK}^{\approx}$, where K is the smallest propositional modal logic.

THEOREM 5.6.3 (Soundness for $\mathrm{QK}^{\approx}$). *If $\vdash_{\mathrm{QK}^{\approx}} \varphi$, then φ is valid in all model structures.*

PROOF. $\mathrm{QK}^{\approx}$ is the intersection of all logics with identity, so for any model structure $\mathcal{S}$, $\mathrm{QK}^{\approx}$ is included in the logic $\mathrm{L}_{\mathcal{S}}$ of the previous Theorem. Hence $\mathcal{S}$ validates $\mathrm{QK}^{\approx}$. ⊣

Here are some useful facts about identity that are derivable in our logics.

LEMMA 5.6.4. *Any logic with identity contains all instances of the following schemes.*

(1) $\tau \approx \tau' \to \tau' \approx \tau$.
(2) $\tau \approx \tau' \to (\tau' \approx \tau'' \to \tau \approx \tau'')$.
(3) $\tau \approx \tau' \to \tau' \approx \tau'$.

PROOF. (1) An instance of AtSI is $\tau \approx \tau' \to (\tau \approx \tau \to \tau' \approx \tau)$, which by axiom WR and PC leads to (1).

(2) Another instance of AtSI is $\tau' \approx \tau \to (\tau' \approx \tau'' \to \tau \approx \tau'')$, which yields (2) by (1) and PC.

(3) Yet another instance of AtSI is $\tau \approx \tau' \to (\tau \approx \tau' \to \tau' \approx \tau')$, which directly yields (3) by PC. ⊣

Now fix a logic with identity L for a *countable* signature $\mathcal{L}$ whose sets Con and ObCon are both infinite and have Con − ObCon infinite as well. As a first step towards building a canonical model for L, we define a set Σ of $\mathcal{L}$-formulas to be *object rich in* L if, for all templates ρ and terms τ,

$$\text{if } \Sigma \vdash_{\mathrm{L}} \rho(\tau \not\approx \mathsf{c}) \text{ for all } \mathsf{c} \in \mathsf{ObCon}, \text{ then } \Sigma \vdash_{\mathrm{L}} \rho(\neg\mathcal{E}\tau).$$

When Σ is deductively closed, i.e. $\Sigma \vdash_{\mathrm{L}} \varphi$ implies $\varphi \in \Sigma$, this condition has the equivalent form that

$$\{\rho(\tau \not\approx \mathsf{c}) : \mathsf{c} \in \mathsf{ObCon}\} \subseteq \Sigma \text{ implies } \rho(\neg\mathcal{E}\tau) \in \Sigma.$$

We sometimes say "L-rich", or just "rich", for "object rich in L", since we will not consider any other kind of richness. The ultimate importance of this notion is that any existence assertion $\mathcal{E}\tau$ belonging to an object rich L-*maximal* set is "witnessed" by an object constant, in the following sense.

LEMMA 5.6.5. *If Σ is* L*-maximal and object rich, then for any term τ, $\mathcal{E}\tau \in \Sigma$ iff $\tau \approx \mathsf{c} \in \Sigma$ for some object constant* c.

PROOF. Taking $\rho = \#$, by richness $\{\tau \not\approx \mathsf{c} : \mathsf{c} : \mathsf{Ob}\} \subseteq \Sigma$ implies $\neg\mathcal{E}\tau \in \Sigma$. Hence if $\mathcal{E}\tau \in \Sigma$, then $\neg\mathcal{E}\tau \notin \Sigma$ as Σ is L-consistent, so $\tau \not\approx \mathsf{c} \notin \Sigma$ for some $\mathsf{c} \in \mathsf{ObCon}$, and then $\tau \approx \mathsf{c} \in \Sigma$ by negation completeness of Σ.

Conversely, if $\tau \approx \mathsf{c} \in \Sigma$, then since $(\tau \approx \mathsf{c} \to \mathcal{E}\tau) \in \Sigma$ by axiom WR, we get $\mathcal{E}\tau \in \Sigma$. ⊣

Countability of the signature $\mathcal{L}$ is needed to prove that there are sufficiently many rich L-maximal sets. This analysis, which we now present, is exactly parallel to the use of templates to construct $\mathcal{E}\forall$-complete L-maximal sets in Lemma 3.2.5 and Theorem 3.2.6.

LEMMA 5.6.6. *Let Σ be* L*-rich. Then*

(1) *$\Sigma \cup \Gamma$ is* L*-rich for every finite set Γ of formulas.*
(2) *$\Box^{-\mathrm{L}}\Sigma = \{\varphi : \Sigma \vdash_{\mathrm{L}} \Box\varphi\}$ is* L*-rich.*

PROOF. (1): Let $\Sigma \cup \Gamma \vdash_{\mathrm{L}} \rho(\tau \not\approx \mathsf{c})$ for all $\mathsf{c} \in \mathsf{ObCon}$. If ψ is the conjunction of the members of Γ, then $\Sigma \vdash_{\mathrm{L}} \psi \to \rho(\tau \not\approx \mathsf{c})$ for all $\mathsf{c} \in \mathsf{ObCon}$. Applying the richness of Σ to the template $\psi \to \rho$ then gives $\Sigma \vdash_L \psi \to \rho(\neg\mathcal{E}\tau)$. Hence $\Sigma \cup \Gamma \vdash_L \rho(\neg\mathcal{E}\tau)$.

(2): Let $\Box^{-L}\Sigma \vdash_L \rho(\tau \not\approx \mathsf{c})$ for all $\mathsf{c} \in \mathsf{ObCon}$. Then by the relationship (2.5.1), $\Sigma \vdash_L \Box\rho(\tau \not\approx \mathsf{c})$ for all $\mathsf{c} \in \mathsf{ObCon}$. Applying the richness of Σ to the template $\Box\rho$ then gives $\Sigma \vdash_L \Box\rho(\neg\mathcal{E}\tau)$, hence $\Box^{-L}\Sigma \vdash_L \rho(\neg\mathcal{E}\tau)$. ⊣

The version of Lindenbaum's Lemma we need here is

THEOREM 5.6.7. *Every object rich* L*-consistent set of formulas has an object rich* L*-maximal extension.*

PROOF. This parallels Theorem 3.2.6, which was itself is an adaptation of Theorem 2.5.2.

Let Σ_0 be L-consistent and object rich. Since the signature is countable, there is an enumeration $\{\psi_n : n \in \omega\}$ of the set of all formulas of the form $\rho(\neg\mathcal{E}\tau)$. Suppose inductively that Σ_n has been defined to be L-consistent, with $\Sigma_n - \Sigma_0$ finite. If $\Sigma_n \vdash_L \psi_n$, put $\Sigma_{n+1} = \Sigma_n \cup \{\psi_n\}$. Otherwise, where $\psi_n = \rho(\neg\mathcal{E}\tau)$, put

$$\Sigma_{n+1} = \Sigma_n \cup \{\neg\rho(\tau \not\approx \mathsf{c})\}$$

for some constant $\mathsf{c} \in \mathsf{ObCon}$ with $\Sigma_n \nvdash_L \rho(\tau \not\approx \mathsf{c})$. Such a c exists because Σ_n is rich by the above Lemma. In both cases we get that Σ_{n+1} is L-consistent with $\Sigma_{n+1} - \Sigma_0$ finite.

Then $\Sigma = \bigcup_{n\in\omega} \Sigma_n$ is L-consistent, and any L-maximal extension of Σ is object rich and extends Σ_0. ⊣

COROLLARY 5.6.8. $\vdash_L \varphi$ *iff* φ *belongs to every object rich* L*-maximal set.*

PROOF. The result holds from left to right because an L-theorem belongs to every L-maximal set.

For the converse, if $\nvdash_L \varphi$, then $\{\neg\varphi\}$ is L-consistent. But $\{\neg\varphi\}$ is also L-rich. For if $\{\neg\varphi\} \vdash_L \rho(\tau \not\approx \mathsf{c})$ for all $\mathsf{c} \in \mathsf{ObCon}$, then choosing some constant c not in φ or ρ we have $\vdash_L \neg\varphi \to \rho(\tau \not\approx \mathsf{c})$, hence $\vdash_L \neg\varphi \to \rho(\neg\mathcal{E}\tau)$ by the rule TNE for the case of the template $\neg\varphi \to \rho$, so $\{\neg\varphi\} \vdash_L \rho(\neg\mathcal{E}\tau)$. Hence by Theorem 5.6.7 there is a rich L-maximal set Σ extending $\{\neg\varphi\}$. Then $\varphi \notin \Sigma$, by L-consistency. ⊣

Let W_L be the set of all object rich L-maximal sets of formulas.[29] A general frame $\mathcal{G}_L = (W_L, R_L, Prop_L)$ on this W_L is given by familiar definitions:

- wR_Lu iff $\{\varphi : \Box\varphi \in w\} \subseteq u$.
- $Prop_L = \{\, |\varphi|_L : \varphi$ is an $\mathcal{L}$-sentence$\}$, with $|\varphi|_L = \{w \in W_L : \varphi \in w\}$.

The closure of $Prop_L$ under the operation $[R_L]$ in this case is given by the following result.

LEMMA 5.6.9. $[R_L]|\varphi|_L = |\Box\varphi|_L \in Prop_L$.

[29]For the logics of Chapter 1, W_L denoted the set of *all* L-maximal sets. For our language with identity, we will only use the object rich ones.

PROOF. The inclusion $|\Box\varphi|_L \subseteq [R_L]|\varphi|_L$ holds by definition of R_L.

For the converse inclusion, if $w \in W_L$ has $w \notin |\Box\varphi|_L$, then $\Box\varphi \notin w$, so $\Box^- w \nvdash_L \varphi$ (1.9.1), hence $\Box^- w \cup \{\neg\varphi\}$ is L-consistent.

Now since w is L-maximal, $\Box^- w$ is identical to $\Box^{-L} w = \{\psi : w \vdash_L \Box\psi\}$, which is rich by Lemma 5.6.6(2), because w is rich by definition of W_L.

Since $\Box^- w$ is rich, so too is $\Box^- w \cup \{\neg\varphi\}$ by Lemma 5.6.6(1). Therefore by Theorem 5.6.7, $\Box^- w \cup \{\neg\varphi\}$ has a rich L-maximal extension u. Then $u \in W_L$, so wR_Lu as $\Box^- w \subseteq u$. But $\neg\varphi \in u$, so $u \notin |\varphi|_L$, showing that $w \notin [R_L]|\varphi|_L$. ⊣

An important feature of the set of object rich L-maximal sets is its representation of material implication:

LEMMA 5.6.10. *$\vdash_L \varphi \to \psi$ iff $|\varphi|_L \subseteq |\psi|_L$.*

PROOF. If $\vdash_L \varphi \to \psi$, then this L-theorem belongs to all members of W_L, from which it follows that all such members containing φ must contain ψ, by closure of maximal sets under Modus Ponens.

For the converse, if $\nvdash_L \varphi \to \psi$, then by Corollary 5.6.8 there is some rich L-maximal w with $\varphi \to \psi \notin w$. Then $w \in W_L$, $\varphi \in w$ and $\psi \notin w$, so w shows $|\varphi|_L \nsubseteq |\psi|_L$. ⊣

To extend $\mathcal{G}_L = (W_L, R_L, Prop_L)$ to a model structure $\mathcal{S}_L$ we use the constants, with object constants playing a special role in defining the domains $D_L(w)$. First, for each $w \in W_L$, define a relation $\sim_w$ on the set Con of all constants by putting

$$\mathsf{c} \sim_w \mathsf{d} \quad \text{iff} \quad \mathsf{c} \approx \mathsf{d} \in w.$$

Now by Lemma 5.6.4(1), $(\mathsf{c} \approx \mathsf{d} \to \mathsf{d} \approx \mathsf{c}) \in w$, so $\mathsf{c} \approx \mathsf{d} \in w$ implies $\mathsf{d} \approx \mathsf{c} \in w$, i.e. $\sim_w$ is symmetric. Similarly, using Lemma 5.6.4(2), $\sim_w$ is transitive. It is not however reflexive, since we do not have $\vdash_L \mathsf{c} \approx \mathsf{c}$, and may not have $\mathsf{c} \approx \mathsf{c} \in w$. What we do have, by symmetry and transitivity, is that $\mathsf{c} \sim_w \mathsf{c}$ iff there is some d such that $\mathsf{c} \sim_w \mathsf{d}$. This can also be seen directly from the instance $(\mathsf{c} \approx \mathsf{d} \to \mathsf{c} \approx \mathsf{c})$ of axiom WR.

Next, for each *object* constant c, let $|\mathsf{c}|_w = \{\mathsf{d} \in \mathsf{ObCon} : \mathsf{c} \sim_w \mathsf{d}\}$, and define the domain at w to be

$$D_L(w) = \{|\mathsf{c}|_w : \mathsf{c} : \mathsf{Ob} \text{ and } \mathsf{c} \sim_w \mathsf{c}\}.$$

Note that c may not be a member of $|\mathsf{c}|_w$, and indeed

$$\mathsf{c} \in |\mathsf{c}|_w \quad \text{iff} \quad \mathsf{c} \sim_w \mathsf{c} \quad \text{iff} \quad \mathcal{E}\mathsf{c} \in w.$$

We do have that

$$\mathsf{c} \sim_w \mathsf{d} \quad \text{implies} \quad |\mathsf{c}|_w = |\mathsf{d}|_w,$$

by symmetry and transitivity, and the converse is also true if $|\mathsf{c}|_w \neq \emptyset$ (in fact $|\mathsf{c}|_w \cap |\mathsf{d}|_w \neq \emptyset$ alone suffices to ensure that $\mathsf{c} \sim_w \mathsf{d}$).

The "universe" of $\mathcal{S}_{\mathrm{L}}$ is specified by putting

$$U_{\mathrm{L}} = \bigcup_{w \in W_{\mathrm{L}}} D_{\mathrm{L}}(w).$$

Now for an arbitrary constant $\mathsf{k} \in \mathsf{Con}$ we define a partial function $|\mathsf{k}|_{\mathrm{L}}$ from W_{L} to U_{L}. The domain of this function is given by putting

$$w \in E|\mathsf{k}|_{\mathrm{L}} \quad \text{iff} \quad \text{there is some } \mathsf{c} \in \mathsf{ObCon} \text{ with } \mathsf{k} \sim_w \mathsf{c}.$$

By Lemma 5.6.5 we then have

$$w \in E|\mathsf{k}|_{\mathrm{L}} \quad \text{iff} \quad \mathcal{E}\mathsf{k} \in w \quad \text{iff} \quad \mathsf{k} \sim_w \mathsf{k}, \tag{5.6.1}$$

and in particular

$$E|\mathsf{k}|_{\mathrm{L}} = |\mathcal{E}\mathsf{k}|_{\mathrm{L}}. \tag{5.6.2}$$

Note that if $\mathsf{k} \sim_w \mathsf{c}$ and $\mathsf{k} \sim_w \mathsf{d}$ with $\mathsf{c}, \mathsf{d} : \mathsf{Ob}$, then $\mathsf{c} \sim_w \mathsf{d}$, so $|\mathsf{c}|_w = |\mathsf{d}|_w$. Thus a well-defined function $|\mathsf{k}|_{\mathrm{L}}$ is given by putting

$$|\mathsf{k}|_{\mathrm{L}}(w) = |\mathsf{c}|_w \quad \text{iff} \quad \mathsf{k} \sim_w \mathsf{c}$$

for all $w \in E|\mathsf{k}|_{\mathrm{L}}$. Moreover, if $\mathsf{k} \sim_w \mathsf{c}$ here, then $\mathsf{c} \sim_w \mathsf{c}$ and so $|\mathsf{c}|_w \in D_{\mathrm{L}}(w)$. Hence $|\mathsf{k}|_{\mathrm{L}}(w) \in D_{\mathrm{L}}(w)$ for all $w \in E|\mathsf{k}|_{\mathrm{L}}$, a property required of the members of $\mathbb{C}$ in any model structure.

Also, if it happens that k itself is of sort Ob, and $|\mathsf{k}|_{\mathrm{L}}(w)$ is defined, then as $\mathsf{k} \sim_w \mathsf{k}$ we have $|\mathsf{k}|_{\mathrm{L}}(w) = |\mathsf{k}|_w$ in this case.

The *canonical* model structure for L can now be defined as

$$\mathcal{S}_{\mathrm{L}} = (W_{\mathrm{L}}, R_{\mathrm{L}}, \mathit{Prop}_{\mathrm{L}}, U_{\mathrm{L}}, D_{\mathrm{L}}, \mathbb{C}_{\mathrm{L}}, \mathbb{O}_{\mathrm{L}}),$$

where

- $\mathbb{C}_{\mathrm{L}} = \{|\mathsf{k}|_{\mathrm{L}} : \mathsf{k} \in \mathsf{Con}\}$, and
- $\mathbb{O}_{\mathrm{L}} = \{|\mathsf{k}|_{\mathrm{L}} : \mathsf{k} \in \mathsf{ObCon}\} \subseteq \mathbb{C}_{\mathrm{L}}$.

It is immediate from the above analysis that $\mathcal{S}_{\mathrm{L}}$ is *full of objects*. For if $a \in D_{\mathrm{L}}(w)$, then $a = |\mathsf{k}|_w$ for some k of sort Ob with $\mathsf{k} \sim_w \mathsf{k}$. But then $|\mathsf{k}|_{\mathrm{L}} \in \mathbb{O}_{\mathrm{L}}$ and $|\mathsf{k}|_{\mathrm{L}}(w) \simeq |\mathsf{k}|_w = a$.

To verify that $\mathcal{S}_{\mathrm{L}}$ is a model structure, it remains to show each $|\mathsf{k}|_{\mathrm{L}} \in \mathsf{Ob}_{\mathrm{L}}$ is rigid. Since $\mathsf{k} : \mathsf{Ob}$ in this case, in general $|\mathsf{k}|_{\mathrm{L}}(w)$ is defined iff $\mathsf{k} \sim_w \mathsf{k}$, with $|\mathsf{k}|_{\mathrm{L}}(w) = |\mathsf{k}|_w$ when it is defined, as noted above. Now suppose $wR_{\mathrm{L}}u$. Then in fact $|\mathsf{k}|_w = |\mathsf{k}|_u$. To show this we use the axioms NI and NNI. For if $\mathsf{c} \in |\mathsf{k}|_w$, then $\mathsf{k} \approx \mathsf{c} \in w$, hence $\Box(\mathsf{k} \approx \mathsf{c}) \in w$ by NI as $\mathsf{k}, \mathsf{c} : \mathsf{Ob}$, so $\mathsf{k} \approx \mathsf{c} \in u$ and therefore $\mathsf{c} \in |\mathsf{k}|_u$. But if $\mathsf{c} \notin |\mathsf{k}|_w$, then $\mathsf{k} \approx \mathsf{c} \notin w$, hence $\mathsf{k} \not\approx \mathsf{c} \in w$, so $\Box(\mathsf{k} \not\approx \mathsf{c}) \in w$ by NNI, giving $\mathsf{k} \not\approx \mathsf{c} \in u$ and so $\mathsf{k} \approx \mathsf{c} \notin u$ and thus $\mathsf{c} \notin |\mathsf{k}|_u$.

In particular, since $|\mathsf{k}|_w = |\mathsf{k}|_u$ we have $\mathsf{k} \in |\mathsf{k}|_w$ iff $\mathsf{k} \in |\mathsf{k}|_u$, i.e. $\mathsf{k} \sim_w \mathsf{k}$ iff $\mathsf{k} \sim_u \mathsf{k}$. This means that $|\mathsf{k}|_{\mathrm{L}}(w)$ is defined iff $|\mathsf{k}|_{\mathrm{L}}(u)$ is defined, and if they are both defined, then

$$|\mathsf{k}|_{\mathrm{L}}(w) = |\mathsf{k}|_w = |\mathsf{k}|_u = |\mathsf{k}|_{\mathrm{L}}(u).$$

Altogether we have shown that if $wR_{\mathrm{L}}u$, then either $|\mathsf{k}|_{\mathrm{L}}(w) \simeq |\mathsf{k}|_{\mathrm{L}}(u)$, or both are undefined. That completes the demonstration that $|\mathsf{k}|_{\mathrm{L}}$ is rigid, and that $\mathcal{S}_{\mathrm{L}}$ as defined is a model structure.

LEMMA 5.6.11. *For any* $\mathsf{k}, \mathsf{k}' \in \mathsf{Con}$:

(1) $|\mathsf{k}|_{\mathrm{L}}(w) \simeq |\mathsf{k}'|_{\mathrm{L}}(w)$ *iff* $\mathsf{k} \approx \mathsf{k}' \in w$.

(2) $[\![\, |\mathsf{k}|_{\mathrm{L}} \simeq |\mathsf{k}'|_{\mathrm{L}} \,]\!] = |\mathsf{k} \approx \mathsf{k}'|_{\mathrm{L}}$.

PROOF. (1) If $|\mathsf{k}|_{\mathrm{L}}(w) \simeq |\mathsf{k}'|_{\mathrm{L}}(w)$, with $|\mathsf{k}|_{\mathrm{L}}(w) \simeq |\mathsf{c}|_w$ and $|\mathsf{k}'|_{\mathrm{L}}(w) \simeq |\mathsf{c}'|_w$ for some $\mathsf{c}, \mathsf{c}' : \mathsf{Ob}$, then $\mathsf{k} \sim_w \mathsf{c}$, $\mathsf{k}' \sim_w \mathsf{c}'$, and $|\mathsf{c}|_w = |\mathsf{c}'|_w$. Now $\mathsf{k} \sim_w \mathsf{c}$ implies $\mathsf{c} \sim_w \mathsf{c}$, hence $\mathsf{c} \in |\mathsf{c}|_w = |\mathsf{c}'|_w$ and therefore $\mathsf{c} \sim_w \mathsf{c}'$. These relations together imply $\mathsf{k} \sim_w \mathsf{k}'$, i.e. $\mathsf{k} \approx \mathsf{k}' \in w$.

Conversely, if $\mathsf{k} \sim_w \mathsf{k}'$, then $\mathsf{k} \sim_w \mathsf{k}$ and $\mathsf{k}' \sim_w \mathsf{k}'$, so both $|\mathsf{k}|_{\mathrm{L}}(w)$ and $|\mathsf{k}'|_{\mathrm{L}}(w)$ are defined, with $|\mathsf{k}|_{\mathrm{L}}(w) = |\mathsf{c}|_w$ and $|\mathsf{k}'|_{\mathrm{L}}(w) = |\mathsf{c}'|_w$ for some $\mathsf{c}, \mathsf{c}' : \mathsf{Ob}$. Then $\mathsf{k} \sim_w \mathsf{c}$ and $\mathsf{k}' \sim_w \mathsf{c}'$. These relations imply $\mathsf{c} \sim_w \mathsf{c}'$, so $|\mathsf{c}|_w = |\mathsf{c}'|_w$, hence $|\mathsf{k}|_{\mathrm{L}}(w) \simeq |\mathsf{k}'|_{\mathrm{L}}(w)$.

(2) By definition, $[\![\, |\mathsf{k}|_{\mathrm{L}} \simeq |\mathsf{k}'|_{\mathrm{L}} \,]\!] = \{w \in W_{\mathrm{L}} : |\mathsf{k}|_{\mathrm{L}}(w) \simeq |\mathsf{k}'|_{\mathrm{L}}(w)\}$. By (1), this is equal to $\{w : \mathsf{k} \approx \mathsf{k}' \in w\}$, which is $|\mathsf{k} \approx \mathsf{k}'|_{\mathrm{L}}$. ⊣

The *canonical premodel* $\mathcal{M}_{\mathrm{L}} = (\mathcal{S}_{\mathrm{L}}, |-|^{\mathcal{M}_{\mathrm{L}}})$ on $\mathcal{S}_{\mathrm{L}}$ is defined as follows:

- $|\mathsf{k}|^{\mathcal{M}_{\mathrm{L}}} = |\mathsf{k}|_{\mathrm{L}} \in \mathbb{C}_{\mathrm{L}}$, for all constants $\mathsf{k} \in \mathsf{Con}$.
  If $\mathsf{k} \in \mathsf{ObCon}$, this automatically makes $|\mathsf{k}|^{\mathcal{M}_{\mathrm{L}}} \in \mathbb{O}_{\mathrm{L}}$ as required.
- $|P|^{\mathcal{M}_{\mathrm{L}}} : \mathbb{C}^n_{\mathrm{L}} \to Prop_{\mathrm{L}}$ is given by

$$|P|^{\mathcal{M}_{\mathrm{L}}}(|\mathsf{k}_1|_{\mathrm{L}}, \dots, |\mathsf{k}_n|_{\mathrm{L}}) = |P\mathsf{k}_1 \cdots \mathsf{k}_n|_{\mathrm{L}}, \quad \text{for all } \mathsf{k}_1, \dots, \mathsf{k}_n \in \mathsf{Con}.$$

As well as showing that $|P|^{\mathcal{M}_{\mathrm{L}}}$ satisfies the Extensionality and Existence properties, we have to verify that the definition of $|P|^{\mathcal{M}_{\mathrm{L}}}$ is proper, by showing that it does not depend on what constants k_i are used to name the arguments $|\mathsf{k}_i|_{\mathrm{L}}$. First we show

LEMMA 5.6.12. *If* $|\mathsf{k}_i|_{\mathrm{L}}(w) \simeq |\mathsf{k}'_i|_{\mathrm{L}}(w)$ *for all* $i \leq n$, *then* $w \in |P\mathsf{k}_1 \cdots \mathsf{k}_n|_{\mathrm{L}}$ *iff* $w \in |P\mathsf{k}'_1 \cdots \mathsf{k}'_n|_{\mathrm{L}}$.

PROOF. For all $i \leq n$, let $|\mathsf{k}_i|_{\mathrm{L}}(w) \simeq |\mathsf{k}'_i|_{\mathrm{L}}(w)$, so that $\mathsf{k}_i \approx \mathsf{k}'_i \in w$ by Lemma 5.6.11(1). Now we can show that

$$\vdash_{\mathrm{L}} \textstyle\bigwedge_{i \leq n} \mathsf{k}_i \approx \mathsf{k}'_i \to (P\mathsf{k}_1 \cdots \mathsf{k}_n \to P\mathsf{k}'_1 \cdots \mathsf{k}'_n),$$

by repeated use of AtSI and PC. So if $w \in |P\mathsf{k}_1 \cdots \mathsf{k}_n|_{\mathrm{L}}$, then $P\mathsf{k}_1 \cdots \mathsf{k}_n \in w$, hence by this L-theorem $P\mathsf{k}'_1 \cdots \mathsf{k}'_n \in w$, hence $w \in |P\mathsf{k}'_1 \cdots \mathsf{k}'_n|_{\mathrm{L}}$.

Interchanging k_i and k'_i gives the converse. ⊣

COROLLARY 5.6.13. *If* $|\mathsf{k}_i|_{\mathrm{L}} = |\mathsf{k}'_i|_{\mathrm{L}}$ *for all* $i \leq n$, *then* $|P\mathsf{k}_1 \cdots \mathsf{k}_n|_{\mathrm{L}} = |P\mathsf{k}'_1 \cdots \mathsf{k}'_n|_{\mathrm{L}}$.

PROOF. Let $w \in |P\mathsf{k}_1 \cdots \mathsf{k}_n|_{\mathrm{L}}$, i.e. $P\mathsf{k}_1 \cdots \mathsf{k}_n \in w$. Then for each $i \leq n$, axiom EX gives $\mathcal{E}\mathsf{k}_i \in w$, hence $w \in E|\mathsf{k}_i|_{\mathrm{L}}$ (5.6.2). Thus if $|\mathsf{k}_i|_{\mathrm{L}} = |\mathsf{k}'_i|_{\mathrm{L}}$, then $|\mathsf{k}_i|_{\mathrm{L}}(w) \simeq |\mathsf{k}'_i|_{\mathrm{L}}(w)$. Hence by Lemma 5.6.12, $w \in |P\mathsf{k}'_1 \cdots \mathsf{k}'_n|_{\mathrm{L}}$.

Again, interchanging the k_i and k'_i gives the converse. ⊣

This Corollary justifies the definition of $|P|^{\mathcal{M}_L}$. Lemma 5.6.12 then gives the required Extensionality property for $|P|^{\mathcal{M}_L}$. The Existence property was in effect shown in the first sentence of the proof of the Corollary. For if $w \in |P|^{\mathcal{M}}(\alpha_1, \dots, \alpha_n)$ with $\alpha_i = |\mathsf{k}_i|_L$ for each $i \leq n$, then $P\mathsf{k}_1 \cdots \mathsf{k}_n \in w$ and hence from EX we get $\mathcal{E}\mathsf{k}_i \in w$, making $w \in E|\mathsf{k}_i|_L$. That completes the verification that $\mathcal{M}_L$ is a premodel.

Now we come to an examination of the role played by the deduction machinery for $\forall$ in reflecting the semantics of the quantifier in the structure of $\mathcal{S}_L$. For the $\mathcal{E}$-logics of Chapter 3, which had no identity predicate $\approx$, we relied on the fact that an $\mathcal{E}$-logic is a logic in the sense of Section 1.2, so the treatment of $\forall$ in the canonical model of an $\mathcal{E}$-logic was reduced to that given in Chapter 1, specifically in Lemma 1.9.2. Here, for logics with identity, we have a different conception of the "E-function" in $\mathcal{S}_L$, and will give a new proof of the property of $\forall$ that corresponds to Lemma 1.9.2. The proof will show how the axiom EI^+ and the rule $\mathcal{E}\forall$-Intro exactly capture this property:

LEMMA 5.6.14. *If $\forall x\varphi$ is a sentence, then in $\mathcal{S}_L$,*

$$|\forall x\varphi|_L = \bigsqcap_{\mathsf{c}\in\mathsf{X}} \left(E|\mathsf{c}|_L \Rightarrow |\varphi(\mathsf{c}/x)|_L\right),$$

where X *is* ObCon *or* Con *according as $x \in$* ObVar *or not.*

PROOF. Let $\mathbb{X}$ be $\mathbb{O}_L$ or $\mathbb{C}_L$, and correspondingly X be ObCon or Con, as appropriate.

Since EI^+ gives $\vdash_L \forall x\varphi \to (\mathcal{E}\mathsf{c} \to \varphi(\mathsf{c}/x))$, we have by Lemma 5.6.10 that $|\forall x\varphi|_L \subseteq |\mathcal{E}\mathsf{c} \to \varphi(\mathsf{c}/x)|_L = |\mathcal{E}\mathsf{c}|_L \Rightarrow |\mathcal{E}\mathsf{c} \to \varphi(\mathsf{c}/x)|_L$. Now $|\mathcal{E}\mathsf{c}|_L = E|\mathsf{c}|_L$ by (5.6.2), so

$$|\forall x\varphi|_L \subseteq E|\mathsf{c}|_L \Rightarrow |\varphi(\mathsf{c}/x)|_L$$

for all $\mathsf{c} \in \mathsf{X}$. But $|\forall x\varphi|_L \in Prop_L$, so $|\forall x\varphi|_L \subseteq \bigsqcap_{\mathsf{c}\in\mathsf{X}} \left(E|\mathsf{c}|_L \Rightarrow |\varphi(\mathsf{c}/x)|_L\right)$.

For the converse inclusion, if $w \in \bigsqcap_{\mathsf{c}\in\mathsf{X}} \left(E|\mathsf{c}|_L \Rightarrow |\varphi(\mathsf{c}/x)|_L\right)$, then there is some $Y \in Prop_L$ with $w \in Y$ and $Y \subseteq E|\mathsf{c}|_L \Rightarrow |\varphi(\mathsf{c}/x)|_L$ for all $\mathsf{c} \in \mathsf{X}$. The definition of $Prop_L$ gives $Y = |\psi|_L$ for some sentence ψ. Now choose a constant $\mathsf{c} \in \mathsf{X}$ that does not occur in φ or ψ. Then as

$$|\psi|_L \subseteq E|\mathsf{c}|_L \Rightarrow |\varphi(\mathsf{c}/x)|_L = |\mathcal{E}\mathsf{c} \to \varphi(\mathsf{c}/x)|_L,$$

we get $\vdash_L \psi \to (\mathcal{E}\mathsf{c} \to \varphi(\mathsf{c}/x))$ by Lemma 5.6.10.

But $\psi \to (\mathcal{E}\mathsf{c} \to \varphi(\mathsf{c}/x)) = (\psi \to (\mathcal{E}x \to \varphi))(\mathsf{c}/x)$, so by the rule GC, $\vdash_L \psi \to (\mathcal{E}x \to \varphi)$. Hence $\vdash_L \psi \to \forall x\varphi$ by $\mathcal{E}\forall$-Intro, and so $|\psi|_L \subseteq |\forall x\varphi)|_L$. Thus $w \in |\forall x\varphi)|_L$ as required. ⊣

COROLLARY 5.6.15. *For any formula φ and any f :* InVar $\to$ Con,

$$|(\forall x\varphi)^f|_L = \bigsqcap_{\mathsf{c}\in\mathsf{X}} \left(E|\mathsf{c}|_L \Rightarrow |\varphi^{f[\mathsf{c}/x]}|_L\right),$$

where X *is as defined in the Lemma.*

PROOF. Replace φ by $\varphi^{f\backslash x}$ in Lemma 5.6.14, and use the equations $(\forall x\varphi)^f = \forall x(\varphi^{f\backslash x})$ and $\varphi^{f\backslash x}(\mathsf{c}/x) = \varphi^{f[\mathsf{c}/x]}$ from Lemma 1.9.3 (cf. the proof of part (5) of that Lemma). ⊣

A new approach to the treatment of variable assignments is needed, as we want to retain the use of functions of the form $f : \mathsf{InVar} \to \mathsf{Con}$ to construct a sentence φ^f from any formula φ by replacing each free occurrence of any variable x in φ by the constant fx. Given any such f, we can lift it to a function $|f|_{\mathrm{L}} : \mathsf{InVar} \to \mathbb{C}_{\mathrm{L}}$ by composing it with the natural map $\mathsf{c} \mapsto |\mathsf{c}|_{\mathrm{L}}$ from Con to $\mathbb{C}_{\mathrm{L}}$. In other words, we put

$$|f|_{\mathrm{L}}(x) = |fx|_{\mathrm{L}} \in \mathbb{C}_{\mathrm{L}} \text{ for all } x \in \mathsf{InVar}.$$

But any function $h : \mathsf{InVar} \to \mathbb{C}_{\mathrm{L}}$ is equal to $|f|_{\mathrm{L}}$ for some suitable choice of an $f : \mathsf{InVar} \to \mathsf{Con}$. To see this, just choose fx to be any c such that $hx = |\mathsf{c}|_{\mathrm{L}}$. Moreover, if $hx \in \mathbb{O}_{\mathrm{L}}$, we can choose fx to be a c of sort Ob with $hx = |\mathsf{c}|_{\mathrm{L}}$. Then $|f|_{\mathrm{L}}(x) \in \mathbb{O}_{\mathrm{L}}$ for all $x \in \mathsf{ObVar}$. Thus we can describe the set of all valuations in $\mathcal{S}_{\mathrm{L}}$ as

$$Val_{\mathcal{S}_{\mathrm{L}}} = \{|f|_{\mathrm{L}} : fx : \mathsf{Ob} \text{ for all } x : \mathsf{Ob}\}.$$

LEMMA 5.6.16. (1) *Updating*: $|f|_{\mathrm{L}}[\,|\mathsf{c}|_{\mathrm{L}}/x] = |f[\mathsf{c}/x]|_{\mathrm{L}}$.
(2) *Term Evaluation*: $|\tau|^{\mathcal{M}_{\mathrm{L}}}|f|_{\mathrm{L}} = |\tau^f|_{\mathrm{L}}$.

PROOF. (1) We have to show that for all variables y,

$$|f|_{\mathrm{L}}[\,|\mathsf{c}|_{\mathrm{L}}/x](y) = |f[\mathsf{c}/x]|_{\mathrm{L}}(y).$$

If $y \neq x$, $|f|_{\mathrm{L}}[\,|\mathsf{c}|_{\mathrm{L}}/x](y) = |f|_{\mathrm{L}}(y) = |fy|_{\mathrm{L}} = |f[\mathsf{c}/x]y|_{\mathrm{L}} = |f[\mathsf{c}/x]|_{\mathrm{L}}(y)$.

If $y = x$, then $|f|_{\mathrm{L}}[\,|\mathsf{c}|_{\mathrm{L}}/x](y) = |\mathsf{c}|_{\mathrm{L}} = |f[\mathsf{c}/x]x|_{\mathrm{L}} = |f[\mathsf{c}/x]|_{\mathrm{L}}(x) = |f[\mathsf{c}/x]|_{\mathrm{L}}(y)$.

(2) If τ is a variable x, then $|x|^{\mathcal{M}_{\mathrm{L}}}|f|_{\mathrm{L}} = |f|_{\mathrm{L}}(x) = |fx|_{\mathrm{L}} = |x^f|_{\mathrm{L}}$.

If τ is a constant c, then $|\mathsf{c}|^{\mathcal{M}_{\mathrm{L}}}|f| = |\mathsf{c}|^{\mathcal{M}_{\mathrm{L}}} = |\mathsf{c}|_{\mathrm{L}} = |\mathsf{c}^f|_{\mathrm{L}}$. ⊣

We are now in a position to prove a suitably formulated Truth Lemma for the premodel $\mathcal{M}_{\mathrm{L}}$.

THEOREM 5.6.17 (Truth Is Membership). *Let φ be any formula. Then for all $f : \mathsf{InVar} \to \mathbb{C}_{\mathrm{L}}$,*

$$|\varphi|^{\mathcal{M}_{\mathrm{L}}}|f|_{\mathrm{L}} = |\varphi^f|_{\mathrm{L}},$$

and hence for all $w \in W_{\mathrm{L}}$,

$$\mathcal{M}_{\mathrm{L}}, w, |f|_{\mathrm{L}} \models \varphi \quad \textit{iff} \quad \varphi^f \in w.$$

PROOF. As usual, by induction on the formation of φ.

If φ is the atomic $P\tau_1 \cdots \tau_n$, then

$$\begin{aligned} |\varphi|^{\mathcal{M}_L}|f|_L &= |P|^{\mathcal{M}_L}(|\tau_1|^{\mathcal{M}_L}|f|_L, \dots, |\tau_n|^{\mathcal{M}_L}|f|_L) && \text{definition of } |\varphi|^{\mathcal{M}} \\ &= |P|^{\mathcal{M}_L}(|\tau_1^f|_L, \dots, |\tau_n^f|_L) && \text{Lemma 5.6.16(2)} \\ &= |P(\tau_1^f) \cdots (\tau_n^f)|_L && \text{definition of } |P|^{\mathcal{M}_L} \\ &= |\varphi^f|_L. \end{aligned}$$

If φ is the identity $\tau \approx \tau'$, then

$$\begin{aligned} |\varphi|^{\mathcal{M}_L}|f|_L &= [\![\, |\tau|^{\mathcal{M}_L}|f|_L \simeq |\tau'|^{\mathcal{M}_L}|f|_L \,]\!] && \text{definition of } |\varphi|^{\mathcal{M}} \\ &= [\![\, |\tau^f|_L \simeq |\tau'^f|_L \,]\!] && \text{Lemma 5.6.16(2)} \\ &= |\tau^f \approx \tau'^f|_L && \text{Lemma 5.6.11(2)} \\ &= |\varphi^f|_L. \end{aligned}$$

The case of F and the inductive cases for conjunction and negation involve nothing new. The inductive case for $\Box$ is taken care of by Lemma 5.6.9 and the induction hypothesis, which allows us to reason that if the result holds for φ, then

$$|\Box\varphi|^{\mathcal{M}_L}|f|_L = [R_L]|\varphi|^{\mathcal{M}_L}|f|_L = [R_L]|\varphi^f|_L = |\Box(\varphi^f)|_L = |(\Box\varphi)^f|_L,$$

so the result holds for $\Box\varphi$.

Finally, for the case of $\forall$, assume that the result holds for φ. Let $\mathbb{X}$ be $\mathbb{O}_L$ or $\mathbb{C}_L$, and correspondingly X be ObCon or Con, according as $x \in \mathsf{ObVar}$ or not. Then

$$\begin{aligned} |\forall x\varphi|^{\mathcal{M}_L}|f|_L &= \textstyle\bigsqcap_{\alpha\in\mathbb{X}} \left(E\alpha \Rightarrow |\varphi|^{\mathcal{M}_L}|f|_L[\alpha/x]\right) && \text{semantics of } \forall \\ &= \textstyle\bigsqcap_{\mathsf{c}\in\mathsf{X}} \left(E|\mathsf{c}|_L \Rightarrow |\varphi|^{\mathcal{M}_L}|f|_L[\,|\mathsf{c}|_L/x]\right) && \text{definition of } \mathbb{X} \\ &= \textstyle\bigsqcap_{\mathsf{c}\in\mathsf{X}} \left(E|\mathsf{c}|_L \Rightarrow |\varphi|^{\mathcal{M}_L}|f[\mathsf{c}/x]|_L\right) && \text{Lemma 5.6.16(1)} \\ &= \textstyle\bigsqcap_{\mathsf{c}\in\mathsf{X}} \left(E|\mathsf{c}|_L \Rightarrow |\varphi^{f[\mathsf{c}/x]}|_L\right) && \text{hypothesis on } \varphi \\ &= |(\forall x\varphi)^f|_L && \text{Corollary 5.6.15,} \end{aligned}$$

so the result holds for $\forall x\varphi$. ⊣

THEOREM 5.6.18. *$\mathcal{M}_L$ is a model that characterises* L.

PROOF. Each truth set $|\varphi|^{\mathcal{M}_L}|f|_L$ of $\mathcal{M}_L$ is admissible, because it is equal to $|\varphi^f|_L \in \mathit{Prop}_L$. Hence $\mathcal{M}_L$ is a model.

The proof that $\mathcal{M}_L$ characterises L is parallel to that of Theorem 1.9.6. Here, if $\vdash_L \varphi$, then for any $f : \mathit{Val} \to \mathsf{Con}$, we have $\vdash_L \varphi^f$ by the extended Term Instantiation rule TI*, hence $|\varphi^f|_L = W_L$, i.e. φ^f belongs to every object rich L-maximal set (Corollary 5.6.8). So $|\varphi|^{\mathcal{M}_L}|f|_L = W_L$ by Theorem 5.6.17. This shows that φ is valid in $\mathcal{M}_L$.

For the converse, if $\nvdash_L \varphi$, then using the rule GC* and a suitable choice of fresh constants as in Theorem 1.9.6, we can find an $f : \mathit{Val} \to \mathsf{Con}$ with

$\nvdash_{\mathrm{L}} \varphi^f$. Hence by Corollary 5.6.8, there is some rich L-maximal set $w \in W_{\mathrm{L}}$ with $\varphi^f \notin w$. Therefore $\mathcal{M}_{\mathrm{L}}, w, |f|_{\mathrm{L}} \not\models \varphi$ by Theorem 5.6.17, so φ is not valid in $\mathcal{M}_{\mathrm{L}}$. ⊣

This Theorem immediately implies that L is complete for $\mathcal{S}_{\mathrm{L}}$, i.e. $\mathcal{S}_{\mathrm{L}} \models \varphi$ implies $\vdash_{\mathrm{L}} \varphi$. From that we get

THEOREM 5.6.19 (Completeness for $\mathrm{QK}^{\approx}$). *The canonical model structure* $\mathcal{S}_{\mathrm{QK}^{\approx}}$ *characterises* $\mathrm{QK}^{\approx}$:

$$\vdash_{\mathrm{QK}^{\approx}} \varphi \quad \textit{iff} \quad \mathcal{S}_{\mathrm{QK}^{\approx}} \models \varphi.$$

Hence $\vdash_{\mathrm{QK}^{\approx}} \varphi$ *iff* φ *is valid in all model structures for the language with identity.* ⊣

We can then go on to show that every logic of the form $\mathrm{QS}^{\approx}$ is characterised by validity in its model structures, and indeed by validity in model structures whose underlying general frame validates the propositional formulas in S. This adapts the analysis of Section 1.10, and uses the general frame $\mathcal{G}_L = (W_{\mathrm{L}}, R_{\mathrm{L}}, Prop_{\mathrm{L}})$ underlying the canonical model structure of a logic with identity L that includes $\mathrm{QS}^{\approx}$. For any such L, $\mathcal{G}_{\mathrm{L}}$ validates the set S of propositional modal formulas, and $\mathcal{S}_{\mathrm{L}}$ validates $\mathrm{QS}^{\approx}$. These facts are shown by the methods used in the proofs of Lemma 1.10.1 and Theorem 1.7.11. From this, similarly to Theorem 1.10.2, we obtain

THEOREM 5.6.20 (Completeness for $\mathrm{QS}^{\approx}$). *Let* S *be any set of propositional modal formulas.*

(1) $\mathrm{QS}^{\approx}$ *is characterised by validity in* $\mathcal{S}_{\mathrm{QS}^{\approx}}$, *i.e.* $\vdash_{\mathrm{QS}^{\approx}} \varphi$ *iff* $\mathcal{S}_{\mathrm{QS}^{\approx}} \models \varphi$.
(2) $\mathrm{QS}^{\approx}$ *is characterised by validity in all model structures whose underlying general frame validates* S. ⊣

Unlike the situation with the identity-free logics of Chapter 1, we are unable to show in general that if S is a canonical propositional modal logic, then the *Kripke frame* underlying $\mathcal{S}_{\mathrm{QS}^{\approx}}$ validates S. The reason is that for a logic L extending $\mathrm{QS}^{\approx}$, we can no longer show that the Kripke frame $\mathcal{F}_{\mathrm{L}}$ underlying $\mathcal{S}_{\mathrm{L}}$ is isomorphic to an inner subframe of the canonical propositional S-frame $\mathcal{F}_{\mathrm{S}}$ (see Theorem 1.10.5). To show this we would need $\mathcal{F}_{\mathrm{L}}$ to contain all the L-maximal sets, but now it only contains the *object rich* ones. All we know is that $\mathcal{F}_{\mathrm{L}}$ is isomorphic to a subframe of $\mathcal{F}_{\mathrm{S}}$, giving us the following conclusion (see Theorem 2.8.3 and the discussion preceding it):

THEOREM 5.6.21. *Let* S *be a canonical propositional modal logic that is preserved by subframes. Then* $\mathrm{QS}^{\approx}$ *is characterised by the class of all model structures whose underlying Kripke frame validates* S. ⊣

5.7. Kripkean Models

A (pre)model $\mathcal{M}$ for the language with identity will be called *Kripkean* if

$$|\forall x\varphi|^{\mathcal{M}} f = \bigcap_{\alpha \in \mathbb{X}} \left(E\alpha \Rightarrow |\varphi|^{\mathcal{M}} f[\alpha/x]\right), \tag{5.7.1}$$

where $\mathbb{X}$ is $\mathbb{O}$ or $\mathbb{C}$ according to the sort of x. In terms of the satisfaction relation, this means that

$$\mathcal{M}, w, f \models \forall x\varphi \text{ iff for all } \alpha \in \mathbb{X} \text{ with } \alpha(w) \text{ defined, } \mathcal{M}, w, f[\alpha/x] \models \varphi.$$

This condition should be compared with the Kripkean model condition of (1.6.6) for varying domain models of the language without identity.

The corresponding semantics for the existential quantifier is

$$\mathcal{M}, w, f \models \exists x\varphi \text{ iff for some } \alpha \in \mathbb{X} \text{ with } \alpha(w) \text{ defined, } \mathcal{M}, w, f[\alpha/x] \models \varphi;$$

or equivalently,

$$|\exists x\varphi|^{\mathcal{M}} f = \bigcup_{\alpha \in \mathbb{X}} \left(E\alpha \Rightarrow |\varphi|^{\mathcal{M}} f[\alpha/x]\right).$$

To axiomatise the logic with identity characterised by Kripkean models, we use again the Template $\forall$-Introduction rule. This was introduced in Section 3.2 to axiomatise the $\mathcal{E}$-logic of Kripkean models in the language without identity, and is

$$\text{T}\forall\text{-Intro:} \quad \frac{\rho(\mathcal{E}x \to \psi)}{\rho(\forall x\psi)}, \quad \text{if } \rho \in \mathsf{Tem} \text{ and } x \text{ is not free in } \rho.$$

THEOREM 5.7.1. T$\forall$-Intro *preserves validity in all Kripkean models for the language with identity.*

PROOF. As for Lemma 3.2.3. Let $\mathcal{M}$ be Kripkean. It suffices to show that if $\mathcal{M}, w, f \not\models \rho(\forall x\psi)$, then there is some $\alpha \in \mathbb{C}$, with $\alpha \in \mathbb{O}$ if $x : \mathsf{Ob}$, such that $\mathcal{M}, w, f[\alpha/x] \not\models \rho(\mathcal{E}x \to \psi)$. This is done by induction on the formation of ρ.

For the case $\rho = \#$, if $\mathcal{M}, w, f \not\models \forall x\psi$, then since $\mathcal{M}$ is Kripkean, there is some $\alpha \in \mathbb{C}$, with $\alpha \in \mathbb{O}$ if $x : \mathsf{Ob}$, such that $w \in E\alpha$ and $\mathcal{M}, w, f[\alpha/x] \not\models \psi$. Then $\mathcal{M}, w, f[\alpha/x] \models \mathcal{E}x$, hence $\mathcal{M}, w, f[\alpha/x] \not\models \mathcal{E}x \to \psi$, so the result holds in this case.

The cases of templates $\varphi \to \rho$ and $\Box\rho$, on the inductive assumption that the result holds for ρ, proceed just as in Lemma 3.2.3. ⊣

The rule T$\forall$-Intro allows us to built characteristic canonical models out of $\mathcal{E}\forall$-complete sets of formulas. For the present language, we define Σ to be $\mathcal{E}\forall$-*complete* in a logic L if, for any formula φ, any variable x, and any template ρ, if $\Sigma \vdash_{\mathrm{L}} \rho(\mathcal{E}\mathsf{c} \to \varphi(\mathsf{c}/x))$ for all $\mathsf{c} \in \mathsf{X}$, then $\Sigma \vdash_{\mathrm{L}} \rho(\forall x\varphi)$. Here, X is ObCon or Con according to the sort of the variable x.

LEMMA 5.7.2. *Let* L *be a logic with identity, and let* Σ *be* $\mathcal{E}\forall$*-complete.*

(1) $\Sigma \cup \Gamma$ *is* $\mathcal{E}\forall$*-complete for every finite set* Γ *of formulas.*
(2) $\Box^{-L}\Sigma = \{\varphi : \Sigma \vdash_L \Box\varphi\}$ *is* $\mathcal{E}\forall$*-complete.*
(3) *If* Σ *is also* L*-consistent, then it has an* L*-maximal* $\mathcal{E}\forall$*-complete extension.*

PROOF. As for Lemma 3.2.5 and Theorem 3.2.6. ⊣

Working with T$\forall$-Intro means that we no longer need the rule TNE of Template Non-Existence, but can replace it by the simple formula scheme

$$\text{FNE:} \quad \forall x(\tau \not\approx x) \to \neg\mathcal{E}\tau, \quad \text{with } x : \mathsf{Ob} \text{ and } x \text{ not in } \tau.$$

This may be more readily understood in its equivalent dual form

$$\mathcal{E}\tau \to \exists x(\tau \approx x), \quad \text{with } x : \mathsf{Ob} \text{ and } x \text{ not in } \tau,$$

which asserts that if τ exists them it is identical to some object.

THEOREM 5.7.3. FNE *is valid in all premodels, not just the Kripkean ones.*

PROOF. Suppose $\mathcal{M}, w, f \not\models \neg\mathcal{E}\tau$. Then $|\tau|^{\mathcal{M}} f.w$ is defined. Since the structure underlying $\mathcal{M}$ is full of objects, there is some $\alpha \in \mathbb{O}$ such that $\alpha(w) \simeq |\tau|^{\mathcal{M}} f.w$. If x : Ob and x is not in τ, then $|\tau|^{\mathcal{M}} f[\alpha/x] = |\tau|^{\mathcal{M}} f$, so $|\tau|^{\mathcal{M}} f[\alpha/x].w \simeq \alpha(w)$. Hence $\mathcal{M}, w, f[\alpha/x] \models \tau \approx x$. By the semantics of $\forall$,

$$|\forall x(\tau \not\approx x)|^{\mathcal{M}} \subseteq E\alpha \Rightarrow |\tau \not\approx x|^{\mathcal{M}} f[\alpha/x].$$

But $w \in E\alpha$ and $w \notin |\tau \not\approx x|^{\mathcal{M}} f[\alpha/x]$, so $w \notin E\alpha \Rightarrow |\tau \not\approx x|^{\mathcal{M}} f[\alpha/x]$. Hence $w \notin |\forall x(\tau \not\approx x)|^{\mathcal{M}}$, i.e. $\mathcal{M}, w, f \not\models \forall x(\tau \not\approx x)$. ⊣

By the characterisations of the previous section, this Theorem shows that FNE must be derivable in any logic L with identity. Indeed it is derivable using the instance of TNE in which ρ is the template $\forall x(\tau \not\approx x) \to \#$, with x : Ob and x not in τ. Taking a constant c : Ob that does not occur in τ, as an instance of EI^+ we have $\forall x(\tau \not\approx x) \to (\mathcal{E}\mathsf{c} \to \tau \not\approx \mathsf{c})$. But $\vdash_L \neg\mathcal{E}\mathsf{c} \to \tau \not\approx \mathsf{c}$ from Lemma 5.6.4(3) by contraposition, so from these we get $\vdash_L \forall x(\tau \not\approx x) \to \tau \not\approx \mathsf{c}$ by PC. Since c does not occur in $\forall x(\tau \not\approx x)$, TNE then gives $\vdash_L \forall x(\tau \not\approx x) \to \neg\mathcal{E}\tau$, which is FNE.

Note that in FNE, τ can be any term, not just an object term. In fact if τ is of sort Ob, then the scheme is derivable just by using EI^+. For, taking φ to be $(\tau \not\approx x)$, if x : Ob we obtain $\neg\mathcal{E}\tau$ as $\varphi(\tau/x)$ and, *provided* τ : Ob, we get the EI^+-instance

$$\forall x(\tau \not\approx x) \to (\mathcal{E}\tau \to \neg\mathcal{E}\tau),$$

which is tautologically equivalent to FNE.

Since we are now interested in Kripkean models, we will define a *Kripkean identity logic* to be any set of formulas that has all the properties of a logic with identity *except for* the rule TNE, and in place of that has the rule T$\forall$-Intro and the scheme FNE. The sense in which FNE can replace TNE relative to T$\forall$-Intro is conveyed by the following result.

THEOREM 5.7.4. *Every Kripkean identity logic is a logic with identity.*

PROOF. Let L be a Kripkean identity logic. What is required is to show that L is closed under TNE. So suppose $\vdash_{\mathrm{L}} \rho(\tau \not\approx \mathsf{c})$, with c an object constant not occurring in the term τ or the template ρ. Since $\tau \not\approx \mathsf{c} \to (\mathcal{E}\mathsf{c} \to \tau \not\approx \mathsf{c})$ is a tautology, $\vdash_{\mathrm{L}} \rho(\tau \not\approx \mathsf{c}) \to \rho(\mathcal{E}\mathsf{c} \to \tau \not\approx \mathsf{c})$ by Lemma 3.2.1. Therefore $\vdash_{\mathrm{L}} \rho(\mathcal{E}\mathsf{c} \to \tau \not\approx \mathsf{c})$. Now choose a new variable x of sort Ob that does not occur in τ or ρ. Then $\vdash_{\mathrm{L}} \rho(\mathcal{E}x \to \tau \not\approx x)$ by rule GC. Hence $\vdash_{\mathrm{L}} \rho(\forall x(\tau \not\approx x))$ by T$\forall$-Intro. But $\vdash_{\mathrm{L}} \rho(\forall x(\tau \not\approx x)) \to \rho(\neg\mathcal{E}\tau)$ by Lemma 3.2.1, since $\vdash_{\mathrm{L}} \forall x(\tau \not\approx x) \to \neg\mathcal{E}\tau$ by FNE. Hence $\vdash_{\mathrm{L}} \rho(\neg\mathcal{E}\tau)$, as required for closure under TNE. ⊣

It is noteworthy that from the WR-instance $\tau \approx x \to \mathcal{E}\tau$ there follows $\neg\mathcal{E}\tau \to (\mathcal{E}x \to \tau \not\approx x)$ by PC, and from this second formula there follows $\neg\mathcal{E}\tau \to \forall x(\tau \not\approx x)$ by T$\forall$-Intro, *provided that* x is not in τ. Thus for any Kripkean identity logic L we have the biconditional

$$\vdash_{\mathrm{L}} \forall x(\tau \not\approx x) \leftrightarrow \neg\mathcal{E}\tau,$$

and dually $\vdash_{\mathrm{L}} \mathcal{E}\tau \leftrightarrow \exists x(\tau \approx x)$, whenever $x : \mathsf{Ob}$ and x is not in τ.

To build a canonical Kripkean model for identity, the work of the previous section and Section 3.2 suggests that we need to use maximal sets that are both $\mathcal{E}\forall$-complete and object rich. But in fact:

THEOREM 5.7.5. *In a Kripkean identity logic every $\mathcal{E}\forall$-complete set is object rich.*

PROOF. Let L be a Kripkean, and Σ be $\mathcal{E}\forall$-complete in L. Suppose that $\Sigma \vdash_{\mathrm{L}} \rho(\tau \not\approx \mathsf{c})$ for all $\mathsf{c} \in \mathsf{ObCon}$. Then we have to show that $\Sigma \vdash_{\mathrm{L}} \rho(\neg\mathcal{E}\tau)$. But $\vdash_{\mathrm{L}} \rho(\tau \not\approx \mathsf{c}) \to \rho(\mathcal{E}\mathsf{c} \to \tau \not\approx \mathsf{c})$, as was shown in the proof of the last Theorem, so from $\Sigma \vdash_{\mathrm{L}} \rho(\tau \not\approx \mathsf{c})$ we get $\Sigma \vdash_{\mathrm{L}} \rho(\mathcal{E}\mathsf{c} \to \tau \not\approx \mathsf{c})$.

Now choose a new variable x of sort Ob that does not occur in τ or ρ, and let φ be $\tau \not\approx x$. Then we have $\Sigma \vdash_{\mathrm{L}} \rho(\mathcal{E}\mathsf{c} \to \varphi(\mathsf{c}/x))$. Since this holds for each $\mathsf{c} \in \mathsf{ObCon}$, $\mathcal{E}\forall$-completeness then implies $\Sigma \vdash_{\mathrm{L}} \rho(\forall x(\tau \not\approx x))$. But $\vdash_{\mathrm{L}} \rho(\forall x(\tau \not\approx x)) \to \rho(\neg\mathcal{E}\tau)$ by FNE and Lemma 3.2.1, so this yields $\Sigma \vdash_{\mathrm{L}} \rho(\neg\mathcal{E}\tau)$, showing that Σ is rich in L. ⊣

In view of this fact, we are able to define a suitable canonical model structure

$$\mathcal{S}^K_{\mathrm{L}} = (W^K_{\mathrm{L}}, R^K_{\mathrm{L}}, Prop^K_{\mathrm{L}}, U^K_{\mathrm{L}}, D^K_{\mathrm{L}}, \mathbb{C}^K_{\mathrm{L}}, \mathbb{O}^K_{\mathrm{L}}),$$

for a Kripkean identity logic L by taking W^K_{L} to be the set of all $\mathcal{E}\forall$-complete L-maximal sets of formulas, with the usual definitions for R^K_{L} and $Prop^K_{\mathrm{L}}$ based on this W^K_{L}, as in Section 3.2.

The other items $U^K_{\mathrm{L}}, D^K_{\mathrm{L}}, \mathbb{C}^K_{\mathrm{L}}, \mathbb{O}^K_{\mathrm{L}}$ are defined just like $U_{\mathrm{L}}, D_{\mathrm{L}}, \mathbb{C}_{\mathrm{L}}, \mathbb{O}_{\mathrm{L}}$ in the previous section, but based on W^K_{L} rather than W_{L}, and using the fact that the members of W^K_{L} are object rich, by the Theorem just proved. The members of $\mathbb{C}^K_{\mathrm{L}}$ are the partial functions $|\mathsf{c}|^K_{\mathrm{L}}$ on W^K_{L}, with domain $E|\mathsf{c}|^K_{\mathrm{L}}$ equal to $|\mathcal{E}\mathsf{c}|^K_{\mathrm{L}}$, i.e. to $\{w \in W^K_{\mathrm{L}} : \mathcal{E}\mathsf{c} \in w\}$.

In place of Lemma 5.6.14, the $\mathcal{E}\forall$-completeness of the members of W^K_{L} ensures that for a sentence $\forall x\varphi$,

$$|\forall x\varphi|^K_{\mathrm{L}} = \bigcap_{\mathsf{c}\in\mathsf{X}} E|\mathsf{c}|^K_{\mathrm{L}} \Rightarrow |\varphi(\mathsf{c}/x)|^K_{\mathrm{L}} = \prod_{\mathsf{c}\in\mathsf{X}} E|\mathsf{c}|^K_{\mathrm{L}} \Rightarrow |\varphi(\mathsf{c}/x)|^K_{\mathrm{L}}$$

(see Lemma 3.2.8). This makes $\mathcal{M}^K_{\mathrm{L}}$, defined similarly to $\mathcal{M}_{\mathrm{L}}$, into a Kripkean model, one that characterises L.

Ultimately this analysis allows us to show the following.

THEOREM 5.7.6. *Let* S *be any set of propositional modal formulas. Then the logic* $\mathrm{QS}^{\approx} + \mathrm{T}\forall\text{-Intro} + \mathrm{FNE}$ *is characterised by validity in all Kripkean models on model structures whose underlying general frame validates* S. ⊣

The full details of this are left to the interested reader to work through by adapting the material of Section 3.2 and Section 5.6 as outlined. We will however return to this construction, and give more details, later on in the next section, where we apply it to an expanded language.

One advantage of working with the notion of Kripkean model that we have described is that we can use it, as in the non-Kripkean case, to build a single model that characterises the logic. But if the model is Kripkean, we can simplify its description by decomposing it into its R-connected components, and then replacing the rigid partial functions making up the admissible objects in each component by actual individuals or the empty function. This works because the evaluation of a formula $\Box\varphi$ at world w takes place within the component of w, and in a *Kripkean* model, the evaluation of $\forall x\varphi$ at w refers only to w and not to any other worlds or admissible sets of worlds. Thus in evaluating formulas at w we can ignore the structure outside of the component of w.

But on the connected component of w, each world has its domain of individuals equal to Dw (see (5.4.1)), and each rigid partial function in $\mathbb{O}$ is either total and constant, and hence is identifiable with a single individual from Dw, or is the empty function on the component, which we will denote by $\emptyset^w$ (see the analysis in Section 5.3). Thus $\mathbb{O}$ becomes identifiable on this component with either Dw or $Dw \cup \{\emptyset^w\}$. This relates to a kind of model developed by Scott [1967] for one-sorted non-modal quantificational logic without existence assumptions. Such a model has a domain of individuals D that is the range of the variable x when evaluating the quantifier $\forall x$; and an element $\star_D$ not in D, thought of as a "null entity", with certain terms assigned $\star_D$ as their value to indicate that they are not legitimately defined, hence do not exist in the model. Evidently $\star_D$ performs a similar role to $\emptyset^w$.

To see how the decomposition works, let

$$\mathcal{M} = (W, R, Prop, U, D, \mathbb{C}, \mathbb{O}, |-|^{\mathcal{M}})$$

be a Kripkean model for identity, and w a member of W. Then we define a new type of structure

$$\mathcal{M}^w = (W^w, R^w, Prop^w, U^w, D^w, \mathbb{C}^w, \mathbb{O}^w, |-|^{\mathcal{M}^w})$$

as follows. First, W^w is the underlying set of the R-connected component of W in the frame (W, R), and R^w is the restriction of R to W^w (see Section 2.4 for the theory of components). (W^w, R^w) is an *inner* subframe of (W, R), i.e. if $u \in W^w$ and uRu', then $u' \in W^w$. The set of admissible propositions of $\mathcal{M}^w$ is defined to be

$$Prop^w = \{X \cap W^w : X \in Prop\}.$$

$Prop^w$ is a Boolean subalgebra of the powerset algebra $\wp(W^w)$, and is closed under the operator $[R^w]$. D^w is the restriction of the domain function D to W^w, and in fact since $Du = Dw$ for all $u \in W^w$ we could just identify D^w with the set Dw. In any case we take the universe U^w to be Dw.

Now for each $\alpha \in \mathbb{C}$ we define an entity α^w. If $\alpha \notin \mathbb{O}$, α^w is the restriction of the partial function α to W^w. If $\alpha \in \mathbb{O}$, and the restriction of α to W^w is the empty function $\emptyset^w$ on W^w, then $\alpha^w = \emptyset^w$. Finally, if $\alpha \in \mathbb{O}$ and the restriction of α to W^w is not empty, then $\alpha^w = \alpha(w) \in Dw$, with $\alpha(w)$ being the constant value that α takes on the whole of W^w. Let $\mathbb{C}^w = \{\alpha^w : \alpha \in \mathbb{C} - \mathbb{O}\}$ and $\mathbb{O}^w = \{\alpha^w : \alpha \in \mathbb{O}\}$. Now if $a \in Dw$ then there is some $\alpha \in \mathbb{O}$ with $\alpha(w) \simeq a$, so that $a = \alpha^w \in \mathbb{O}^w$. But every member of $\mathbb{O}^w$ is either in Dw or is $\emptyset^w$, so altogether

$$Dw \subseteq \mathbb{O}^w \subseteq Dw \cup \{\emptyset^w\},$$

and in a model having $\emptyset_U^W$ in $\mathbb{O}$ we always get $\mathbb{O}^w = Dw \cup \{\emptyset^w\}$.

To define $\mathcal{M}^w$, we put $|\mathsf{c}|^{\mathcal{M}^w} = (|\mathsf{c}|^{\mathcal{M}})^w$ for any constant $\mathsf{c} \in \mathsf{Con}$. The interpretation of predicate symbols by $\mathcal{M}^w$ is complicated by the fact that in this new kind of structure, $\mathbb{O}^w$ is not a subset of $\mathbb{C}^w$, but rather $\mathbb{C}^w$ is a collection of partial functions whose values belong to Dw. So for an n-ary predicate symbol P, we take $|P|^{\mathcal{M}^w}$ to be an n-ary function on $\mathbb{C}^w \cup \mathbb{O}^w$, defined by

$$|P|^{\mathcal{M}^w}(\alpha_1^w, \dots, \alpha_n^w) = |P|^{\mathcal{M}}(\alpha_1, \dots, \alpha_n) \cap W^w$$

for all $\alpha_1, \dots, \alpha_n \in \mathbb{C}$. We need to check that this is well-defined, i.e. that if $\alpha_i^w = \beta_i^w$ for all $i \leq n$, then

$$|P|^{\mathcal{M}}(\alpha_1, \dots, \alpha_n) \cap W^w = |P|^{\mathcal{M}}(\beta_1, \dots, \beta_n) \cap W^w. \tag{5.7.2}$$

But if $u \in |P|^{\mathcal{M}}(\alpha_1, \dots, \alpha_n) \cap W^w$, then for all $i \leq n$ we have that $\alpha_i(u)$ is defined, by the Existence condition on $|P|^{\mathcal{M}}$ in $\mathcal{M}$ as a premodel, and so if $\alpha_i^w = \beta_i^w$, then $\alpha_i(u) \simeq \beta_i(u)$ as the restrictions of α_i and β_i to W^w agree. Therefore the Extensionality condition on $|P|^{\mathcal{M}}$ implies that $u \in |P|^{\mathcal{M}}(\beta_1, \dots, \beta_n) \cap W^w$. The converse inclusion of (5.7.2) holds likewise by interchange of the α_i and β_i.

It is readily checked that $\mathcal{M}^w$ itself satisfies the Extensionality and Existence conditions for $|P|^{\mathcal{M}^w}$, and so is a premodel. To deal with variable-assignments, we associate with each $f : \mathsf{InVar} \to \mathbb{C}$ the function $f^w : \mathsf{InVar} \to \mathbb{C}^w \cup \mathbb{O}^w$ defined by putting $f^w(x) = (f(x))^w$. Routine calculations then show that in general

$$(f[\alpha/x])^w = f^w[\alpha^w/x]. \tag{5.7.3}$$

Moreover, any function from InVar to $\mathbb{C}^w \cup \mathbb{O}^w$ is equal to f^w for a suitable choice of valuation f on $\mathcal{M}$.

Term functions $|\tau|^{\mathcal{M}^w}$ are defined in the usual way by putting $|x|^{\mathcal{M}^w} f^w = f^w(x)$ and $|\mathsf{c}|^{\mathcal{M}^w} f^w = |\mathsf{c}|^{\mathcal{M}^w}$. These satisfy

$$|\tau|^{\mathcal{M}^w} f^w = (|\tau|^{\mathcal{M}} f)^w. \tag{5.7.4}$$

The satisfaction relation $\mathcal{M}^w, u, f^w \models \varphi$ is defined as usual, except for the case that φ is of the form $\forall x\psi$. Then if $x \notin \mathsf{ObVar}$, the condition is that

$$\text{for all } \alpha^w \in \mathbb{C}^w \text{ with } \alpha^w(u) \text{ defined, } \mathcal{M}^w, u, f^w[\alpha^w/x] \models \varphi;$$

while if $x : \mathsf{Ob}$ then it can be taken simply as

$$\text{for all } a \in Dw, \quad \mathcal{M}^w, u, f^w[a/x] \models \varphi.$$

With this construction set out, we are in a position to formalise and confirm the above claim that "in evaluating formulas at w we can ignore the structure outside of the component of w". In fact for any formula φ, any valuation f on $\mathcal{M}$, and any member u of W^w, we have

$$\mathcal{M}^w, u, f^w \models \varphi \quad \text{iff} \quad \mathcal{M}, u, f \models \varphi. \tag{5.7.5}$$

In particular, $\mathcal{M}, w, f \models \varphi$ iff $\mathcal{M}^w, w, f^w \models \varphi$, which fulfils the claim.

(5.7.5) is proven by induction on the formation of φ, with (5.7.4) assisting in the base case that φ is $P\tau_1 \cdots \tau_n$, and (5.7.3) being used in the inductive case that φ is $\forall x\psi$.

Another way of expressing the result of (5.7.5) is that, in terms of truth sets,

$$|\varphi|^{\mathcal{M}^w} f^w = |\varphi|^{\mathcal{M}} f \cap W^u.$$

Now since $\mathcal{M}$ is a model, $|\varphi|^{\mathcal{M}} f$ is admissible in $\mathcal{M}$, so $|\varphi|^{\mathcal{M}} f \cap W^u \in \mathit{Prop}^w$, hence this last equation shows that $|\varphi|^{\mathcal{M}^w} f^w$ is admissible in $\mathcal{M}^w$. Thus $\mathcal{M}^w$ is also a model.

If a formula φ is falsifiable in $\mathcal{M}$ at some point w under some valuation f, then it is falsified in $\mathcal{M}^w$ at w under f^w. On the other hand (5.7.5) also implies that if φ is valid in $\mathcal{M}$, then it is valid in $\mathcal{M}^w$, for any w. So altogether

$$\mathcal{M} \models \varphi \quad \text{iff} \quad \text{for all } w \in W, \mathcal{M}^w \models \varphi.$$

Thus if a logic is characterised by a class of Kripkean models, then it is characterised by a class of models of the form $\mathcal{M}^w$. These are models in which the set of admissible rigid objects is identifiable with either Dw or

$Dw \cup \{\emptyset^w\}$, where Dw is the universe of the model, while the admissible individual concepts are partial functions from W^w to Dw.

5.8. Definite Descriptions

Russell's famous theory of definite descriptions involves formation of terms of the form $\imath x.\varphi$, where φ is a formula. Such a term maybe read "the φ", as in "the present King of France". $\imath x.\varphi$ denotes the unique object satisfying φ, if there is such.

Our analysis of terms as partial functions lends itself naturally to the interpretation of definite description terms. The intension of $\imath x.\varphi$ can be taken as the function that assigns to each world the unique object satisfying φ at that world, if there is one, and is undefined at that world otherwise. We now show how such a theory can be developed formally.

Syntactically, we allow formation of $\imath x.\varphi$ for any formula φ and any *object* variable x. Although x is of sort Ob, the term $\imath x.\varphi$ itself is not: it may well define a non-rigid individual concept. The variable x is regarded as bound in $\imath x.\varphi$, in the same way that it is in the formula $\forall x\varphi$. Thus we will say that $\imath x.\varphi$ is a *closed* term if it has no free variables, which amounts to saying that φ has at most x free. The formula φ will be called the *body* of $\imath x.\varphi$. To provide every term with a body, we take a constant or a variable to be its own body.

In a premodel $\mathcal{M}$ for the language with identity, we can specify a term function $|\imath x.\varphi|^{\mathcal{M}}$ on the set Val_S of valuations by defining a partial function $|\imath x.\varphi|^{\mathcal{M}} f$ on W for each $f \in Val_S$ as follows.

DEFINITION 5.8.1. For each $w \in W$, declare

$$|\imath x.\varphi|^{\mathcal{M}} f.w$$

to be defined iff there exists some $\alpha \in \mathbb{O}$ with $\alpha(w)$ defined, such that

(i) $\mathcal{M}, w, f[\alpha/x] \models \varphi$; and
(ii) for any $\beta \in \mathbb{O}$, if $w \in E\beta$ and $\mathcal{M}, w, f[\beta/x] \models \varphi$, then $\beta(w) \simeq \alpha(w)$.

For such an α, put $|\imath x.\varphi|^{\mathcal{M}} f.w \simeq \alpha(w)$. ⊣

Notice that this definition of the term function $|\imath x.\varphi|^{\mathcal{M}}$ is based on the inductive assumption that the propositional function $|\varphi|^{\mathcal{M}}$ has been defined.

To see that $|\imath x.\varphi|^{\mathcal{M}} f.w$ is indeed uniquely determined, given α as above suppose there is also some other $\alpha' \in \mathbb{O}$ with $\alpha'(w)$ defined and (i) and (ii) holding with α' in place of α. Then by (i) for α', we can take β to be α' in (ii) for α, to conclude that $\alpha'(w) \simeq \alpha(w)$.

Thus we are taking $|\imath x.\varphi|^{\mathcal{M}} f.w$ to be the unique interpretation of x at w as a rigid concept α satisfying φ, unique in the sense that any other such concept is identical to α at w. Since our model structures are full of objects, every member of Dw is the value $\alpha(w)$ of some $\alpha \in \mathbb{O}$, so it is quite natural to define $|\imath x.\varphi|^{\mathcal{M}} f.w$ as a value of this kind.

The definition of "model" will now require, not only that the truth sets of formulas are admissible propositions, i.e. $|\varphi|^{\mathcal{M}} f \in \mathit{Prop}$, but also that the partial functions interpreting definite descriptions are admissible concepts, i.e. $|\imath x.\varphi|^{\mathcal{M}} f \in \mathbb{C}$.

To illustrate this semantics of definite description terms consider the scheme

$$\text{DD:} \quad \imath x.\varphi \approx \tau \leftrightarrow \big(\mathcal{E}\tau \wedge \forall x(\varphi \leftrightarrow x \approx \tau)\big),$$
$$\text{where } \tau : \mathsf{Ob} \text{ and } x \text{ is not free in the body of } \tau.$$

This says that an object τ is the value of a definite description iff it exists and is the unique object satisfying that description.

DD is valid in any model of the kind just described, even the ones that are not Kripkean. That suggests that our definition of $|\imath x.\varphi|^{\mathcal{M}} f.w$ has captured the right notion in this context.

To show the validity of DD we need some preliminary facts. First since x is not free in the body of τ, we have $|\tau|^{\mathcal{M}} f = |\tau|^{\mathcal{M}} f[\alpha/x]$ for any $\alpha \in \mathbb{O}$. If τ is a constant or variable, this is immediate as $\tau \neq x$. But the fact holds also if τ is some description term $\imath y.\psi$, since, as a second fact, we have that $|\psi|^{\mathcal{M}} f = |\psi|^{\mathcal{M}} f[\alpha/x]$ because x is not free in the body ψ of τ. To show that, we have to allow the possibility that ψ contains further description terms, for which we need the first fact. So these two facts are interdependent, and need to be proven together by simultaneous induction. A third fact we need is part (1) of the Indistinguishability Lemma 5.5.2, which provides that if $\alpha(w) \equiv \beta(w)$ with $\alpha, \beta \in \mathbb{O}$, then

$$\mathcal{M}, w, f[\alpha/x] \models \varphi \quad \text{iff} \quad \mathcal{M}, w, f[\beta/x] \models \varphi.$$

This was proved for any φ in the language with identity, but needs to be verified also for formulas containing description terms. We leave these preliminary facts as exercises for the reader.

THEOREM 5.8.2. *The scheme* DD *is valid in all models for the language with identity and descriptions.*

PROOF. Suppose that τ is an object term that does not have free x in its body. Let $\mathcal{M}$ be a model as described above. Assume that

$$\mathcal{M}, w, f \models \mathcal{E}\tau \wedge \forall x(\varphi \leftrightarrow x \approx \tau).$$

Let $\alpha = |\tau|^{\mathcal{M}} f$. Then $\alpha \in \mathbb{O}$, and $w \in E\alpha$ as $\mathcal{M}, w, f \models \mathcal{E}\tau$. Hence $\alpha(w) \simeq |\tau|^{\mathcal{M}} f.w$. So to show $\mathcal{M}, w, f \models \imath x.\varphi \approx \tau$, it will suffice to show that $|\imath x.\varphi|^{\mathcal{M}} f.w \simeq \alpha(w)$. For this we use our definition of $|\imath x.\varphi|^{\mathcal{M}} f.w$, requiring us to prove (i) and (ii) for α.

Since $|x|^{\mathcal{M}} f[\alpha/x] = \alpha$ and $\alpha(w) \simeq |\tau|^{\mathcal{M}} f.w$, the semantics of identity gives $\mathcal{M}, w, f[\alpha/x] \models x \approx \tau$. But as $\mathcal{M}, w, f \models \forall x(\varphi \leftrightarrow x \approx \tau)$ and $w \in E\alpha$, the semantics of $\forall$ (in any premodel, Kripkean or not) yields

$\mathcal{M}, w, f[\alpha/x] \models \varphi \leftrightarrow x \approx \tau$. Hence $\mathcal{M}, w, f[\alpha/x] \models \varphi$, which proves (i) for this α.

For (ii), suppose $\beta \in \mathbb{O}$ has $w \in E\beta$ and $\mathcal{M}, w, f[\beta/x] \models \varphi$. Then from $\mathcal{M}, w, f \models \forall x(\varphi \leftrightarrow x \approx \tau)$ we get $\mathcal{M}, w, f[\beta/x] \models \varphi \leftrightarrow x \approx \tau$, and so $\mathcal{M}, w, f[\beta/x] \models x \approx \tau$. Hence $\beta(w) \simeq |\tau|^{\mathcal{M}} f[\beta/x].w$. But then $\beta(w) \simeq |\tau|^{\mathcal{M}} f.w$, as x is not free in the body of τ, so $\beta(w) \simeq \alpha(w)$, as required to prove (ii) for α. This completes the proof that $\mathcal{M}, w, f \models \imath x.\varphi \approx \tau$.

For the converse, suppose $\mathcal{M}, w, f \models \imath x.\varphi \approx \tau$. Then $|\imath x.\varphi|^{\mathcal{M}} f.w \simeq |\tau|^{\mathcal{M}} f.w$, so $|\tau|^{\mathcal{M}} f.w$ is defined, hence $\mathcal{M}, w, f \models \mathcal{E}\tau$. So it remains to show $\mathcal{M}, w, f \models \forall x(\varphi \leftrightarrow x \approx \tau)$. For this we first need that for any $\gamma \in \mathbb{O}$,

$$|\imath x.\varphi \approx \tau|^{\mathcal{M}} f \subseteq E\gamma \Rightarrow |\varphi \leftrightarrow x \approx \tau|^{\mathcal{M}} f[\gamma/x]. \tag{5.8.1}$$

To prove this, suppose $u \in |\imath x.\varphi \approx \tau|^{\mathcal{M}} f$. Then $|\imath x.\varphi|^{\mathcal{M}} f.u \simeq |\tau|^{\mathcal{M}} f.u$, so there is some $\alpha \in \mathbb{O}$ such that $\alpha(u) \simeq |\tau|^{\mathcal{M}} f.u$, and (i) and (ii) hold for α with u in place of w. Now let $u \in E\gamma$. Then we have to show that $u \in |\varphi \leftrightarrow x \approx \tau|^{\mathcal{M}} f[\gamma/x]$, i.e. that $\mathcal{M}, u, f[\gamma/x] \models \varphi \leftrightarrow x \approx \tau$. Now if $\mathcal{M}, u, f[\gamma/x] \models \varphi$, then we can put γ for β in (ii) for u and conclude that $\gamma(u) \simeq \alpha(u) \simeq |\tau|^{\mathcal{M}} f.u$. But $\gamma(u) = |x|^{\mathcal{M}} f[\gamma/x].u$, and $|\tau|^{\mathcal{M}} f.u = |\tau|^{\mathcal{M}} f[\gamma/x].u$ as x is not free in the body of τ, so this all yields $\mathcal{M}, u, f[\gamma/x] \models x \approx \tau$ as required.

In the opposite direction, let $\mathcal{M}, u, f[\gamma/x] \models x \approx \tau$. Hence $\gamma(u) \simeq |\tau|^{\mathcal{M}} f[\gamma/x].u$. But $|\tau|^{\mathcal{M}} f[\gamma/x].u = |\tau|^{\mathcal{M}} f.u \simeq \alpha(u)$ as in the previous paragraph, so this implies $\alpha(u) \simeq \gamma(u)$. Since we have $\mathcal{M}, u, f[\alpha/x] \models \varphi$ by (i), and $\alpha, \gamma \in \mathbb{O}$, we then get $\mathcal{M}, u, f[\gamma/x] \models \varphi$ by the Indistinguishability Lemma. Hence indeed $\mathcal{M}, u, f[\gamma/x] \models \varphi \leftrightarrow x \approx \tau$, and that completes the proof of (5.8.1).

Now as $\mathcal{M}$ is a model, $|\imath x.\varphi \approx \tau|^{\mathcal{M}} f$ belongs to *Prop*. Since (5.8.1) holds for every $\gamma \in \mathbb{O}$, it follows that

$$|\imath x.\varphi \approx \tau|^{\mathcal{M}} f \subseteq \bigsqcap_{\gamma \in \mathbb{O}} \left(E\gamma \Rightarrow |\varphi \leftrightarrow x \approx \tau|^{\mathcal{M}} f[\gamma/x]\right) = |\forall x(\varphi \leftrightarrow x \approx \tau)|^{\mathcal{M}} f.$$

But $w \in |\imath x.\varphi \approx \tau|^{\mathcal{M}} f$ by assumption, so this finally gives $\mathcal{M}, w, f \models \forall x(\varphi \leftrightarrow x \approx \tau)$, completing the proof that DD is valid in $\mathcal{M}$. ⊣

Schemes like DD are used in theories of *elimination* of definite descriptions. The idea is that for suitable choices of τ, the right side of the biconditional DD contains no occurrence of $\imath x.\varphi$, and might be usable in some way in place of the definite description. This happens for instance whenever $\imath x.\varphi$ does not occur in τ. Then the right side of DD is a $\imath x.\varphi$-free equivalent of the identity $\imath x.\varphi \approx \tau$.

For a less immediate example, we will show how to obtain a $\imath x.\varphi$-free formula equivalent to the existence assertion $\mathcal{E}\imath x.\varphi$. Taking τ to be an object variable y not occurring in $\imath x.\varphi$, the formula

$$\imath x.\varphi \approx y \leftrightarrow (\mathcal{E}y \wedge \forall x(\varphi \leftrightarrow x \approx y))$$

is an instance of DD, so is valid. Hence

$$\exists y(\imath x.\varphi \approx y) \leftrightarrow \exists y(\mathcal{E}y \wedge \forall x(\varphi \leftrightarrow x \approx y))$$

is valid (see the $\exists$-Equivalence rule, Lemma 1.2.1(1)). Now $\exists y(\imath x.\varphi \approx y)$ is equivalent to $\mathcal{E}(\imath x.\varphi)$. Indeed $\exists y(\sigma \approx y)$ is equivalent to $\mathcal{E}\sigma$ whenever y is not in σ.[30] Also $\exists y(\mathcal{E}y \wedge \forall x(\varphi \leftrightarrow x \approx y))$ is equivalent to $\exists y \forall x(\varphi \leftrightarrow x \approx y)$. Indeed $\exists y(\mathcal{E}y \wedge \chi) \leftrightarrow \exists y\chi$ is derivable in any $\mathcal{E}$-logic (see Theorem 3.1.1(2)). So we conclude that

$$\mathcal{E}(\imath x.\varphi) \leftrightarrow \exists y \forall x(\varphi \leftrightarrow x \approx y) \tag{5.8.2}$$

is valid. The right side of (5.8.2) can then be substituted for occurrences of the formula $\mathcal{E}(\imath x.\varphi)$, thereby eliminating the definite description.

To see whether this might work more generally, take a formula containing occurrences of $\imath x.\varphi$, and create a new formula ψ by replacing all those occurrences by some completely new object variable y. Then the original formula is $\psi(\imath x.\varphi/y)$. Now if we could show that

$$\psi(\imath x.\varphi/y) \leftrightarrow \exists y(\imath x.\varphi \approx y \wedge \psi) \tag{5.8.3}$$

is valid, then we could use DD to replace the subformula $\imath x.\varphi \approx y$ on the right here, to get

$$\psi(\imath x.\varphi/y) \leftrightarrow \exists y(\mathcal{E}y \wedge \forall x(\varphi \leftrightarrow x \approx y) \wedge \psi),$$

which is itself equivalent to

$$\psi(\imath x.\varphi/y) \leftrightarrow \exists y(\forall x(\varphi \leftrightarrow x \approx y) \wedge \psi), \tag{5.8.4}$$

providing an equivalent to $\psi(\imath x.\varphi/y)$ on the right that eliminates $\imath x.\varphi$.

However this strategy is impeded by the fact that (5.8.3) is not in general valid. Its right side entails $\exists y(\imath x.\varphi \approx y)$ and hence $\mathcal{E}(\imath x.\varphi)$, whereas $\psi(\imath x.\varphi/y)$ need not entail $\mathcal{E}(\imath x.\varphi)$, e.g. when $\psi(\imath x.\varphi/y)$ is $\neg\mathcal{E}(\imath x.\varphi)$ and $\mathcal{E}(\imath x.\varphi)$ is not valid.

The implication within (5.8.3) from right to left *is* valid, but the one from left to right is not. What is missing from this left-to-right implication is precisely the assertion that $\imath x.\varphi$ exists: what we can show in general is the validity of

$$\psi(\imath x.\varphi/y) \wedge \mathcal{E}(\imath x.\varphi) \leftrightarrow \exists y(\imath x.\varphi \approx y \wedge \psi),$$

and hence of

$$\psi(\imath x.\varphi/y) \wedge \mathcal{E}(\imath x.\varphi) \leftrightarrow \exists y(\forall x(\varphi \leftrightarrow x \approx y) \wedge \psi). \tag{5.8.5}$$

So it is only in cases where

$$\psi(\imath x.\varphi/y) \rightarrow \mathcal{E}(\imath x.\varphi)$$

---

[30] $\sigma \approx y \rightarrow \mathcal{E}\sigma$ is valid (an instance of axiom WR) and entails $\exists y(\sigma \approx y) \rightarrow \mathcal{E}\sigma$ when y is not in σ. The converse formula $\mathcal{E}\sigma \rightarrow \exists y(\sigma \approx y)$ is (equivalent to) an instance of the valid axiom FNE (Theorem 5.7.3).

is valid, and hence $\psi(\imath x.\varphi/y) \wedge \mathcal{E}(\imath x.\varphi)$ is equivalent to $\psi(\imath x.\varphi/y)$, that we get the validity of (5.8.4) and the desired elimination of $\imath x.\varphi$.

Thus it appears that for our present language with identity and actualist quantifiers ranging over partially-defined individual concepts, the addition of definite descriptions provides an increase in expressive power that cannot be "defined away".

There is a simple addition that can be made to our language that will allow descriptions to be eliminated if they define a rigid object. The addition is a special object constant n with a fixed meaning in all models $\mathcal{M}$. This constant is to be thought of as the *null object*, or the "non-entity", namely the (rigid) individual concept that is defined nowhere. Thus we stipulate that $|\mathsf{n}|^{\mathcal{M}} = \emptyset_U^W$, the empty function from W to U, in any model. So n may be viewed as an "impossible" object, one whose description cannot be satisfied, like a square circle, or the least prime number divisible by 4. The definition of a model structure has to be extended to require that $\emptyset_U^W \in \mathbb{O}$, to ensure that n denotes an admissible object of the structure.

Since $|\mathsf{n}|^{\mathcal{M}}$ is everywhere undefined, $\neg\mathcal{E}\mathsf{n}$ is valid in all models, as is $\tau \not\approx \mathsf{n}$ for any term τ. The simple sentence $\neg\mathcal{E}\mathsf{n}$ suffices to axiomatise the resulting logic, since if $\vdash_{\mathrm{L}} \neg\mathcal{E}\mathsf{n}$, then in the canonical structure $\mathcal{S}_{\mathrm{L}}$, for each $w \in W_{\mathrm{L}}$ we have $\mathcal{E}\mathsf{n} \notin w$, and so w is not in the domain of the function $|\mathsf{n}|_{\mathrm{L}}$ (see (5.6.1)). This ensures that $|\mathsf{n}|_{\mathrm{L}}$, i.e. $|\mathsf{n}|^{\mathcal{M}_{\mathrm{L}}}$, is the empty function.

Now in a situation $\mathcal{M}, w, f$ in which $\mathcal{E}(\imath x.\varphi)$ is satisfied, from (5.8.5) we get that (5.8.4) is satisfied, providing an $\imath x.\varphi$-free equivalent of $\psi(\imath x.\varphi/y)$. But if $\mathcal{E}(\imath x.\varphi)$ is not satisfied, then $|\imath x.\varphi|^{\mathcal{M}} f$ is undefined at w, hence is indistinguishable from $|\mathsf{n}|^{\mathcal{M}}$ at w. This indistinguishability does not imply that the identity $\imath x.\varphi \approx \mathsf{n}$ is satisfied. On the contrary, $\imath x.\varphi \not\approx \mathsf{n}$ is valid. But it does imply that the two terms are semantically indistinguishable in the sense that n can be substituted for $\imath x.\varphi$ salva veritate. So in this case we get that $\psi(\imath x.\varphi/y)$ is equivalent to $\psi(\mathsf{n}/y)$. We formalise this observation as follows.

LEMMA 5.8.3. *In any model $\mathcal{M}$, if $|\imath x.\varphi|^{\mathcal{M}} f \in \mathbb{O}$, then for all $w \in W$, and any y :* Ob,

$$\mathcal{M}, w, f \models \neg\mathcal{E}(\imath x.\varphi) \to (\psi(\imath x.\varphi/y) \leftrightarrow \psi(\mathsf{n}/y)).$$

PROOF. If $\mathcal{M}, w, f \models \neg\mathcal{E}(\imath x.\varphi)$, then $|\imath x.\varphi|^{\mathcal{M}} f.w$ is undefined, and therefore $|\imath x.\varphi|^{\mathcal{M}} f.w \equiv |\mathsf{n}|^{\mathcal{M}} f.w$. Hence

$$\mathcal{M}, w, f[\,|\imath x.\varphi|^{\mathcal{M}} f/x] \models \psi \quad \text{iff} \quad \mathcal{M}, w, f[\,|\mathsf{n}|^{\mathcal{M}} f/x] \models \psi$$

by the Indistinguishability Lemma, as $|\imath x.\varphi|^{\mathcal{M}} f$ and $|\mathsf{n}|^{\mathcal{M}} f$ are in $\mathbb{O}$. Hence by the Substitution Lemma,

$$\mathcal{M}, w, f \models \psi(\imath x.\varphi/y) \quad \text{iff} \quad \mathcal{M}, w, \models \psi(\mathsf{n}/y),$$

giving $\mathcal{M}, w, f \models \psi(\imath x.\varphi/y) \leftrightarrow \psi(\mathsf{n}/y)$. ⊣

From this result and (5.8.5) it follows that in a model in which $|\imath x.\varphi|^{\mathcal{M}} f \in \mathbb{O}$, the formula

$$\psi(\imath x.\varphi/y) \leftrightarrow \exists y(\forall x(\varphi \leftrightarrow x \approx y) \wedge \psi) \vee (\neg\mathcal{E}(\imath x.\varphi) \wedge \psi(\mathsf{n}/y)).$$

is true at every world under f. By (5.8.2), this formula is equivalent to

$$\psi(\imath x.\varphi/y) \leftrightarrow \exists y(\forall x(\varphi \leftrightarrow x \approx y) \wedge \psi) \vee (\neg\exists y \forall x(\varphi \leftrightarrow x \approx y) \wedge \psi(\mathsf{n}/y)),$$

with no occurrence of $\imath x.\varphi$ on the right of the biconditional.

But this works only for definite descriptions that are rigid, like "the number 8", or "the least prime number". If anything, the analysis serves to clarify the limitations on our ability to eliminate descriptions, and supports an approach in which the $\imath x$ construct is taken as a primitive term-forming operator of our language. In that case, the question arises of how to axiomatise the resulting logic.

We will now give an answer to this question for the logic of Kripkean models, showing that the scheme DD is the only new axiom needed. In the present language with the description operator, let L be any Kripkean identity logic, as defined in Section 5.7, and suppose L includes DD. To obtain a completeness proof we use the construction of the Kripkean canonical model $\mathcal{M}_{\mathrm{L}}^{K}$ on the structure $\mathcal{S}_{\mathrm{L}}^{K}$ described briefly in that previous section. This needs some adaptation in view of the new class of terms we now have. We focus on describing the most important features of the adaptation, especially the new definition of the set $\mathbb{C}_{\mathrm{L}}^{K}$ of admissible concepts.

Recall that by a *closed term* we mean either an individual constant, i.e. a member of Con, or a description term $\imath x.\varphi$ that has no free variables, i.e. φ has at most x free. Let Cterm be the set of closed terms. Cterm will take over the role previously played by its subset Con.

Recall that $\mathcal{M}_{\mathrm{L}}^{K}$ is based on the set W_{L}^{K} of all $\mathcal{E}\forall$-complete L-maximal sets of formulas. $\mathcal{E}\forall$-completeness of a set Σ now means that if $\Sigma \vdash_{\mathrm{L}} \rho(\mathcal{E}\tau \to \varphi(\tau/x))$ for all $\tau \in \mathsf{X}$, then $\Sigma \vdash_{\mathrm{L}} \rho(\forall x\varphi)$, where, X is ObCon or Cterm according to whether $x : \mathsf{Ob}$ or not. The proof that every $\mathcal{E}\forall$-complete L-maximal set is object rich still holds here as in Theorem 5.7.5.

Each closed term τ determines a partial function $|\tau|_{\mathrm{L}}^{K}$ on W_{L}^{K}. We already saw how to define this for τ in Con, and will use the same definition now for all members of Cterm. Recall that for each $\mathsf{c} \in \mathsf{ObCon}$ we have $|\mathsf{c}|_w = \{\mathsf{d} \in \mathsf{ObCon} : \mathsf{c} \approx \mathsf{d} \in w\}$, and $D_{\mathrm{L}}^{K}(w) = \{|\mathsf{c}|_w : \mathcal{E}\mathsf{c} \in w\}$. The function $|\tau|_{\mathrm{L}}^{K}$ is defined by

$$|\tau|_{\mathrm{L}}^{K}(w) \simeq |\mathsf{c}|_w \quad \text{iff} \quad \tau \sim_w \mathsf{c} \quad \text{iff} \quad \tau \approx \mathsf{c} \in w.$$

Then $\mathbb{C}_{\mathrm{L}}^{K} = \{|\tau|_{\mathrm{L}}^{K} : \tau \in \mathsf{Cterm}\}$, and $\mathbb{O}_{\mathrm{L}}^{K} = \{|\mathsf{c}|_{\mathrm{L}}^{K} : \mathsf{c} \in \mathsf{ObCon}\} \subseteq \mathbb{C}_{\mathrm{L}}^{K}$.

To deal with valuations, we now take each function $f : \mathsf{InVar} \to \mathsf{Cterm}$ and lift it to $|f|_{\mathrm{L}}^{K} : \mathsf{InVar} \to \mathbb{C}_{\mathrm{L}}^{K}$ by putting

$$|f|_{\mathrm{L}}^{K}(x) = |fx|_{\mathrm{L}}^{K} \in \mathbb{C}_{\mathrm{L}}^{K} \text{ for all } x \in \mathsf{InVar}.$$

Any function $h : \mathsf{InVar} \to \mathbb{C}^K_{\mathrm{L}}$ is equal to $|f|^K_{\mathrm{L}}$ for a suitable choice of an $f : \mathsf{InVar} \to \mathsf{Cterm}$, with $fx : \mathsf{Ob}$ whenever $hx \in \mathbb{O}_{\mathrm{L}}$. So the set of all valuations in $\mathcal{S}^K_{\mathrm{L}}$ is now

$$Val_{\mathcal{S}^K_{\mathrm{L}}} = \{|f|^K_{\mathrm{L}} : fx : \mathsf{Ob} \text{ for all } x : \mathsf{Ob}\}.$$

The premodel $\mathcal{M}^K_{\mathrm{L}}$ is specified as follows:

- $|\mathsf{k}|^{\mathcal{M}^K_{\mathrm{L}}} = |\mathsf{k}|^K_{\mathrm{L}} \in \mathbb{C}^K_{\mathrm{L}}$, for all $\mathsf{k} \in \mathsf{Con}$.
  If $\mathsf{k} \in \mathsf{ObCon}$, this automatically makes $|\mathsf{k}|^{\mathcal{M}^K_{\mathrm{L}}} \in \mathbb{O}^K_{\mathrm{L}}$ as required.
- $|P|^{\mathcal{M}^K_{\mathrm{L}}} : (\mathbb{C}^K_{\mathrm{L}})^n \to Prop^K_{\mathrm{L}}$ is given by

$$|P|^{\mathcal{M}^K_{\mathrm{L}}}(|\tau_1|^K_{\mathrm{L}}, \dots, |\tau_n|^K_{\mathrm{L}}) = |P\tau_1 \cdots \tau_n|^K_{\mathrm{L}}, \quad \text{for all } \tau_1, \dots, \tau_n \in \mathsf{Cterm}.$$

We emphasise that if $\tau \in \mathsf{Cterm}$ is a definite description $\imath x.\varphi$, then $|\tau|^{\mathcal{M}^K_{\mathrm{L}}}$ is not specified as part of the definition of $\mathcal{M}^K_{\mathrm{L}}$ itself, but rather is determined by Definition 5.8.1 once $\mathcal{M}^K_{\mathrm{L}}$ has been given.

To show that $\mathcal{M}^K_{\mathrm{L}}$ is a Kripkean model that characterises L in this expanded language, we need to establish that the Truth Lemma

$$\mathcal{M}^K_{\mathrm{L}}, w, |f|^K_{\mathrm{L}} \models \varphi \quad \text{iff} \quad \varphi^f \in w \tag{5.8.6}$$

holds for all formulas, some of which will contain definite descriptions. Reviewing the proof for $\mathcal{M}_{\mathrm{L}}$ in Theorem 5.6.17, we see that what is required now is to show that the Term Evaluation equation of Lemma 5.6.16(2) continues to hold in this context. This means that

$$|\tau|^{\mathcal{M}^K_{\mathrm{L}}}|f|^K_{\mathrm{L}} = |\tau^f|^K_{\mathrm{L}} \tag{5.8.7}$$

not just when τ is an individual variable or constant, but also when it has the form $\imath x.\varphi$. In that case $(\imath x.\varphi)^f$ is equal to $\imath x.(\varphi^{f\setminus x})$, parallel to the formula for $(\forall x\varphi)^f$ in Lemma 1.9.3(2).

For the purpose of showing (5.8.7), we will say that a set Σ of formulas is *definite in* L if the following holds:

> For any constant $\mathsf{d} : \mathsf{Ob}$, variable $x : \mathsf{Ob}$, and formula φ, if $\Sigma \vdash_{\mathrm{L}} \mathcal{E}\mathsf{d}$ and $\Sigma \vdash_{\mathrm{L}} \mathcal{E}\mathsf{c} \to (\varphi(\mathsf{c}/x) \leftrightarrow \mathsf{c} \approx \mathsf{d})$ for every $\mathsf{c} : \mathsf{Ob}$, then $\Sigma \vdash_{\mathrm{L}} \imath x.\varphi \approx \mathsf{d}$.

Here is where we use the right-to-left direction of DD as an axiom:

THEOREM 5.8.4. *If Σ is $\mathcal{E}\forall$-complete in* L, *then it is definite in* L.

PROOF. If $\Sigma \vdash_{\mathrm{L}} \mathcal{E}\mathsf{c} \to (\varphi(\mathsf{c}/x) \leftrightarrow \mathsf{c} \approx \mathsf{d})$ for every $\mathsf{c} : \mathsf{Ob}$, then by $\mathcal{E}\forall$-completeness of Σ we get $\Sigma \vdash_{\mathrm{L}} \forall x(\varphi \leftrightarrow x \approx \mathsf{d})$. So if $\Sigma \vdash_{\mathrm{L}} \mathcal{E}\mathsf{d}$, we then have $\Sigma \vdash_{\mathrm{L}} \mathcal{E}\mathsf{d} \wedge \forall x(\varphi \leftrightarrow x \approx \mathsf{d})$, from which $\Sigma \vdash_{\mathrm{L}} \imath x.\varphi \approx \mathsf{d}$ follows since L includes DD. ⊣

The other direction of DD is involved in proving that (5.8.7) holds for definite descriptions, i.e. that we have the partial-function equality

$$|\imath x.\varphi|^{\mathcal{M}^K_{\mathrm{L}}}|f|^K_{\mathrm{L}} = |(\imath x.\varphi)^f|^K_{\mathrm{L}}. \tag{5.8.8}$$

Since $(\imath x.\varphi)^f$ is a closed term, this equality implies that every value of the term function $|\imath x.\varphi|^{\mathcal{M}^K_{\mathrm{L}}}$ is admissible, i.e. belongs to $\mathbb{C}^K_{\mathrm{L}}$.

The proof of (5.8.8) is quite intricate, and uses the induction hypothesis that the body φ of the description term satisfies the Truth Lemma (5.8.6) for all $f : \mathsf{InVar} \to \mathsf{Cterm}$. To begin the proof, suppose that $|(\imath x.\varphi)^f|^K_{\mathrm{L}}$ is defined at $w \in W^K_{\mathrm{L}}$, and takes the value $|\mathsf{d}|_w$ there for some $\mathsf{d} : \mathsf{Ob}$. But $(\imath x.\varphi)^f = \imath x.(\varphi^{f\backslash x})$, so we have

$$|\imath x.(\varphi^{f\backslash x})|^K_{\mathrm{L}}.w \simeq |\mathsf{d}|_w,$$

implying that the identity $\imath x.(\varphi^{f\backslash x}) \approx \mathsf{d}$ belongings to w. Hence $\mathcal{E}\mathsf{d} \in w$, and by DD,

$$\forall x(\varphi^{f\backslash x} \leftrightarrow x \approx \mathsf{d}) \in w. \tag{5.8.9}$$

Now let $\alpha = |\mathsf{d}|^K_{\mathrm{L}} \in \mathbb{O}^K_{\mathrm{L}}$. Then $\alpha(w) \simeq |\mathsf{d}|_w$. We will show that

$$|\imath x.\varphi|^{\mathcal{M}^K_{\mathrm{L}}}|f|^K_{\mathrm{L}}.w \simeq |\mathsf{d}|_w$$

by showing that our choice of α fulfills conditions (i) and (ii) of Definition 5.8.1 for $|\imath x.\varphi|^{\mathcal{M}^K_{\mathrm{L}}}|f|^K_{\mathrm{L}}.w$.

For this we need to know that for any $\mathsf{c} : \mathsf{Ob}$,

$$\mathcal{M}^K_{\mathrm{L}}, w, |f|^K_{\mathrm{L}}[\,|\mathsf{c}|^K_{\mathrm{L}}/x] \models \varphi \quad \text{iff} \quad \varphi^{f\backslash x}(\mathsf{c}/x) \in w. \tag{5.8.10}$$

But $|f|^K_{\mathrm{L}}[\,|\mathsf{c}|^K_{\mathrm{L}}/x] = |f[\mathsf{c}/x]|^K_{\mathrm{L}}$ as in the Term Evaluation Lemma 5.6.16(1), and

$$\mathcal{M}^K_{\mathrm{L}}, w, |f[\mathsf{c}/x]|^K_{\mathrm{L}} \models \varphi \quad \text{iff} \quad \varphi^{f[\mathsf{c}/x]} \in w$$

by the induction hypothesis (5.8.6) on φ. Since $\varphi^{f[\mathsf{c}/x]} = \varphi^{f\backslash x}(\mathsf{c}/x)$ by Lemma 1.9.3(3), this confirms that (5.8.10) holds for all $\mathsf{c} : \mathsf{Ob}$.

Now to prove (i) for $\alpha = |\mathsf{d}|^K_{\mathrm{L}}$, from (5.8.9) and the fact that $\mathcal{E}\mathsf{d} \in w$, by the instantiation axiom EI$^+$ we get $(\varphi^{f\backslash x}(\mathsf{d}/x) \leftrightarrow \mathsf{d} \approx \mathsf{d}) \in w$. As $\mathsf{d} \approx \mathsf{d}$ is just $\mathcal{E}\mathsf{d} \in w$, this then yields $\varphi^{f\backslash x}(\mathsf{d}/x) \in w$. Hence by (5.8.10),

$$\mathcal{M}^K_{\mathrm{L}}, w, |f|^K_{\mathrm{L}}[\,|\mathsf{d}|^K_{\mathrm{L}}/x] \models \varphi, \tag{5.8.11}$$

i.e. $\mathcal{M}^K_{\mathrm{L}}, w, |f|^K_{\mathrm{L}}[\alpha/x] \models \varphi$, which is (i).

For (ii), suppose $\mathcal{M}^K_{\mathrm{L}}, w, |f|^K_{\mathrm{L}}[\beta/x] \models \varphi$ with $\beta \in \mathbb{O}^K_{\mathrm{L}}$ and $w \in E\beta$. We have to show $\beta(w) \simeq \alpha(w)$. Now $\beta = |\mathsf{c}|^K_{\mathrm{L}}$ for some $\mathsf{c} : \mathsf{Ob}$, so by (5.8.10), $\varphi^{f\backslash x}(\mathsf{c}/x) \in w$. But $|\mathsf{c}|^K_{\mathrm{L}}(w)$ is defined, so also $\mathcal{E}\mathsf{c} \in w$. Hence from (5.8.9) and EI$^+$ we get $\mathsf{c} \approx \mathsf{d} \in w$. So $|\mathsf{c}|_w = |\mathsf{d}|_w$, i.e. $\beta(w) \simeq \alpha(w)$ as required for (ii). Altogether this shows that if $|(\imath x.\varphi)^f|^K_{\mathrm{L}}.w$ is defined, then so is $|\imath x.\varphi|^{\mathcal{M}^K_{\mathrm{L}}}|f|^K_{\mathrm{L}}.w$ and the two are equal.

That is half the story about (5.8.8). We now need to run this in reverse, starting with the assumption that $|\imath x.\varphi|^{\mathcal{M}^K_{\mathrm{L}}}|f|^K_{\mathrm{L}}.w$ is defined. Here we will use the fact that w is a definite set of formulas. By Definition 5.8.1 there is some $\alpha = |\mathsf{d}|^K_{\mathrm{L}}$ with $\mathsf{d} : \mathsf{Ob}$ such that $|\imath x.\varphi|^{\mathcal{M}^K_{\mathrm{L}}}|f|^K_{\mathrm{L}}.w \simeq \alpha(w)$ and conditions (i) and (ii) of that definition hold for α in $\mathcal{M}_{\mathsf{L}}^{K}$. Since $|\mathsf{d}|_{\mathsf{L}}^{K}(w)$ is defined we have $\mathcal{E}\mathsf{d} \in w$. By (i) we have (5.8.11), which by (5.8.10) implies $\varphi^{f\setminus x}(\mathsf{d}/x) \in w$. Now we show that

$$\mathcal{E}\mathsf{c} \to (\varphi^{f\setminus x}(\mathsf{c}/x) \leftrightarrow \mathsf{c} \approx \mathsf{d}) \in w \tag{5.8.12}$$

for every $\mathsf{c} \in \mathsf{ObCon}$. To see this, let $\mathcal{E}\mathsf{c} \in w$. Then we show that

$$(\varphi^{f\setminus x}(\mathsf{c}/x) \leftrightarrow \mathsf{c} \approx \mathsf{d}) \in w.$$

For if $\varphi^{f\setminus x}(\mathsf{c}/x) \in w$, then $\mathcal{M}_{\mathsf{L}}^{K}, w, |f|_{\mathsf{L}}^{K}[\,|\mathsf{c}|_{\mathsf{L}}^{K}/x] \models \varphi$ by (5.8.10), hence $|\mathsf{c}|_{\mathsf{L}}^{K}(w) \simeq \alpha(w) = |\mathsf{d}|_{\mathsf{L}}^{K}(w)$ by (ii), so $\mathsf{c} \approx \mathsf{d} \in w$. Conversely, if $\mathsf{c} \approx \mathsf{d} \in w$, then $|\mathsf{c}|_{w} = |\mathsf{d}|_{w}$ and $|\mathsf{c}|_{\mathsf{L}}^{K}(w) \simeq |\mathsf{d}|_{\mathsf{L}}^{K}(w)$, hence as (5.8.11) holds we get that $\mathcal{M}_{\mathsf{L}}^{K}, w, |f|_{\mathsf{L}}^{K}[\,|\mathsf{c}|_{\mathsf{L}}^{K}/x] \models \varphi$ by Indistinguishability, and so $\varphi^{f\setminus x}(\mathsf{c}/x) \in w$ by (5.8.10).

This completes the proof that (5.8.12) holds for all $\mathsf{c} : \mathsf{Ob}$. But $\mathcal{E}\mathsf{d} \in w$, and w is definite by Theorem 5.8.4, so applying the definition of definiteness to the formula $\varphi^{f\setminus x}$, it follows that the identity $\imath x.(\varphi^{f\setminus x}) \approx \mathsf{d}$ belongs to w. Since $\imath x.(\varphi^{f\setminus x})$ is $(\imath x.\varphi)^{f}$, we now have

$$|(\imath x.\varphi)^{f}|_{\mathsf{L}}^{K}.w \simeq |\mathsf{d}|_{w} = |\imath x.\varphi|^{\mathcal{M}_{\mathsf{L}}^{K}}|f|_{\mathsf{L}}^{K}.w.$$

Altogether we have shown that at each $w \in W_{\mathsf{L}}^{K}$, $|\imath x.\varphi|^{\mathcal{M}_{\mathsf{L}}^{K}}|f|_{\mathsf{L}}^{K}$ is defined iff $|(\imath x.\varphi)^{f}|_{\mathsf{L}}^{K}$ is defined, and when they are defined they are equal. This establishes (5.8.8), which is the main new feature required to build a canonical model for a language with definite descriptions.

The rest of the analysis of $\mathcal{M}_{\mathsf{L}}^{K}$ as a Kripkean model characterising L proceeds much as in previous cases, and leads us to the following result.

THEOREM 5.8.5. *Let* S *be any set of propositional modal formulas. Then for the language with identity and definite descriptions, the logic* $\mathrm{QS}^{\approx}$ + T$\forall$-Intro + FNE + DD *is characterised by the class of all Kripkean models on model structures whose underlying general frame validates* S. ⊣

If we restrict a Kripkean model $\mathcal{M}$ to the connected component of one of its worlds w, to form the model $\mathcal{M}^{w}$ defined at the end of Section 5.7, then we get a model in which the term function $|\imath x.\varphi|^{\mathcal{M}^{w}}$ is specified as follows:

- $|\imath x.\varphi|^{\mathcal{M}^{w}} g.u \simeq b$ if b is the unique member of Dw such that $\mathcal{M}^{w}, u, g[b/x] \models \varphi$.
- $|\imath x.\varphi|^{\mathcal{M}^{w}} g.u$ is undefined if there is no such b.

Then $\mathrm{QS}^{\approx}$ + T$\forall$-Intro + FNE + DD can be shown to be characterised by validity in models of the $\mathcal{M}^{w}$ kind whose general frame $(W^{w}, R^{w}, Prop^{w})$ validates S, since this validation is inherited from $(W, R, Prop)$.

For logics characterised by models that are not in general Kripkean, we no longer have the rule T$\forall$-Intro available, and we cannot confine the points of our canonical models to $\mathcal{E}\forall$-complete maximal sets. So we lose the guarantee of Theorem 5.8.4 that these points are definite. Instead we must directly build models whose points are definite, but may not be $\mathcal{E}\forall$-complete.

The situation is similar to that with object richness. For Kripkean models we needed only the axiom FNE, since this combined with the rule T$\forall$-Intro to ensure that all $\mathcal{E}\forall$-complete maximal sets are object rich (Theorem 5.7.5). For logics with non-Kripkean models, the axiom FNE must be replaced by the template rule TNE, enabling us to build maximal sets that are object rich but possibly not $\mathcal{E}\forall$-complete. Likewise, for non-Kripkean logics the axiom DD for definite descriptions is insufficent for axiomatisation. We retain one half of it, namely

$$\imath x.\varphi \approx \tau \to \big(\mathcal{E}\tau \wedge \forall x(\varphi \leftrightarrow x \approx \tau)\big),$$

but replace the converse implication by the template rule

$$\frac{\rho(\mathcal{E}\tau \wedge (\mathcal{E}\mathsf{c} \to (\varphi(\mathsf{c}/x) \leftrightarrow \mathsf{c} \approx \tau)))}{\rho(\imath x.\varphi \approx \tau)}$$

where $\rho \in \mathsf{Tem}$, $\mathsf{c} : \mathsf{Ob}$ and c is not in ρ, τ or φ. This rule will allow us to construct sufficiently many maximal sets that are definite. Again, the fine details of this are left to the interested reader.

Chapter 6

COVER SEMANTICS FOR RELEVANT LOGIC

The main aim of this chapter is to set out a new kind of admissible model theory for the propositional relevant logic R and its quantified extension RQ. First we review the relational semantics for R of Routley and Meyer [1973], and its adaptation by Mares and Goldblatt [2006] to an admissible semantics for RQ. Then we introduce an alternative kind of structure, called a *cover system*, motivated by topological ideas about "local truth" from the Kripke-Joyal semantics for intuitionistic logic in topos theory. These are combined with a modelling of negation by a binary world-relation of orthogonality, or incompatibility, as in [Goldblatt 1974], and an operation of combination, or "fusion", of worlds to interpret relevant implication. Characteristic model systems for R have an algebra *Prop* of admissible propositions, while those for RQ have a set *PropFun* of admissible propositional functions as well.

We then show that by conservatively adding an intuitionistic implication connective to R it is possible to characterise that logic by models in which all possible propositions are admissible. The prospects for a similar analysis of RQ are considered at the end.

6.1. Routley-Meyer Models for R

The subject of *relevant logic* (also known as *relevance* logic) is based on the view that an implication $A \to B$ can only be true if the meaning of A is relevant to the meaning of B. On this view, such Boolean-tautological schemes as $A \wedge \neg A \to B$, $A \to (B \to A)$, and $A \to B \vee \neg B$ are not valid, since B may have no symbols in common with A and hence have its truth conditions determined independently of those of A.

Anderson and Belnap formulated a propositional logic R that was intended to encapsulate this conception of relevant implication (see Anderson and Belnap [1975], especially §28). Axioms and rules for R are listed in Table 6.1.1. They refer to formulas A, B, C that are generated from a denumerable set PropVar of propositional variables, and a propositional constant t, by the

Axioms

$A \to A$	Self-Implication
$(A \to B) \to ((B \to C) \to (A \to C))$	Suffixing
$A \to ((A \to B) \to B)$	Assertion
$(A \to (A \to B)) \to (A \to B)$	Contraction
$A \wedge B \to A,\ \ A \wedge B \to B$	Conjunction Elimination
$(A \to B) \wedge (A \to C) \to (A \to B \wedge C)$	Conjunction Introduction
$A \to A \vee B,\ \ B \to A \vee B$	Disjunction Introduction
$(A \to C) \wedge (B \to C) \to (A \vee B \to C)$	Disjunction Elimination
$A \wedge (B \vee C) \to (A \wedge B) \vee (A \wedge C)$	$\wedge\vee$-Distribution
$(A \to \neg B) \to (B \to \neg A)$	Contraposition
$\neg\neg A \to A$	Double-Negation Elimination
$A \leftrightarrow (t \to A)$	t-Axiom

Rules

$\dfrac{A, B}{A \wedge B}$	Adjunction
$\dfrac{A, A \to B}{B}$	Modus Ponens

TABLE 6.1.1. Axioms and Rules for R

connectives $\wedge$, $\neg$ and $\to$.[31] Disjunction and the biconditional are introduced by the definitions

$$A \vee B = \neg(\neg A \wedge \neg B),$$
$$A \leftrightarrow B = (A \to B) \wedge (B \to A).$$

R has the property that if $A \to B$ is derivable, then the formulas A and B must have some propositional variable in common. This variable-sharing is regarded as a necessary condition for the relevance of A to B [Anderson and Belnap 1975, p. 33].

The constant t may be thought of as a strongest logical truth, one that implies all others. A binary "intensional conjunction" $\circ$ can be introduced by the definition

$$A \circ B = \neg(A \to \neg B).$$

[31] R is sometimes formulated without the constant t, in which case our logic is called R^t. But we will always assume t is present.

This connective was called "co-tenability" in [Anderson and Belnap 1975, p. 344], suggesting a kind of compatibility in the sense that "A does not preclude B". The connective is now commonly known as "fusion". Its behaviour in R is illustrated in Table 6.1.2, which lists a number of R-theorem schemes:

$(A \to B) \to ((C \to A) \to (C \to B))$	Prefixing
$(A \to (B \to C)) \to (B \to (A \to C))$	Permutation
$A \to \neg\neg A$	Double-Negation Introduction
$(A \to \neg A) \to \neg A$	Reductio
$\neg(A \vee B) \leftrightarrow \neg A \wedge \neg B$	De Morgan
$\neg(A \wedge B) \leftrightarrow \neg A \vee \neg B$	De Morgan
t	Logical Truth
$((A \circ B) \to C) \leftrightarrow (A \to (B \to C))$	Residuation
$A \circ (B \circ C) \leftrightarrow (A \circ B) \circ C$	Associative Fusion
$A \circ B \leftrightarrow B \circ A$	Commutative Fusion
$(t \circ A) \leftrightarrow A$	
$A \circ (B \vee C) \leftrightarrow (A \circ B) \vee (A \circ C)$	
$A \wedge B \to A \circ B$	
$A \to A \circ A$	

TABLE 6.1.2. Some Theorems of R

Anderson and Belnap also introduced a quantified extension of R, known as RQ, which we will define in the next section. But the metalogical analysis of these systems proved to be difficult. For one thing, R turned out to be undecidable [Urquhart 1984]. A possible-worlds style semantical characterisation for R was given by Routley and Meyer [1973], using frames that carry a ternary relation R for modelling implication. We will now review this model theory in some detail.

First we describe R-*frames*, which are based on structures of the form $\mathcal{F} = (W, 0, R, {}^*)$, where:

- W is a set, whose members may be thought of as worlds (or states of affairs, or situations ...).
- 0 is a subset of W, whose members are the *base* worlds.
- R is a ternary relation on the set W.
- * is a unary function on W.

The set 0 of base worlds, which will provide the interpretation of the logical constant t, is used to define a binary relation $\leq$ on W by

$$u \leq v \quad \text{iff} \quad \text{there exists } w \in 0 \text{ with } Rwuv.$$

This relation is called a *preorder* if it is reflexive and transitive, and a *partial order* if it is an antisymmetric preorder. A subset X of W is called an *up-set* if it is closed upwards under $\leq$, which means that

$$\text{if } u \in X \text{ and } u \leq v, \text{ then } v \in X.$$

If X and Y are up-sets, then so are $X \cap Y$ and $X \cup Y$. We write $Up(\mathcal{F})$ for the collection of all up-sets of $\mathcal{F}$.

The structure $\mathcal{F} = (W, 0, R, ^*)$ is called an R-*frame* if it satisfies the following conditions.

- $\leq$ is a preorder.
- 0 is an up-set.
- $Ruvw$ implies $Rvuw$.
- $\exists w(Ruvw \;\&\; Rwu'v')$ implies $\exists w(Ruu'w \;\&\; Rwvv')$.
- $Rwww$.
- $Rwuv$, implies Rwv^*u^*.
- $w^{**} = w$.
- if $Rwuv$ and $w' \leq w$, then $Rw'uv$.

The ternary relation R induces a binary operation $\Rightarrow$ on the powerset $\wp W$, defined for all $X, Y \subseteq W$ by

$$X \Rightarrow Y = \{w : \forall u \forall v(Rwuv \text{ and } u \in X \text{ implies } v \in Y)\}.$$

The last condition on an R-frame ensures that if X and Y are up-sets, then $X \Rightarrow Y$ is an up-set.[32]

The unary operation * on W lifts to $\wp W$ by putting, for all $X \subseteq W$,

$$X^* = \{w : w^* \notin X\}.$$

This operation satisfies the De Morgan laws $(X \cap Y)^* = X^* \cup Y^*$ and $(X \cup Y)^* = X^* \cap Y^*$, and is inclusion-reversing, i.e. $X \subseteq Y$ implies $Y^* \subseteq X^*$. Those facts require no special properties of *. When * is involutary, i.e. $w^{**} = w$, then so is its lifting, i.e. $X^{**} = X$. Also, in any R-frame, if X is an up-set then so is X^*.

In summary, the collection $Up(\mathcal{F})$ of up-sets of an R-frame $\mathcal{F}$ contains 0 and is closed under the operations $\cap$, $\cup$, *, and $\Rightarrow$.

A *model* $\mathcal{M}$ on an R-frame is given by a function $|-|^{\mathcal{M}}$ that assigns to each propositional variable p an *up-set* $|p|^{\mathcal{M}} \subseteq W$, the *truth set of* p. This assignment is then extended inductively to define truth sets $|A|^{\mathcal{M}}$ for all propositional

[32]When $Rwuv$ iff $w = u = v$, we get $X \Rightarrow Y = (W - X) \cup Y$, so the present definition of $\Rightarrow$ generalises the Boolean set implication operation.

formulas, as follows:

$$\begin{aligned} |t|^{\mathcal{M}} &= 0, \\ |A \wedge B|^{\mathcal{M}} &= |A|^{\mathcal{M}} \cap |B|^{\mathcal{M}}, \\ |\neg A|^{\mathcal{M}} &= (|A|^{\mathcal{M}})^*, \\ |A \to B|^{\mathcal{M}} &= |A|^{\mathcal{M}} \Rightarrow |B|^{\mathcal{M}}. \end{aligned} \tag{6.1.1}$$

The closure properties of $Up(\mathcal{F})$ that we have mentioned then ensure that $|A|^{\mathcal{M}}$ is an up-set for every formula A.

Writing $\mathcal{M}, w \models A$ as usual to mean that $w \in |A|^{\mathcal{M}}$, we have the following description of the truth/satisfaction relation in $\mathcal{M}$:

- $\mathcal{M}, w \models p$ iff $w \in |p|^{\mathcal{M}}$.
- $\mathcal{M}, w \models t$ iff $w \in 0$.
- $\mathcal{M}, w \models A \wedge B$ iff $\mathcal{M}, w \models A$ and $\mathcal{M}, w \models B$.
- $\mathcal{M}, w \models \neg A$ iff $\mathcal{M}, w^* \not\models A$.
- $\mathcal{M}, w \models A \to B$ iff for all u, v such that $Rwuv$, if $\mathcal{M}, u \models A$ then $\mathcal{M}, v \models B$.

A formula A is *true in the model* $\mathcal{M}$, symbolised $\mathcal{M} \models A$, if A is true at every *base* world, i.e. if $0 \subseteq |A|^{\mathcal{M}}$. A is *valid in* $\mathcal{F}$, symbolised $\mathcal{F} \models A$, if A is true in every model based on $\mathcal{F}$.

A entails B in $\mathcal{M}$ if $|A|^{\mathcal{M}} \subseteq |B|^{\mathcal{M}}$, i.e. if B is true at every world that A is in $\mathcal{M}$. An important fact relating this notion of entailment to relevant implication is that A entails B in $\mathcal{M}$ iff $A \to B$ is true in $\mathcal{M}$, i.e.

$$|A|^{\mathcal{M}} \subseteq |B|^{\mathcal{M}} \quad \text{iff} \quad \mathcal{M} \models A \to B \tag{6.1.2}$$

[Routley and Meyer 1973, Theorem 1].

It can be shown that the logic R is *sound* for validity in R-frames: if $\mathcal{F}$ is any R-frame, then in general $\vdash_{\mathrm{R}} A$ implies $\mathcal{F} \models A$. The proof uses all the conditions defining R-frames, as well as (6.1.2), and depends also on the fact that truth-sets are up-sets, which amounts to saying that

$$\text{if } \mathcal{M}, u \models A \text{ and } u \leq v, \text{ then } \mathcal{M}, v \models A$$

[Routley and Meyer 1973, Lemma 1].

In any model on an R-frame, the disjunction connective gets the standard interpretation

$$\mathcal{M}, w \models A \vee B \ \text{ iff } \ \mathcal{M}, w \models A \text{ or } \mathcal{M}, w \models B;$$

while fusion has the semantics

$$\mathcal{M}, w \models A \circ B \ \text{ iff } \text{ there exists } u, v \text{ such that } Ruvw \text{ and } \mathcal{M}, u \models A \text{ and } \mathcal{M}, v \models B.$$

To illustrate the nature of the truth condition for relevant implication, take a binary operation $\cdot$ on W and let R be its ternary graph, i.e.

$$R = \{(w, u, v) : \ v = w \cdot u\}.$$

For this R the truth condition for $\to$ amounts to:

$$\mathcal{M}, w \models A \to B \text{ iff } \mathcal{M}, u \models A \text{ implies } \mathcal{M}, w \cdot u \models B. \tag{6.1.3}$$

We may view $w \cdot u$ as a situation whose information content consists of combinations of co-tenable pieces of information from w and u (together with whatever they entail). Then (6.1.3) states that A implies B in w if every situation satisfying A combines with w to give a situation satisfying B. This kind of "operational" interpretation of implication was introduced to relevant logic by Urquhart [1972] (a revised version appeared as §47 of Anderson, Belnap, and Dunn [1992]). For a general R-frame, we may view $Rwuv$ as holding when the information content of v *includes* all combinations of co-tenable pieces of information from w and u. In Section 6.4 we will give a different kind of semantics for R in which the truth condition for $\to$ is exactly given by the simpler (6.1.3) for some binary operation.

The logic R is complete for validity in R-frames. This can be shown by a canonical model construction that we now outline. A formula A is said to be *deducible* from a set of formulas Γ, which is symbolised $\Gamma \vdash_R A$, if there are finitely many formulas $A_0, \ldots, A_n \in \Gamma$ such that $\vdash_R A_0 \wedge \cdots \wedge A_n \to A$. We call Γ an R-*theory* if it is closed under deducibility, i.e. if $\Gamma \vdash_R A$ implies $A \in \Gamma$. A theory Γ is *prime* if $A \vee B \in \Gamma$ implies $A \in \Gamma$ or $B \in \Gamma$, and is *regular* if it contains all R-theorems (which is equivalent to it containing t). The relevant version of Lindenbaum's Lemma in this context states that:

> *If Δ is a set of formulas that is closed under disjunction, and Γ is an* R-*theory disjoint from Δ, then Γ can be extended to a prime* R-*theory disjoint from Δ.*

The *canonical* R-*frame* is

$$\mathcal{F}_R = (W_R, 0_R, R, {}^*),$$

where:

- W_R is the set of all prime R-theories.
- $0_R = \{w \in W_R : t \in w\}$, the set of all regular prime R-theories.
- $Rwuv$ iff for all A and B, if $A \to B \in w$ and $A \in u$, then $B \in v$.
- $w^* = \{A : \neg A \notin w\}$.

This structure is an R-frame (Routley and Meyer [1973] or Routley with others [1982, Chapter 4]). The *canonical* model $\mathcal{M}$ on $\mathcal{F}_R$ is defined by putting $|p|^{\mathcal{M}} = \{w \in W_R : p \in w\}$. It satisfies the Truth Lemma

$$\mathcal{M}, w \models A \quad \text{iff} \quad A \in w$$

for all formulas A and all $w \in W_R$. Thus if $\nvdash_R A$, then $\nvdash_R t \to A$, so by Lindenbaum's Lemma there is a $w \in W_R$ with $t \in w$ and $A \notin w$. Then $w \in 0_R$ and $\mathcal{M}, w \not\models A$, showing that A is not true in $\mathcal{M}$, hence not valid in the R-frame $\mathcal{F}_R$. Contrapositively, if a formula is valid in all R-frames then it is a R-theorem. This construction also shows that R is characterised by validity in the single R-frame $\mathcal{F}_{\mathrm{R}}$, as well as by validity in all R-frames.

6.2. Admissible Semantics for RQ

The quantified extension RQ of R is defined axiomatically as follows:

Axioms

All substitution instances of R-axioms.	(see Table 6.1.1)
$\forall x\varphi \to \varphi(\tau/x)$, with τ free for x in φ.	Universal Instantiation
$\varphi \wedge \exists x\psi \to \exists x(\varphi \wedge \psi)$, with x not free in φ.	$\wedge\exists$-Distribution

Rules

$\dfrac{\varphi, \psi}{\varphi \wedge \psi}$	Adjunction
$\dfrac{\varphi, \varphi \to \psi}{\psi}$	Modus Ponens
$\dfrac{\varphi \to \psi}{\varphi \to \forall x\psi}$ if x is not free in φ.	$\forall$-Introduction

TABLE 6.2.1. Axioms and Rules for RQ

This refers to formulas φ, ψ which, for a given signature $\mathcal{L}$, are generated from atomic formulas $P\tau_1 \cdots \tau_n$ and the constant t by the connectives $\to$, $\wedge$, $\neg$, and the quantifiers $\forall x$. $\mathcal{L}$ will be taken to consist of finitary predicate symbols and individual constants. Hence a term τ is either an individual variable or a constant. Existential quantifiers are introduced by defining $\exists x\varphi$ to be the formula $\neg\forall x\neg\varphi$.

Deletion of the $\wedge\exists$-Distribution axiom from RQ gives the system QR. Over QR, the $\wedge\exists$-Distribution axiom is equivalent to the $\forall\vee$-*Distribution* scheme

$$\forall x(\varphi \vee \psi) \to (\varphi \vee \forall x\psi), \quad \text{with } x \text{ not free in } \varphi.$$

This last scheme is usually taken as an axiom in presentations of RQ, but we find it more convenient to take its dual equivalent, for reasons that will become evident later in discussing cover semantics.

From Universal Instantiation and $\forall$-Introduction, many facts about $\forall$ and $\exists$ that hold for classical Boolean logic can be derived in both QR and RQ, with the help of the De Morgan laws, Contraposition and Double Negation Elimination. Also, deducibility in both QR and RQ is finitary, i.e. is characterised by the existence of proof-sequences of finite length, in the sense described in

Theorem 1.2.5. This means that these logics are closed under the rules of Term Instantiation, Generalisation on Constants, and related rules, by the sort of arguments used at the end of Section 1.2. We record some such facts now. For further details, see Mares and Goldblatt [2006], Anderson, Belnap, and Dunn [1992].

LEMMA 6.2.1. *Let* L *be either* RQ *or* QR.

(1) Existential Generalisation:
$\vdash_{\mathrm{L}} \varphi(\tau/x) \to \exists x\varphi$, *where* τ *free for* x *in* φ.

(2) L *includes the schemes*
UD : $\forall x(\varphi \to \psi) \to (\forall x\varphi \to \forall x\psi)$,
VQ : $\varphi \to \forall x\varphi$, *where* x *is not free in* φ,
CQ : $\forall x\forall y\varphi \to \forall y\forall x\varphi$.

(3) *If* x, y *are distinct variables with* y *not free in* φ *but freely substitutable for* x *in* φ, *then* $\vdash_{\mathrm{L}} \forall x\varphi \leftrightarrow \forall y\varphi(y/x)$ *and* $\vdash_{\mathrm{L}} \exists x\varphi \leftrightarrow \exists y\varphi(y/x)$.

(4) Relettering of Bound Variables:
If φ *and* ψ *differ only in that* φ *has free* x *exactly where* ψ *has free* y, *then* $\vdash_{\mathrm{L}} \forall x\varphi \leftrightarrow \forall y\psi$ *and* $\vdash_{\mathrm{L}} \exists x\varphi \leftrightarrow \exists y\psi$.

(5) L *is closed under the following rules*:

$$\text{TI}^*: \quad \frac{\varphi}{\varphi(\tau_1/x_1, \ldots, \tau_n/x_n)}, \quad \textit{if each } \tau_i \textit{ is free for } x_i \textit{ in } \varphi.$$

$$\text{GC}^*: \quad \frac{\varphi(\mathsf{c}_1/x_1, \ldots, \mathsf{c}_n/x_n)}{\varphi}, \quad \textit{if the } \mathsf{c}_i \textit{ are distinct and not in } \varphi.$$

$$\forall\text{GC}: \quad \frac{\varphi \to \psi(\mathsf{c}/x)}{\varphi \to \forall x\psi}, \quad \textit{if } \mathsf{c} \textit{ is not in } \varphi \textit{ or } \psi.$$

$$\exists\text{GC}: \quad \frac{\psi(\mathsf{c}/x) \to \varphi}{\exists x\psi \to \varphi}, \quad \textit{if } \mathsf{c} \textit{ is not in } \psi \textit{ or } \varphi. \quad \dashv$$

Routley and Meyer [1973] suggested that RQ could be characterised by models that are based on R-frames, have one universal domain U and are Kripkean, so have the quantifier semantics

$$\mathcal{M}, w, f \models \forall x\varphi \;\text{ iff }\; \text{for all } a \in U,\ \mathcal{M}, w, f[a/x] \models \varphi$$

(see Section 2.4). This conjecture turned out to be incorrect: a formula was found by Fine [1989] that is true in all such models, but is not an RQ-theorem. Fine [1988] then devised a complete semantics for RQ using varying-domain models having many new relational constructs, satisfying many intricate conditions, giving a different kind of interpretation to the propositional connectives and defining a quantified sentence $\forall v\varphi$ to be true if $\varphi(v)$ is true for an "arbitrary or generic individual" in a suitable sense.

An alternative approach was developed in [Mares and Goldblatt 2006]. This built an admissible semantics on Routley-Meyer frames for R by introducing algebras *Prop* and *PropFun* of propositions and propositional functions, obtaining a notion of functional model structure. In this context a QR-*model structure* can be defined as a system

$$\mathfrak{S} = (W, 0, R, {}^*, U, Prop, PropFun),$$

where:

- $(W, 0, R, {}^*)$ is an R-frame.
- U is a set, the *universe of* $\mathfrak{S}$.
- *Prop* is a collection of *up-sets* of the R-frame, called the *admissible propositions* of $\mathfrak{S}$. *Prop* contains 0 and is closed under the operations $\Rightarrow$, $\cap$ and *.
- *PropFun* is a set of functions from U^ω to *Prop*, called the *admissible propositional functions* of $\mathfrak{S}$. *PropFun* contains the function α_0 with constant value 0, and is closed under the operations $\cap$, * and $\Rightarrow$ that are defined pointwise by lifting the corresponding operations on *Prop*. For $\alpha, \beta : U^\omega \to \wp W$, the functions $\alpha \cap \beta$, α^*, $\alpha \Rightarrow \beta : U^\omega \to \wp W$ are defined by:

$$(\alpha \cap \beta)f = \alpha f \cap \beta f.$$
$$(\alpha^*)f = -(\alpha f)^*.$$
$$(\alpha \Rightarrow \beta)f = \alpha f \Rightarrow \beta f.$$

- *PropFun* is closed under operations $\forall_x$ for all individual variables x, where, for each $\alpha : U^\omega \to \wp W$, the function $\forall_x \alpha : U^\omega \to \wp W$ is defined by

$$(\forall_x \alpha)f = \bigsqcap_{a \in U} \alpha(f[a/x]).$$

As usual, a *premodel* $\mathcal{M}$ on such a structure $\mathfrak{S}$ assigns to each constant $\mathsf{c} \in \mathcal{L}$ an element $|\mathsf{c}|^{\mathcal{M}} \in U$, and to each n-ary predicate symbol $P \in \mathcal{L}$ a function $|P|^{\mathcal{M}} : U^n \to \wp W$. Then $\mathcal{M}$ assigns a propositional function $|\varphi|^{\mathcal{M}} : U^\omega \to \wp W$ to each $\mathcal{L}$-formula φ. When φ is the atomic $P\tau_1 \cdots \tau_n$, this propositional function is given, for each $f \in U^\omega$, by the familiar

$$|P\tau_1 \cdots \tau_n|^{\mathcal{M}} f = |P|^{\mathcal{M}}(|\tau_1|^{\mathcal{M}} f, \ldots, |\tau_n|^{\mathcal{M}} f). \tag{6.2.1}$$

The cases for the other forms of φ are given by these equations:

- $|t|^{\mathcal{M}} = \alpha_0$.
- $|\varphi \wedge \psi|^{\mathcal{M}} = |\varphi|^{\mathcal{M}} \cap |\psi|^{\mathcal{M}}$.
- $|\neg\varphi|^{\mathcal{M}} = (|\varphi|^{\mathcal{M}})^*$.
- $|\varphi \to \psi|^{\mathcal{M}} = |\varphi|^{\mathcal{M}} \Rightarrow |\psi|^{\mathcal{M}}$.
- $|\forall x \varphi|^{\mathcal{M}} = \forall_x |\varphi|^{\mathcal{M}}$.

$\mathcal{M}$ is called a *model* if $|\varphi|^{\mathcal{M}} \in PropFun$ for all *atomic* formulas φ. The closure conditions on *PropFun* ensure that if $\mathcal{M}$ is a model, then $|\varphi|^{\mathcal{M}} \in PropFun$

for every formula φ, and hence that $|\varphi|^{\mathcal{M}} f \in Prop$ for every φ and every $f \in U^{\omega}$.

A formula φ is *valid in* $\mathcal{M}$ if $0 \subseteq |\varphi|^{\mathcal{M}} f$ for all $f \in U^{\omega}$, which means that φ is true at every *base* world in $\mathcal{M}$ under every variable assignment. Then QR is *sound* for validity in these functional QR-model structures, i.e. the QR-theorems are all valid in all models on QR-model structures.

EXAMPLE 6.2.2. A Non-Kripkean QR-Model.

Let $\mathbb{Z}$ be the set of all integers, and write $\mathbf{0}$ for the integer zero. Let $W = \mathbb{Z} \cup \{\infty, -\infty\}$, where ∞ and $-\infty$ are two new objects, and extend the natural linear ordering of $\mathbb{Z}$ to W by putting $-\infty < m < \infty$ for all $m \in \mathbb{Z}$. Let $-(-\infty) = \infty$ and define $w^* = -w$ for all $w \in W$. For each $w \in W$, let $[w) = \{u : w \leq u\}$. Then $[w)^* = \{u \in W : w^* < u\} = [w^*) - \{w^*\}$.

To make this into an R-frame we define a binary operation $\cdot$ on W by

$$w \cdot u = \begin{cases} \max\{w, u\} & \text{if } u^* < w, \\ \min\{w, u\} & \text{otherwise,} \end{cases}$$

and put $Rwuv$ iff $w \cdot u \leq v$. Then the structure $(W, [\mathbf{0}), R, {}^*)$ proves to be an R-frame whose preorder, defined by the condition "for some $w \in [\mathbf{0})$, $Rwuv$", is just the arithmetical ordering $\leq$ of W. The operation $\Rightarrow$ induced by R has

$$X \Rightarrow Y = \begin{cases} X^* \cup Y & \text{if } X \subseteq Y, \\ X^* \cap Y & \text{otherwise.} \end{cases}$$

The fact that $\leq$ is linear ensures that up-sets are linearly ordered by inclusion, i.e. if X and Y are up-sets then $X \subseteq Y$ or $Y \subseteq X$. The collection

$$Prop = \{W, \emptyset\} \cup \{[m) : m \in \mathbb{Z}\}$$

is closed under $\cap$, $\cup$ and *, hence under $\Rightarrow$. Note that *Prop* does not include all up-sets, since it contains neither $\{\infty\}$ nor $\mathbb{Z} \cup \{\infty\}$. An important feature of *Prop* is that it is $\bigsqcap$-complete, i.e. it is closed under the operation $\bigsqcap$ on $\wp W$ that is induced by *Prop*. This is because if $\mathcal{Z} \subseteq Prop$, then $\bigcap \mathcal{Z}$ is one of W, $\emptyset$, $[m)$ for some $m \in \mathbb{Z}$, or $\{\infty\}$; in which case $\bigsqcap \mathcal{Z}$ is W, $\emptyset$, $[m)$, or $\emptyset$, respectively.

Finally, to make this into a model structure, we take the universe U to be ω, and *PropFun* to be the set of *all* functions from U^{ω} to *Prop*. The closure of this *PropFun* under $\Rightarrow$, $\cap$ and * is automatic from the closure of *Prop* under the corresponding operations, and its closure under $\forall_x$ follows from the $\bigsqcap$-completeness of *Prop*.

Now take a unary predicate symbol P and define $|P|^{\mathcal{M}} : U \to Prop$ by putting $|P|^{\mathcal{M}} n = [n)$ for all $n \in \omega$. Let φ be Px for some variable x. Then for any $f \in U^{\omega}$ and any $n \in \omega$,

$$|\varphi|^{\mathcal{M}} f[n/x] = |P|^{\mathcal{M}}(|x|^{\mathcal{M}} f[n/x]) = [n).$$

Since $\infty \in [n)$, this gives $\mathcal{M}, \infty, f[n/x] \models \varphi$. But it also gives

$$\bigcap_{n \in U} |\varphi|^{\mathcal{M}} f[n/x] = \bigcap_{n \in U} [n) = \{\infty\},$$

and so $|\forall x\varphi|^{\mathcal{M}} f = \bigsqcap_{n \in U} |\varphi|^{\mathcal{M}} f[n/x] = \emptyset$.

Altogether we have $\mathcal{M}, \infty, f[n/x] \models \varphi$ for all $n \in U$, but $\mathcal{M}, \infty, f \not\models \forall x\varphi$, hence $\mathcal{M}$ is not Kripkean.

This example, and others like it, are analysed in more detail in [Mares and Goldblatt 2006, §5]. ⊣

To prove that QR is complete for validity in QR-model structures, a canonical model structure $\mathfrak{S}_{\mathrm{QR}}$ is constructed. Its underlying R-frame is defined by repeating in the present language the canonical frame construction of the previous section for R itself. The universe of $\mathfrak{S}_{\mathrm{QR}}$ is the set $\mathsf{Con}_{\mathcal{L}}$ of individual constants of $\mathcal{L}$, and is assumed to be infinite. $\mathit{Prop}_{\mathrm{QR}}$ and $\mathit{PropFun}_{\mathrm{QR}}$ are defined just as we did for quantified modal logics:

$$\mathit{Prop}_{\mathrm{QR}} = \{\, |\varphi|_{\mathrm{QR}} : \varphi \text{ is an } \mathcal{L}\text{-sentence}\},$$

where $|\varphi|_{\mathrm{QR}} = \{w : \varphi \in w\}$; and

$$\mathit{PropFun}_{\mathrm{QR}} = \{\alpha_\varphi : \varphi \text{ is any } \mathcal{L}\text{-formula}\},$$

where $\alpha_\varphi f = |\varphi^f|_{\mathrm{QR}}$ for all $f \in \mathsf{Con}_{\mathcal{L}}{}^{\omega}$. The canonical premodel $\mathfrak{M}_{\mathrm{QR}}$ on $\mathfrak{S}_{\mathrm{QR}}$ is defined in the now familiar way by putting $|\mathsf{c}|^{\mathfrak{M}_{\mathrm{QR}}} = \mathsf{c}$, and

$$|P|^{\mathfrak{M}_{\mathrm{QR}}}(\mathsf{c}_1, \ldots, \mathsf{c}_n) = |P\mathsf{c}_1 \cdots \mathsf{c}_n|_{\mathrm{QR}} \in \mathit{Prop}_{\mathrm{QR}}.$$

This premodel satisfies the Truth Lemma: $|\varphi|^{\mathfrak{M}_{\mathrm{QR}}} f = |\varphi^f|_{\mathrm{QR}}$ for all φ and f. From this it can be shown that $\mathfrak{M}_{\mathrm{QR}}$ is a model that characterises QR. Hence $\mathfrak{S}_{\mathrm{QR}}$ characterises QR, as does the class of all functional QR-model structures.

It is not really necessary to use *functional* model structures to characterise QR. We could drop *PropFun* from the above discussion to conclude that QR is characterised by validity in models on structures of the form $(W, 0, R, {}^*, U, \mathit{Prop})$ that are based on R-frames. But the use of admissible propositional functions does appear to be essential in dealing with the stronger logic RQ. We define an RQ-*model structure* to be a QR-model structure in which

$$X - Y \subseteq \bigcap_{a \in U} \alpha(f[a/x]) \quad \text{implies} \quad X - Y \subseteq (\forall_x \alpha) f \tag{6.2.2}$$

for all $\alpha \in \mathit{PropFun}$, all $x \in \mathsf{InVar}$, all $X, Y \in \mathit{Prop}$, and all $f \in U^\omega$. This condition ensures the validity of the $\forall\vee$-Distribution axiom, and hence the validity of all RQ-theorems. Conversely, a canonical model structure for RQ, defined as for QR above, satisfies (6.2.2) and so validates RQ. It follows that RQ is characterised by validity in all models on RQ-model structures. The details of all this are set out in [Mares and Goldblatt 2006].

Neither RQ nor QR is characterised by *Kripkean* models with one universal domain based on R-frames, so neither logic has a Kripkean canonical model of this kind. The counter-example to Kripkean completeness of RQ given in [Fine 1989] applies also to QR. But for QR a simpler example can be given, by using our present semantics to show that the Kripkean-valid scheme of $\forall\vee$-Distribution is not derivable in QR, confirming that QR $\neq$ RQ. This is done by modifying the structure of Example 6.2.2. That structure in fact satisfies condition (6.2.2) so is not itself suitable. We modify it by deleting the points ∞ and $-\infty$, and adding a new isolated point θ with $\theta^* = \theta$ and θ being $\leq$-incomparable to the members of $\mathbb{Z}$. New admissible propositions are created by adding θ to each $[m)$ while retaining $[m)$ itself, and without adding $\{\theta\}$ to *Prop*. The resulting structure violates (6.2.2).

If, instead of adding θ, we adjoin to $\mathbb{Z}$ a disjoint copy of the three-element structure $\{-1, \mathbf{0}, 1\}$, a QR-model can be constructed that falsifies not just $\forall\vee$-Distribution, but also its consequence

$$\neg\forall x(\varphi \vee \psi) \vee \varphi \vee \forall x\psi, \quad \text{where } x \text{ is not free in } \varphi.$$

For details see [Mares and Goldblatt 2006, §11].

These examples show that admissible models built on Routley-Meyer frames can be usefully applied to the metatheory of quantified relevant logic. But it might be argued that a condition like (6.2.2) is essentially a functional-algebraic translation of the corresponding axiom scheme of $\forall\vee$-Distribution, and contributes little to clarifying its "meaning". In the rest of this chapter we will develop a different semantic characterisation of RQ that is based on intuitive structural ideas of independent interest.

6.3. Local Truth and Covers

Our new semantics for R and RQ is based on the idea of a property being determined "locally". This is a notion that originally came from topology, as we now explain.

A topological space comprises a set with a collection of distinguished subsets, called the *open* sets of the space, satisfying certain axioms. Intuitively, a subset is open if, whenever it contains a point p, then it also contains every point that is "near" to p. So if p belongs to an open set whose members have a particular property, then we may think of this property as holding "near" p. But in this abstract setting there is no explicit relation of "nearness" provided with the space: all that we have is the given collection of open sets satisfying the axioms for a topology: the intersection of two open sets is open, the union of any collection of open sets is open, etc.

A property is said to hold *locally of X*, where X may be the whole space or an open subset of it, if X has an *open cover* consisting of sets that all have the property in question. An open cover of X is a collection $Z = \{O_i : i \in I\}$ of open sets such that every point of X belongs to some O_i, i.e. $X \subseteq \bigcup\{O_i : i \in I\}$. So if Z is an open cover of X, and each member O_i of Z has a certain property π, then each point of X belongs to an open set having π. Intuitively this means that π holds near each point of X.

Note that if X is a proper open subset of the ambient space, it may be that $X \subsetneq \bigcup Z$, so Z covers more than X. But then we can replace Z by $Z' = \{X \cap O : O \in Z\}$, which is a new open cover of X that exactly covers X, i.e. $X = \bigcup Z'$. Such a Z' is known as a *refinement* of Z, obtained by refining each $O \in Z$ to its open subset $X \cap O$.

To illustrate the notion of a property holding locally of X, consider the property of a function f being constant. If X has an open cover such that f takes a constant value on each member of the cover, then f is said to be *locally constant* on X. In other words, the assertion "f is constant" is *locally true* of X. The intuitive idea is that f is constant near each point of X. This is compatible with there being distinct open sets O_i and O_j in the cover such that f assigns a different constant value on O_i to the one it assigns on O_j. So f can be locally constant on X without being actually constant on X.

By contrast, consider the property of functions f and g being equal. If X has an open cover with f and g agreeing on each cover set O_i, then f and g are said to be *locally equal* on X. In other words, the assertion "$f = g$" is *locally true* of X. But in that case, each point p of X belongs to some set O_i on which f and g agree, so $f(p) = g(p)$. Hence f is actually equal to g on the whole of X.

These examples show that there can be properties that are locally true of X without being true of X, and other properties for which their being locally true of X is sufficient to ensure that they are actually true of X. The latter kind of property is said to be *locally determined*, or *of local character*. Thus a property is locally determined if a set must have the property whenever it has a cover consisting of open sets that themselves have the property.

We now move to develop these ideas in a more general context that abstracts away from topological spaces. The concept of a cover allows us to formulate notions of locally true property and locally determined property, and these can be developed in a context in which all that we have are axiomatically given covers. We just need to be able to say when an object has a "cover" consisting of other such objects, without taking those objects to be sets of points, and without taking the relation "Z is a cover of X" to be set-theoretically definable. Rather we take it to be a primitive relation that is subject to certain postulates. Forgetting the topology, we might think of the cover relation as saying that Z is some kind of "decomposition" of X, whatever sort of object X is.[33]

---

[33]An information-theoretic interpretation of $\rhd$ is given on page 217.

Our particular purpose is to take the objects to be the members of the set W in a model $\mathcal{M}$ for the logic R, in such away that the truth relation $\mathcal{M}, w \models \varphi$ is locally determined. This will mean that if w has a cover Z with φ true at all members of Z, then φ is true at w itself. That requirement will ensure that certain R-axioms are valid in our models, as we will see.

With this motivation in mind, we now work with structures $(W, \leq, \rhd)$, for which:

- $\leq$ is a preorder relation on set W, called the *refinement* relation.
- $\rhd$ is a binary relation from the powerset $\wp W$ to W, called the *cover relation*.

Thus $\rhd$ is some set of pairs (Z, w) with $Z \subseteq W$ and $w \in W$. We write $Z \rhd w$ to mean that (Z, w) belongs to $\rhd$, and say that Z *is a cover of* w, or that Z *covers* w, or that Z *is a* w*-cover*. The inverse relation to $\rhd$ will be denoted $\lhd$. Thus when $Z \rhd w$, we may also write $w \lhd Z$, and say that w *is covered by* Z.

When $w \leq u$ we may say that u *refines* w (in the topological case this means $w \supseteq u$—see Example 6.3.2 below). Recall that an *up-set* is a subset of W that is closed upwards under the preorder $\leq$. For an arbitrary $X \subseteq W$, define

$$\uparrow X = \{u \in W : \text{ for some } w \in X,\ w \leq u\}.$$

Then $\uparrow X$ is an up-set, the *smallest* up-set including X, i.e. if Y is an up-set including X, then $\uparrow X \subseteq Y$. Thus X is an up-set iff $\uparrow X = X$.

We write $\uparrow w$ for $\uparrow\{w\} = \{u : w \leq u\}$, the smallest up-set containing w. This will sometimes be called the *cone* at w.

Now for each subset X of W, define

$$jX = \{w \in W : \text{ for some } Z,\ w \lhd Z \subseteq X\}.$$

The condition "$w \lhd Z \subseteq X$" states that Z is a w-cover that consists of members of X. So w belongs to jX just when the property of *being a member of* X holds *locally* of w, i.e. when w is covered by a set of members of X. Thus we may think of jX as the collection of "local members" of X. X is called *localised* if $jX \subseteq X$, i.e. if every local member of X is an actual member of X. Thus X is a localised set when membership of X is locally determined in the sense described above.

We will call $\mathcal{S} = (W, \leq, \rhd)$ a *cover system* if it satisfies the following postulates:

- *Covers are refining*: every w-cover is included in the cone of w, i.e. $Z \rhd w$ implies $Z \subseteq \uparrow w$.
- *Covers exist*: every member of W has a cover, i.e. for all w there exists a Z with $Z \rhd w$.

LEMMA 6.3.1. *In a cover system, the operation* j *is inflationary on up-sets, i.e. if* X *is any up-set, then* $X \subseteq jX$.

PROOF. Let X be an up-set. If $w \in X$, then $\uparrow w \subseteq X$ as X is upwardly closed under refinement. Since *covers exist*, w has a cover Z. Then Z is included in $\uparrow w$ as *covers are refining*. Altogether, $w \lhd Z \subseteq \uparrow w \subseteq X$, showing that $w \in jX$. ⊣

An up-set X in a cover system will be called a *proposition* if it is localised, i.e. if $jX \subseteq X$. By this Lemma, an up-set X is a proposition iff $jX = X$. In general, a set X is a proposition iff $X = \uparrow X = jX$. We write $Prop(\mathcal{S})$ for the set of all localised up-sets of a cover system $\mathcal{S}$.

EXAMPLE 6.3.2. Let W be the set of open subsets of some topological space, with $w \leq u$ iff $w \supseteq u$, and $Z \rhd w$ iff $w = \bigcup Z$. Then $\rhd$ captures the topological notion of open cover, and $(W, \leq, \rhd)$ is a cover system: if $w = \bigcup Z$ then $Z \subseteq \{u \in W : w \supseteq u\} = \uparrow w$, hence *covers are refining*; and w has at least the covers $\{w\}$ and $\uparrow w$, so *covers exist*.

In this cover system, each cone $\uparrow w$ is a proposition, for if $u \lhd Z \subseteq \uparrow w$, then $u = \bigcup Z \subseteq w$, hence $u \in \uparrow w$. If s is any subset of the ambient space, then $\{w \in W : s \cap w = \emptyset\}$ is a proposition. But a union $\uparrow w \cup \uparrow u$ of cones need not be localised.

This example can be generalised to any set W that has a partial order $\sqsubseteq$ that is *complete*, i.e. every subset X has a join, or least upper bound, $\bigsqcup X$ (such a $(W, \sqsubseteq)$ is usually called a *complete lattice*). Then a cover system on W is defined by putting $w \leq u$ iff $u \sqsubseteq w$, and $Z \rhd w$ iff $w = \bigsqcup Z$. In this case the propositions are exactly the cones [Goldblatt 2006a, Theorem 6]: if X is any localised up-set, put $w = \bigsqcup X$ to get $w \lhd X$, hence $w \in jX \subseteq X$, and so $X = \{u : u \sqsubseteq w\} = \uparrow w$. ⊣

One way to interpret a cover system is to think of W as a collection of situations or states that each contain certain information, with $w \leq u$ when the information content of u includes that of w. So $\leq$ is a relation of refinement in the sense of increase of information. For the cover relation, view $Z \rhd w$ as meaning that the information content of w comprises that information which is common to all the states in Z. This idea will be formally represented in a canonical model construction for R to be given in Section 6.5, where states are sets of formulas, and $Z \rhd w$ is defined to mean that $w = \bigcap Z$.

The idea of a semantics based on cover systems originated in topos theory, in the logic of categories of sheaves, where it was introduced by André Joyal. There it became known as *Kripke-Joyal semantics*, because its interpretation of some connectives and quantifiers is reminiscent of Kripke's modelling of intuitionistic logic (see Lambek and Scott [1986], Bell [1988] and Mac Lane and Moerdijk [1992]). But, as we will see, this approach gives a non-classical interpretation to disjunction and existential quantification, and in that sense is more reminscent of the intuitionistic semantics of Beth [1956]. See [Bell 2005] for a discussion of intuitionistic cover semantics, including the connection with Beth models. For the history of Beth's work in this area, see Troelstra and van Ulsen [1999].

6.4. Relevant Cover Systems

To model relevant implication, we are going to replace the Routley-Meyer ternary relation by a binary operation; and to model negation we will replace the unary operation * by a binary relation. We now expand our notion of structure to one of the form

$$\mathcal{S} = (W, \leq, \rhd, \cdot, \varepsilon, \perp),$$

with $\cdot$ a binary operation on W, called *fusion*; ε an element of W, the *identity*; and $\perp$ a *symmetric* binary relation on W, called *orthogonality*.

The fusion operation will be used to model implication in the manner of (6.1.3). The element ε will be an identity for fusion, and may be thought of as the state whose information content is exactly that of the constant formula t. The orthogonality relation may be thought of as a relation of incompatibility between information states, with $w \perp u$ when the two states contain conflicting information.[34] If $w \perp u$ we say that *w is orthogonal to u*, or that *u is orthogonal to w*. The fact that $\perp$ is symmetric ensures that this usage is unambiguous. A negated formula $\neg A$ will be defined to be true at w when A is only true at points orthogonal to (i.e. incompatible with) w. This kind of modelling of negation was introduced in [Goldblatt 1974]. Similar ideas occur in linear logic [Girard 1987].

The fusion operation $\cdot$ is lifted to an operation on subsets of W by putting

$$X \cdot Y = \{w \cdot u : w \in X \text{ and } u \in Y\}.$$

Then $w \cdot Y$ is defined to be $\{w\} \cdot Y$, and $X \cdot u$ is $X \cdot \{u\}$.

An *implication* operation $\Rightarrow$ on sets is defined by

$$X \Rightarrow Y = \{w : w \cdot X \subseteq Y\}. \tag{6.4.1}$$

Thus $w \in X \Rightarrow Y$ iff $u \in X$ implies $w \cdot u \in Y$, so $X \Rightarrow Y$ embodies the semantics for implication of (6.1.3). This operation is *residual* to the fusion operation $X \cdot Y$ on sets, i.e. for all $X, Y, Z \subseteq W$ it has

$$Z \subseteq X \Rightarrow Y \quad \text{iff} \quad Z \cdot X \subseteq Y. \tag{6.4.2}$$

The orthogonality relation is lifted to a relation between sets, and between sets and points, by putting:

[34] The description of $\perp$ as an "incompatibility" relation is due to Dunn [1993], and is discussed further by Mares [2004, Chapter 5]. It is manifest in the canonical model construction of Goldblatt [1974], where states are certain sets of formulas, and $w \perp u$ is defined to hold iff u contains some formula whose negation belongs to w. We use a variant of that definition below in our canonical models for R and RQ.

- $X \perp Y$ iff $w \perp u$ for all $w \in X$ and $u \in Y$.
- $w \perp Y$ iff $w \perp u$ for all $u \in Y$.
- $X \perp u$ iff $w \perp u$ for all $w \in X$.

The set $X^{\perp} = \{w : w \perp X\}$ is the *orthocomplement* of X, and $\{u\}^{\perp} = \{w : w \perp u\}$ may be written $u^{\perp}$.

Observe that if $w \perp (X \cup Y)$ then both $w \perp X$ and $w \perp Y$, and conversely, so we have

$$(X \cup Y)^{\perp} = X^{\perp} \cap Y^{\perp} \tag{6.4.3}$$

for any $X, Y \subseteq W$. In fact for an arbitrary collection $\{X_i : i \in I\}$ of subsets of W we have

$$\Big(\bigcup_{i \in I} X_i\Big)^{\perp} = \bigcap_{i \in I} X_i^{\perp}. \tag{6.4.4}$$

Orthocomplementation reverses inclusion: $X \subseteq Y$ implies $Y^{\perp} \subseteq X^{\perp}$. Hence double-orthocomplementation preserves inclusion: $X \subseteq Y$ implies $X^{\perp\perp} \subseteq Y^{\perp\perp}$. The symmetry of $\perp$ ensures that $X \subseteq X^{\perp\perp}$ for any X. X is called *orthoclosed* when $X^{\perp\perp} \subseteq X$ and so $X = X^{\perp\perp}$. Note that $X \subseteq X^{\perp\perp}$ implies $X^{\perp\perp\perp} \subseteq X^{\perp}$. Hence $X^{\perp}$ is always orthoclosed: $X^{\perp\perp\perp} = X^{\perp}$. In particular, any set of the form $X^{\perp\perp}$ is orthoclosed, and is the smallest orthoclosed superset of X. For if Y is orthoclosed and $X \subseteq Y$, then $X^{\perp\perp} \subseteq Y^{\perp\perp} = Y$. $X^{\perp\perp}$ is called the *orthoclosure* of X.

The intersection of two orthoclosed sets is orthoclosed, for in general we have $(X \cap Y)^{\perp\perp} \subseteq X^{\perp\perp} \cap Y^{\perp\perp}$ as double-orthocomplementation preserves inclusion, so if X and Y are orthoclosed this becomes $(X \cap Y)^{\perp\perp} \subseteq X \cap Y$. However, the union of orthoclosed sets need not be orthoclosed, so we introduce a new operation $X + Y$, called the *orthojoin* of X and Y, defined by putting

$$X + Y = (X \cup Y)^{\perp\perp},$$

the smallest orthoclosed superset of X and Y. More generally we can define an infinitary orthojoin of any collection $\{X_i : i \in I\}$ by putting

$$\sum_{i \in I} X_i = \Big(\bigcup_{i \in I} X_i\Big)^{\perp\perp}. \tag{6.4.5}$$

The collection of all orthoclosed sets is a complete lattice under the partial ordering of set inclusion, with joins (least upper bounds) given by orthojoins $\sum$, and meets (greatest lower bounds) given by set intersections $\bigcap$. This lattice also satisfies the *De Morgan laws* for distribution of orthocomplements over joins and meets. First, from (6.4.4) we obtain

$$\Big(\sum_{i \in I} X_i\Big)^{\perp} = \bigcap_{i \in I} X_i^{\perp} \tag{6.4.6}$$

for any collection of sets X_i; and if the X_i are orthoclosed we also have

$$\Big(\bigcap_{i\in I} X_i\Big)^{\perp} = \sum_{i\in I} X_i^{\perp},$$

as the reader may care to check. From (6.4.6) and the orthoclosure of $\sum_{i\in I} X_i$ we derive

$$\sum_{i\in I} X_i = \Big(\bigcap_{i\in I} X_i^{\perp}\Big)^{\perp}. \tag{6.4.7}$$

All of these results about orthocomplementation flow from the assumption that $\perp$ is symmetric. In general the lattice of orthoclosed sets is not distributive, and may fail to satisfy

$$X \cap (Y + Y') \subseteq (X \cap Y) + (X \cap Y').$$

But by adding further conditions concerning the cover system, and then restricting to a suitable set of admissible propositions, we will obtain a kind of structure that not only satisfies distributivity, but also validates all the other axioms of the relevant logic R. First we define the kind of system we need, and later add the admissibility restrictions on propositions.

A *relevant cover system* is defined to be a structure of the form

$$\mathcal{S} = (W, \leq, \rhd, \cdot, \varepsilon, \perp),$$

that satisfies the following postulates:

- $(W, \leq, \rhd)$ is a cover system.
- $(W, \leq, \cdot, \varepsilon)$ is a *square-decreasing commutative ordered monoid*; i.e. the fusion operation $\cdot$ is:
  - *monotonic under refinement*: if $w \leq w'$ and $u \leq u'$, then $w \cdot u \leq w' \cdot u'$.
  - *associative*: $w \cdot (u \cdot v) = (w \cdot u) \cdot v$.
  - *commutative*: $w \cdot u = u \cdot w$.
  - *with identity* ε: $w \cdot \varepsilon = \varepsilon \cdot w = w$.
  - *square-decreasing*: $w \cdot w \leq w$.
- *Fusion preserves covering*: $w \lhd Z$ implies $w \cdot u \lhd Z \cdot u$.
- *Refinement preserves orthogonality*: if $w \leq w'$ and $u \leq u'$, then $w \perp u$ implies $w' \perp u'$.
- *Orthogonality to* ε *is local*: $w \lhd Z \perp \varepsilon$ implies $w \perp \varepsilon$.
- *Contraposition*: $w \cdot u \perp v$ implies $w \cdot v \perp u$. ⊣

These postulates have been chosen more for their naturalness than their minimality in meeting certain requirements.[35] Here are some fundamental consequences of them.

[35]For example, refinement preserving orthogonality follows from the weaker statement that $w' \geq w \perp u$ implies $w' \perp u$, given the symmetry of $\perp$. Also symmetry of $\perp$ is itself implied by the contraposition postulate and the identity property of ε: if $w \perp u$, then $\varepsilon \cdot w \perp u$, hence $u = \varepsilon \cdot u \perp w$.

LEMMA 6.4.1. *In a relevant cover system, for any* $w, u \in W$ *and any* $X, Y \subseteq W$:

(1) $w \perp u$ *iff* $w \cdot u \perp \varepsilon$.
(2) $X^{\perp}$ *is a localised up-set.*
(3) *If* Y *is an up-set, then so is* $X \Rightarrow Y$.
(4) *If* Y *is localized, then so is* $X \Rightarrow Y$.
(5) $w \perp X$ *implies* $w \perp jX$. *That is,* $X^{\perp} \subseteq (jX)^{\perp}$.
(6) $jX \subseteq (jX)^{\perp\perp} \subseteq X^{\perp\perp}$.
(7) *If* X *is an up-set, then* $w \perp X$ *iff* $w \perp jX$, *i.e.* $X^{\perp} = (jX)^{\perp}$.
(8) *If* X *is an up-set, then* $X \subseteq jX \subseteq (jX)^{\perp\perp} = X^{\perp\perp}$.

PROOF. (1) Since ε is an identity, $w \perp u$ iff $w \cdot \varepsilon \perp u$, iff $w \cdot u \perp \varepsilon$ by the contraposition postulate.

(2) First let $w \perp X$ and $w \leq u$. Then for all $v \in X$ we have $w \perp v$, so $u \perp v$ as refinement preserves orthogonality. Hence $u \perp X$. This shows that $X^{\perp}$ is an up-set.

Next, to show $X^{\perp}$ is localized, suppose $w \lhd Z \subseteq X^{\perp}$. We want $w \in X^{\perp}$. So take any $u \in X$, in order to show $w \perp u$. Now for each $v \in Z$ we have $v \perp u$ as $v \in X^{\perp}$, hence $v \cdot u \perp \varepsilon$ by (1). This shows that $Z \cdot u \perp \varepsilon$. But $w \cdot u \lhd Z \cdot u$ as fusion preserves covering, so then $w \cdot u \perp \varepsilon$ as orthogonality to ε is local, hence $w \perp u$ as required, by (1) again.

(3) Let $u \geq w \in X \Rightarrow Y$. Then $w \cdot X \subseteq Y$ (6.4.1). So for any $v \in X$, using the monotonicity of fusion under refinement we get $u \cdot v \geq w \cdot v \in w \cdot X \subseteq Y$, hence $u \cdot v \in Y$ as Y is an up-set. This shows that $u \cdot X \subseteq Y$, i.e. $u \in X \Rightarrow Y$.

(4) Let $w \lhd Z \subseteq X \Rightarrow Y$. We want $w \in X \Rightarrow Y$, i.e. $w \cdot X \subseteq Y$. We have $Z \cdot X \subseteq (X \Rightarrow Y) \cdot X \subseteq Y$ (see (6.4.2)). But then if $u \in X$, $w \cdot u \lhd Z \cdot u \subseteq Z \cdot X \subseteq Y$, so $w \cdot u \in Y$ as Y is localized.

(5) Let $w \perp X$. Suppose $u \in jX$, so $u \lhd Z \subseteq X$ for some Z. Then $w \cdot u \lhd w \cdot Z$ as fusion preserves covering and $\cdot$ is commutative. But if $v \in Z$ we have $v \in X$, hence $w \perp v$, so then $w \cdot v \perp \varepsilon$ by (1). This shows that $w \cdot u \lhd w \cdot Z \perp \varepsilon$. Therefore $w \cdot u \perp \varepsilon$ as orthogonality to ε is local, hence $w \perp u$ by (1) again. Since this holds for all $u \in jX$, we have $w \perp jX$ as required.

(6) $jX \subseteq (jX)^{\perp\perp}$ holds as double orthocomplementation is inflationary. But (5) gives $X^{\perp} \subseteq (jX)^{\perp}$, which implies $(jX)^{\perp\perp} \subseteq X^{\perp\perp}$.

(7) If X is an up-set, then $X \subseteq jX$ by Lemma 6.3.1, hence $(jX)^{\perp} \subseteq X^{\perp}$. But $X^{\perp} \subseteq (jX)^{\perp}$ by (5).

(8) If X is an up-set, then $X \subseteq jX \subseteq (jX)^{\perp\perp} \subseteq X^{\perp\perp}$ by Lemma 6.3.1 and (6). But also $(jX)^{\perp} = X^{\perp}$ by (7), hence $(jX)^{\perp\perp} = X^{\perp\perp}$. ⊣

COROLLARY 6.4.2. *In a relevant cover system, every orthoclosed up-set is localised, hence is a proposition.*

PROOF. If an up-set X is orthoclosed, i.e. $X = X^{\perp\perp}$, then $X = jX$ by part (8) of the Lemma, so X is localised. But a proposition, by definition, is a localised up-set. ⊣

Now we define an R-*model system* to be a structure

$$\mathcal{S} = (W, \leq, \rhd, \cdot, \varepsilon, \perp, \mathit{Prop}),$$

such that:

- $(W, \leq, \rhd, \cdot, \varepsilon, \perp)$ is a relevant cover system.
- *Prop* is a set of *orthoclosed up-sets* of $(W, \leq)$ such that:
  - the cone $\uparrow\varepsilon$ belongs to *Prop*.
  - $X, Y \in \mathit{Prop}$ implies $X \cap Y$, $X^{\perp}$, $X \Rightarrow Y \in \mathit{Prop}$.
  - $X, Y \in \mathit{Prop}$ implies $X + Y = j(X \cup Y)$.

Here, by the above Corollary, the members of *Prop* are all propositions, so *Prop* is a subset of the set $\mathit{Prop}(\mathcal{S})$ of all localised up-sets of the relevant cover system of $\mathcal{S}$. The members of *Prop* will be called the *admissible* propositions of $\mathcal{S}$. Requiring the cone $\uparrow\varepsilon$ to be admissible implies that it is orthoclosed, which is not a requirement of relevant cover systems in general.

The condition $X + Y = j(X \cup Y)$ on members of *Prop* is used to prove the model condition of (6.4.8) below, which is needed to prove validity of the $\wedge\vee$-Distribution axiom of R (see the proof of Theorem 6.4.4). In the next section we will construct a canonical R-model system and use the $\wedge\vee$-Distribution axiom to prove that its admissible propositions satisfy $X + Y = j(X \cup Y)$ (see Theorem 6.5.4).

A *model* $\mathcal{M}$ on an R-model system is given by a function $|-|^{\mathcal{M}}$ that assigns to each propositional variable p an admissible truth set $|p|^{\mathcal{M}} \in \mathit{Prop}$. This assignment is extended inductively to define truth sets $|A|^{\mathcal{M}}$ for all propositional formulas, as follows:

$$|t|^{\mathcal{M}} = \uparrow\varepsilon.$$

$$|A \wedge B|^{\mathcal{M}} = |A|^{\mathcal{M}} \cap |B|^{\mathcal{M}}.$$

$$|\neg A|^{\mathcal{M}} = (|A|^{\mathcal{M}})^{\perp}.$$

$$|A \to B|^{\mathcal{M}} = |A|^{\mathcal{M}} \Rightarrow |B|^{\mathcal{M}}.$$

The closure properties of *Prop* given in the definition of R-model system then ensure that, for every formula A, $|A|^{\mathcal{M}}$ is a member of *Prop*, and so is an orthoclosed localised up-set. Localisation here means that $j|A|^{\mathcal{M}} \subseteq |A|^{\mathcal{M}}$, which spells out to the condition:

if there exists a w*-cover included in* $\{u : \mathcal{M}, u \models A\}$, *then* $\mathcal{M}, w \models A$.

This fulfills our purpose of making the truth relation locally determined.

Since $A \vee B$ is defined to be $\neg(\neg A \wedge \neg B)$, we have, using (6.4.7), that

$$|A \vee B|^{\mathcal{M}} = ((|A|^{\mathcal{M}})^{\perp} \cap (|B|^{\mathcal{M}})^{\perp})^{\perp} = |A|^{\mathcal{M}} + |B|^{\mathcal{M}},$$

where $|A|^{\mathcal{M}} + |B|^{\mathcal{M}}$ is the orthojoin $(|A|^{\mathcal{M}} \cup |B|^{\mathcal{M}})^{\perp\perp}$. But $|A|^{\mathcal{M}}$ and $|B|^{\mathcal{M}}$ belong to *Prop*, so by the last defining property of *Prop* in an R-model system, their orthojoin is equal to $j(|A|^{\mathcal{M}} \cup |B|^{\mathcal{M}})$. Hence we get

$$|A \vee B|^{\mathcal{M}} = j(|A|^{\mathcal{M}} \cup |B|^{\mathcal{M}}). \tag{6.4.8}$$

Writing $\mathcal{M}, w \models A$ as usual to mean that $w \in |A|^{\mathcal{M}}$, the definition of truth sets becomes the following conditions on the truth relation:

- $\mathcal{M}, w \models p$ iff $w \in |p|^{\mathcal{M}}$.
- $\mathcal{M}, w \models t$ iff $\varepsilon \leq w$.
- $\mathcal{M}, w \models A \wedge B$ iff $\mathcal{M}, w \models A$ and $\mathcal{M}, w \models B$.
- $\mathcal{M}, w \models \neg A$ iff for all u, $\mathcal{M}, u \models A$ implies $w \perp u$.
- $\mathcal{M}, w \models A \to B$ iff for all u, $\mathcal{M}, u \models A$ implies $\mathcal{M}, w \cdot u \models B$.

From (6.4.8) we have

$$\mathcal{M}, w \models A \vee B \text{ iff for some } Z,\ w \lhd Z \subseteq |A|^{\mathcal{M}} \cup |B|^{\mathcal{M}}. \tag{6.4.9}$$

Thus $\mathcal{M}, w \models A \vee B$ when w has a cover that is included in $|A|^{\mathcal{M}} \cup |B|^{\mathcal{M}}$, which is itself the truth set for the classical disjunction of A and B. So we can say that $A \vee B$ is true at w under the cover semantics precisely when the classical disjunction of A and B is *locally* true at w.[36]

A formula A is *true in the model* $\mathcal{M}$, symbolised $\mathcal{M} \models A$, if A is true at the identity element ε, i.e. if $\mathcal{M}, \varepsilon \models A$. Since $|A|^{\mathcal{M}}$ is an up-set, this is equivalent to requiring that $\uparrow\varepsilon \subseteq |A|^{\mathcal{M}}$. We say that A is *valid in* $\mathcal{S}$, symbolised $\mathcal{S} \models A$, if A is true in every model based on $\mathcal{S}$.

LEMMA 6.4.3 (Semantic Entailment). *$\mathcal{M} \models A \to B$ iff $|A|^{\mathcal{M}} \subseteq |B|^{\mathcal{M}}$.*

PROOF. Let $\mathcal{M} \models A \to B$, i.e. $\varepsilon \in |A \to B|^{\mathcal{M}} = |A|^{\mathcal{M}} \Rightarrow |B|^{\mathcal{M}}$. Then if $u \in |A|^{\mathcal{M}}$, by definition of $\Rightarrow$ we get $u = \varepsilon \cdot u \in |B|^{\mathcal{M}}$.

Conversely, let $|A|^{\mathcal{M}} \subseteq |B|^{\mathcal{M}}$. Then for any u, if $\mathcal{M}, u \models A$, then $\varepsilon \cdot u = u \in |A|^{\mathcal{M}} \subseteq |B|^{\mathcal{M}}$, so $\mathcal{M}, \varepsilon \cdot u \models B$. This shows that $\mathcal{M}, \varepsilon \models A \to B$. ⊣

THEOREM 6.4.4 (Soundness for R). *Every* R-*theorem is valid in every* R-*model system.*

PROOF. Let $\mathcal{M}$ be model on an R-model system. We show that all the R-axioms from Table 6.1.1 are true in $\mathcal{M}$, and that the rules of Adjunction and Modus Ponens preserve truth in $\mathcal{M}$. The rules are dealt with first.

Adjunction: If $\mathcal{M} \models A$ and $\mathcal{M} \models B$, then A and B are both true at ε in $\mathcal{M}$, so $A \wedge B$ is true at ε by the semantics of $\wedge$, hence $\mathcal{M} \models A \wedge B$.

Modus Ponens: Let A and $A \to B$ be true at ε. Then by the semantics of $\to$, B is true at $\varepsilon \cdot \varepsilon$, which is just ε.

---

[36] Another kind of non-classical interpretation of disjunction was given for the negation-free fragment of R by Humberstone [1988]. This uses the Urquhart operational semantics of (6.1.3) for $\to$, as we have done here. For disjunction it uses a second operation $+$, with $\mathcal{M}, w \models A \vee B$ iff for some u, v with $w = u + v$, $\mathcal{M}, u \models A$ and $\mathcal{M}, v \models B$.

For the cases of the R-axioms, we make use of the result of the Lemma above, namely that to show an implicative formula $A \to B$ is true in $\mathcal{M}$, it is enough to show that $|A|^{\mathcal{M}} \subseteq |B|^{\mathcal{M}}$, i.e. that $\mathcal{M}, w \models A$ implies $\mathcal{M}, w \models B$ for any $w \in W$. In the proofs we may drop the reference to $\mathcal{M}$ in the truth relation, and just write $w \models A$ instead of $\mathcal{M}, w \models A$.

Self-implication: Since $|A|^{\mathcal{M}} \subseteq |A|^{\mathcal{M}}$, it is immediate that $\mathcal{M} \models A \to A$.

Suffixing: Let $w \models A \to B$. Then we need to show that

$$w \models (B \to C) \to (A \to C).$$

So we have to show that if $u \models B \to C$, then $w \cdot u \models A \to C$. But assuming $u \models B \to C$, then if $v \models A$, we have $w \cdot v \models B$ as $w \models A \to B$, hence $u \cdot (w \cdot v) \models C$ as $u \models B \to C$, so $(w \cdot u) \cdot v \models C$ as fusion is associative and commutative. This proves that $w \cdot u \models A \to C$, as required.

Assertion: Let $w \models A$. Then we need to show $w \models (A \to B) \to B$. But if $u \models A \to B$, then $u \cdot w \models B$ as $w \models A$, hence $w \cdot u \models B$ as fusion is commutative. This shows $w \models (A \to B) \to B$ as required.

Contraction: Let $w \models A \to (A \to B)$. Then we need $w \models A \to B$. But if $u \models A$, then $w \cdot u \models A \to B$, hence $(w \cdot u) \cdot u \models B$. Now

$$(w \cdot u) \cdot u = w \cdot (u \cdot u) \leq w \cdot u$$

as fusion is associative, square-decreasing and monotone under $\leq$. Hence $w \cdot u \models B$ as $|B|^{\mathcal{M}}$ is an up-set. So this show that $u \models A$ implies $w \cdot u \models B$, giving $w \models A \to B$ as required.

Conjunction Elimination: $A \wedge B \to A$ is true in $\mathcal{M}$ as $|A \wedge B|^{\mathcal{M}} \subseteq |A|^{\mathcal{M}}$. Likewise for $A \wedge B \to B$.

Conjunction Introduction: Let $w \models (A \to B) \wedge (A \to C)$. Then if $u \models A$, we have $w \cdot u \models B$ as $w \models A \to B$, and $w \cdot u \models C$ as $w \models A \to C$, hence $w \cdot u \models B \wedge C$. This shows that $w \models A \to B \wedge C$.

Disjunction Introduction: $|A|^{\mathcal{M}} \cup |B|^{\mathcal{M}}$ is an up-set, and by Lemma 6.3.1 the operator j is inflationary on up-sets, so

$$|A|^{\mathcal{M}} \subseteq |A|^{\mathcal{M}} \cup |B|^{\mathcal{M}} \subseteq j(|A|^{\mathcal{M}} \cup |B|^{\mathcal{M}}) = |A \vee B|^{\mathcal{M}}$$

(see (6.4.8)). Hence $\mathcal{M} \models A \to A \vee B$. Likewise $\mathcal{M} \models B \to A \vee B$.

Disjunction Elimination: Let $w \models (A \to C) \wedge (B \to C)$. We require $w \models (A \vee B) \to C$. So let $u \models A \vee B$. Then by (6.4.9), there exists Z with $u \lhd Z \subseteq |A|^{\mathcal{M}} \cup |B|^{\mathcal{M}}$. Since fusion is commutative and preserves covering, we then have $w \cdot u = u \cdot w \lhd Z \cdot w = w \cdot Z$. Moreover, for any $v \in Z$, either $v \models A$, so $w \cdot v \models C$ as $w \models A \to C$; or $v \models B$, so again $w \cdot v \models C$ as $w \models B \to C$. This shows that $w \cdot u \lhd w \cdot Z \subseteq |C|^{\mathcal{M}}$, which implies $w \cdot u \in |C|^{\mathcal{M}}$ as $|C|^{\mathcal{M}}$ is localised.

Altogether, we showed that $u \models A \vee B$ implies $w \cdot u \models C$, giving $w \models (A \vee B) \to C$ as required.

$\wedge\vee$-*Distribution*: Let $w \models A \wedge (B \vee C)$. We need $w \models (A \wedge B) \vee (A \wedge C)$. Since $w \models B \vee C$, there exists Z with $w \lhd Z \subseteq |B|^{\mathcal{M}} \cup |C|^{\mathcal{M}}$. We have $Z \subseteq {\uparrow}w$ as covers are refining, and ${\uparrow}w \subseteq |A|^{\mathcal{M}}$ as $w \models A$. This implies

$$w \lhd Z \subseteq |A|^{\mathcal{M}} \cap (|B|^{\mathcal{M}} \cup |C|^{\mathcal{M}}) = (|A|^{\mathcal{M}} \cap |B|^{\mathcal{M}}) \cup (|A|^{\mathcal{M}} \cap |C|^{\mathcal{M}}).$$

Hence

$$\begin{aligned} w &\in j\big((|A|^{\mathcal{M}} \cap |B|^{\mathcal{M}}) \cup (|A|^{\mathcal{M}} \cap |C|^{\mathcal{M}})\big) \\ &= j\big(\,|A \wedge B|^{\mathcal{M}} \cup |A \wedge C|^{\mathcal{M}}\,\big) && \text{by the semantics of } \wedge, \\ &= |(A \wedge B) \vee (A \wedge C)|^{\mathcal{M}} && \text{by (6.4.8),} \end{aligned}$$

giving $w \models (A \wedge B) \vee (A \wedge C)$ as required.

Contraposition: Let $w \models A \to \neg B$. We need $w \models B \to \neg A$. So let $u \models B$. Then if $v \models A$, we have $w \cdot v \models \neg B$ as $w \models A \to \neg B$, hence $w \cdot v \perp u$ by the semantics of $\neg$, so $w \cdot u \perp v$ by the contraposition postulate of relevant cover systems. This shows that $w \cdot u \models \neg A$.

Altogether we showed that $u \models B$ implies $w \cdot u \models \neg A$, which means that $w \models B \to \neg A$ as required.

Double-Negation Elimination: $|\neg\neg A|^{\mathcal{M}} = (|A|^{\mathcal{M}})^{\perp\perp} = |A|^{\mathcal{M}}$, since $|A|^{\mathcal{M}} \in$ *Prop* and all members of *Prop* are orthoclosed. Hence $\mathcal{M} \models \neg\neg A \to A$.

t-Axiom: Let $w \models A$. Then if $u \models t$, we have $\varepsilon \leq u$, hence $w = w \cdot \varepsilon \leq w \cdot u$, and so $w \cdot u \models A$ as $|A|^{\mathcal{M}}$ is an up-set. This shows that $w \models t \to A$.

Conversely, if $w \models t \to A$, then since $\varepsilon \models t$, we have $w = w \cdot \varepsilon \models A$. Altogether, this proves that $|A|^{\mathcal{M}} = |t \to A|^{\mathcal{M}}$, implying that both $A \to (t \to A)$ and $(t \to A) \to A$ are true in $\mathcal{M}$, hence so is $A \leftrightarrow (t \to A)$. ⊣

6.5. Cover System Completeness for R

The logic R is complete for validity in all R-model systems. This will now be proven by construction of a kind of canonical model, in which the members of W are certain theories, i.e. deductively closed sets of formulas. But since our models on cover-systems use a non-classical interpretation of negation and disjunction, these theories are not required to be prime, as in the Routley-Meyer style canonical frame described at the end of Section 6.1, let alone negation complete as in the constructions for Boolean modal logic of earlier chapters.

In fact we use the simplest kind of theory, of the form $\{B : \vdash_{\mathrm{R}} A \to B\}$ for some formula A. This is called the *principal theory generated by A*. One might ask if it could be replaced by just A itself. However, there may be distinct formulas A, A' that generate the same theory. This happens when (and only when) $\vdash_{\mathrm{R}} A \leftrightarrow A'$.

Recall that the deducibility relation $\Gamma \vdash_{\mathrm{R}} B$ holds when there are finitely many formulas $A_0, \ldots, A_n \in \Gamma$ such that $\vdash_{\mathrm{R}} A_0 \wedge \cdots \wedge A_n \to B$. Since we are dealing with a single logic, we will now drop the subscript from the deducibility relation $\vdash_R$, and take it as understood we are working in R. We put $A \vdash B$ when $\{A\} \vdash B$, i.e. when $\vdash A \to B$; and $A \dashv\vdash B$ when both $A \vdash B$ and $B \vdash A$, or equivalently $\vdash A \leftrightarrow B$. Here are some basic properties of these relations on formulas which follow from the definition of R in Table 6.1.1 and the R-theorems of Table 6.1.2.

LEMMA 6.5.1. • *$\vdash$ is a preorder, i.e.*
- *$A \vdash A$.*
- *If $A \vdash B$ and $B \vdash C$, then $A \vdash C$.*

- *$A \vdash B \wedge C$ iff $A \vdash B$ and $A \vdash C$.*
- *If $A \vdash C$ and $B \vdash C$, then $A \vee B \vdash C$.*
- *If $A \vdash \neg B$, then $B \vdash \neg A$.*
- *$\neg\neg A \dashv\vdash A$.*
- *If $A \vdash B$, then $\neg B \vdash \neg A$.*
- *If $A \vdash B$, then $A \circ C \vdash B \circ C$ and $C \circ A \vdash C \circ B$.*
- *$A \circ B \vdash C$ iff $A \vdash B \to C$.*
- *$A \circ (B \circ C) \dashv\vdash (A \circ B) \circ C$.*
- *$A \circ B \dashv\vdash B \circ A$.*
- *$(t \circ A) \dashv\vdash A$.*
- *$A \vdash A \circ A$.* ⊣

We will use many of these these properties below, often without explicit reference. For each formula A, let

$$A^{\vdash} = \{B : A \vdash B\}.$$

Then $A^{\vdash}$ is a theory: it is closed under conjunction, in that if $B, C \in A^{\vdash}$, then $B \wedge C \in A^{\vdash}$; and is closed under deducibility, in the sense that if $B \in A^{\vdash}$ and $B \vdash C$, then $C \in A^{\vdash}$. Together these facts imply that if $A^{\vdash} \vdash C$, then $C \in A^{\vdash}$. Since $\vdash$ is a preorder, it follows that

$$B^{\vdash} \subseteq A^{\vdash} \quad \text{iff} \quad A \vdash B.$$

Hence

$$A^{\vdash} = B^{\vdash} \quad \text{iff} \quad A \dashv\vdash B.$$

Now define a canonical system

$$\mathcal{S}_{\mathrm{R}} = (W_{\mathrm{R}}, \leq, \rhd, \cdot, \varepsilon, \bot, Prop_{\mathrm{R}}),$$

where

- $W_{\mathrm{R}} = \{A^{\vdash} : A \text{ is a formula}\}$.
- $w \leq u$ iff $w \subseteq u$, for all $w, u \in W_{\mathrm{R}}$.
- $Z \rhd w$ iff $\bigcap Z = w$, for all $Z \subseteq W_{\mathrm{R}}$ and $w \in W_{\mathrm{R}}$.
- $A^{\vdash} \cdot B^{\vdash} = (A \circ B)^{\vdash}$.
- $\varepsilon = t^{\vdash}$.

- $A^{\vdash} \perp B^{\vdash}$ iff $A \vdash \neg B$.
- $Prop_{\mathrm{R}} = \{|A|_{\mathrm{R}} : A \text{ is a formula}\}$, where

$$|A|_{\mathrm{R}} = \{w \in W_{\mathrm{R}} : A \in w\} = \{B^{\vdash} : B \vdash A\}.$$

We need to check that some of these items are well-defined. But if $A_1^{\vdash} = A_2^{\vdash}$ and $B_1^{\vdash} = B_2^{\vdash}$, then $\vdash_{\mathrm{R}} A_1 \circ B_1 \leftrightarrow A_2 \circ B_2$ and so $(A_1 \circ B_1)^{\vdash} = (A_1 \circ B_2)^{\vdash}$; while $A_1 \vdash \neg B_1$ iff $A_2 \vdash \neg B_2$. Hence the fusion operation $\cdot$ and the orthogonality relation $\perp$ are well-defined.

If we view a member w of W_{R} as a state whose information content is determined by the formulas belonging to w, then the condition $\bigcap Z = w$ captures the interpretation of the cover relation $Z \rhd w$ suggested just after Example 6.3.2, namely that the information content of w comprises that information which is common to all the states in Z.

Theorem 6.5.2. *$\mathcal{S}_{\mathrm{R}}$ is based on a relevant cover system.*

Proof. Since $A \vdash \neg B$ implies $B \vdash \neg A$, orthogonality is symmetric. Since $\bigcap Z = w$ implies that every $u \in Z$ has $w \subseteq u$, we have that *covers are refining*. Since $\bigcap\{w\} = w$, i.e. $\{w\} \rhd w$, we have that *covers exist*. Thus $\mathcal{S}_{\mathrm{R}}$ is based on a cover system.

By properties listed in Lemma 6.5.1, the fusion operation $\cdot$ is monotonic under refinement, associative, commutative, and has identity $t^{\vdash}$. For instance, the last of these holds because from $(t \circ A) \dashv\vdash A$ we obtain $t^{\vdash} \cdot A^{\vdash} = A^{\vdash}$. Also, from $A \vdash A \circ A$ we get $A^{\vdash} \cdot A^{\vdash} \subseteq A^{\vdash}$, so $\cdot$ is square-decreasing.

To show that *fusion preserves covering*, suppose $w \lhd Z$, i.e., $\bigcap Z = w$, where $w = A^{\vdash}$. We want to show that $w \cdot u \lhd Z \cdot u$ for any u. Let $u = B^{\vdash}$. Then for any $v \in Z$, with say $v = C^{\vdash}$, we have $w \subseteq v$, and therefore $C \vdash A$. Hence $C \circ B \vdash A \circ B$, so $(A \circ B)^{\vdash} \subseteq (C \circ B)^{\vdash}$, giving $w \cdot u \subseteq v \cdot u$. This shows that $Z \cdot u \subseteq \uparrow(w \cdot u)$, implying $w \cdot u \subseteq \bigcap(Z \cdot u)$. Now we prove the converse of this last inclusion. Take $D \in \bigcap(Z \cdot u)$. Then for each $C^{\vdash} \in Z$, we have $D \in C^{\vdash} \cdot B^{\vdash} = (C \circ B)^{\vdash}$, so $C \circ B \vdash D$, hence $C \vdash B \to D$, and thus $B \to D \in C^{\vdash}$. This shows that $B \to D \in \bigcap Z = A^{\vdash}$, hence $A \vdash B \to D$ and so $A \circ B \vdash D$, giving $D \in (A \circ B)^{\vdash} = w \cdot u$. This proves that $\bigcap(Z \cdot u) = w \cdot u$, i.e. that $w \cdot u \lhd Z \cdot u$ as required.

To show that *refinement preserves orthogonality*, suppose $A_1^{\vdash} \subseteq A_2^{\vdash}$, $B_1^{\vdash} \subseteq B_2^{\vdash}$ and $A_1^{\vdash} \perp B_1^{\vdash}$. Then $A_2 \vdash A_1$, $B_2 \vdash B_1$ and $A_1 \vdash \neg B_1$. From $B_2 \vdash B_1$ we infer $\neg B_1 \vdash \neg B_2$. Transitivity of $\vdash$ then yields $A_2 \vdash \neg B_2$, and hence $A_2^{\vdash} \perp B_2^{\vdash}$ as required.

To show that *orthogonality to ε is local*, suppose $w \lhd Z \perp \varepsilon$. We want $w \perp \varepsilon$. Now if $C^{\vdash} \in Z$, then $C^{\vdash} \perp \varepsilon = t^{\vdash}$, so $C \vdash \neg t$, hence $\neg t \in C^{\vdash}$. This shows that $\neg t \in \bigcap Z = w$. Thus if $w = A^{\vdash}$, then $A \vdash \neg t$, showing $A^{\vdash} \perp t^{\vdash}$ as required.

Finally we prove the *contraposition* postulate. Suppose $A^{\vdash} \cdot B^{\vdash} \perp C^{\vdash}$, i.e. $(A \circ B)^{\vdash} \perp C^{\vdash}$. Then $A \circ B \vdash \neg C$, hence $A \vdash B \to \neg C$, implying

$A \vdash C \to \neg B$ by the Contraposition axiom, hence $A \circ C \vdash \neg B$, and therefore $A^{\vdash} \cdot C^{\vdash} = (A \circ C)^{\vdash} \perp B^{\vdash}$ as required. ⊣

To show that $\mathcal{S}_{\mathrm{R}}$ is an R-model system, we now need to examine the structure of its admissible propositions.

LEMMA 6.5.3. (1) $\uparrow(A^{\vdash}) = |A|_{\mathrm{R}}$.
(2) $|A|_{\mathrm{R}} \cap |B|_{\mathrm{R}} = |A \wedge B|_{\mathrm{R}}$.
(3) $|A|_{\mathrm{R}}{}^{\perp} = |\neg A|_{\mathrm{R}}$.
(4) $|A|_{\mathrm{R}} \Rightarrow |B|_{\mathrm{R}} = |A \to B|_{\mathrm{R}}$.
(5) $|A|_{\mathrm{R}} + |B|_{\mathrm{R}} = |A \vee B|_{\mathrm{R}}$.

PROOF. (1) $B^{\vdash} \in \uparrow(A^{\vdash})$ iff $A^{\vdash} \subseteq B^{\vdash}$ iff $B \vdash A$ iff $B^{\vdash} \in |A|_{\mathrm{R}}$.
(2) $C^{\vdash} \in |A|_{\mathrm{R}} \cap |B|_{\mathrm{R}}$ iff $C \vdash A$ and $C \vdash B$ iff $C \vdash A \wedge B$ iff $C^{\vdash} \in |A \wedge B|_{\mathrm{R}}$.
(3) Let $B^{\vdash} \in |A|_{\mathrm{R}}{}^{\perp}$. Then as $A^{\vdash} \in |A|_{\mathrm{R}}$, we have $B^{\vdash} \perp A^{\vdash}$, hence $B \vdash \neg A$ and so $B^{\vdash} \in |\neg A|_{\mathrm{R}}$. Conversely, let $B^{\vdash} \in |\neg A|_{\mathrm{R}}$. Then if $C^{\vdash} \in |A|_{\mathrm{R}}$, we have $C \vdash A$, hence $\neg A \vdash \neg C$. But $B \vdash \neg A$, so then $B \vdash \neg C$, giving $B^{\vdash} \perp C^{\vdash}$. This shows $B^{\vdash} \in |A|_{\mathrm{R}}{}^{\perp}$.
(4) Let $C^{\vdash} \in |A|_{\mathrm{R}} \Rightarrow |B|_{\mathrm{R}}$. Since $A^{\vdash} \in |A|_{\mathrm{R}}$, it follows that $C^{\vdash} \cdot A^{\vdash} \in |B|_{\mathrm{R}}$. Thus $(C \circ A)^{\vdash} \in |B|_{\mathrm{R}}$, hence $C \circ A \vdash B$, so $C \vdash A \to B$, giving $C^{\vdash} \in |A \to B|_{\mathrm{R}}$.

Conversely, let $C^{\vdash} \in |A \to B|_{\mathrm{R}}$, so $C \vdash A \to B$ and thus $C \circ A \vdash B$. To show that $C^{\vdash} \in |A|_{\mathrm{R}} \Rightarrow |B|_{\mathrm{R}}$, we have to show that if $D^{\vdash} \in |A|_{\mathrm{R}}$, then $C^{\vdash} \cdot D^{\vdash} \in |B|_{\mathrm{R}}$, or equivalently that if $D \vdash A$, then $C \circ D \vdash B$. But if $D \vdash A$, then $C \circ D \vdash C \circ A$, hence as $C \circ A \vdash B$ we get $C \circ D \vdash B$ as required.
(5) $|A|_{\mathrm{R}} + |B|_{\mathrm{R}} = (|A|_{\mathrm{R}} \cup |B|_{\mathrm{R}})^{\perp\perp}$, which is $(|A|_{\mathrm{R}}{}^{\perp} \cap |B|_{\mathrm{R}}{}^{\perp})^{\perp}$ by (6.4.3). But by parts (2) and (3), this is $|\neg(\neg A \wedge \neg B)|_{\mathrm{R}}$, i.e. $|A \vee B|_{\mathrm{R}}$. ⊣

THEOREM 6.5.4. *$\mathcal{S}_{\mathrm{R}}$ is an* R*-model system.*

PROOF. Since $\mathcal{S}_{\mathrm{R}}$ was already shown to be a relevant cover system in Theorem 6.5.2, we now need to show that $Prop_{\mathrm{R}}$ has the properties required by the definition of an R-model system.

It is immediate that each member $|A|_{\mathrm{R}}$ of $Prop_{\mathrm{R}}$ is an up-set for the refinement relation of $\mathcal{S}_{\mathrm{R}}$, which is just set inclusion $\subseteq$. Moreover, $|A|_{\mathrm{R}}$ is orthoclosed, since by part (3) of the Lemma just proved,

$$|A|_{\mathrm{R}}{}^{\perp\perp} = |\neg\neg A|_{\mathrm{R}} = |A|_{\mathrm{R}}$$

as $\neg\neg A \dashv\vdash A$. Also, by part (1) of the Lemma,

$$\uparrow\varepsilon = \uparrow(t^{\vdash}) = |t|_{\mathrm{R}} \in Prop_{\mathrm{R}}. \tag{6.5.1}$$

Parts (2)–(4) of the Lemma imply that $Prop_{\mathrm{R}}$ is closed under the operations $\cap$, $(-)^{\perp}$, and $\Rightarrow$. Thus it remains to show that, in general,

$$|A|_{\mathrm{R}} + |B|_{\mathrm{R}} = j(|A|_{\mathrm{R}} \cup |B|_{\mathrm{R}}).$$

Now $|A|_R + |B|_R$ is $(|A|_R \cup |B|_R)^{\perp\perp}$, which includes $j(|A|_R \cup |B|_R)$ by Lemma 6.4.1(6). So our burden is to show that $|A|_R + |B|_R \subseteq j(|A|_R \cup |B|_R)$, or equivalently by Lemma 6.5.3(5) that

$$|A \vee B|_R \subseteq j(|A|_R \cup |B|_R). \tag{6.5.2}$$

This is precisely the point in our completeness theorem for R at which we need the $\wedge\vee$-Distribution axiom. Note how (6.5.2) corresponds to the model condition (6.4.8), which was used to prove validity of $\wedge\vee$-Distribution in R-model systems.

Suppose $w \in |A \vee B|_R$, and that $w = C^{\vdash}$. Then $\uparrow w = |C|_R$ by Lemma 6.5.3(1). Let

$$Z = (|C|_R \cap |A|_R) \cup (|C|_R \cap |B|_R) = |C \wedge A|_R \cup |C \wedge B|_R.$$

Then $Z \subseteq |A|_R \cup |B|_R$, so if we can show that $w \lhd Z$, this will make $w \in j(|A|_R \cup |B|_R)$, proving (6.5.2).

Thus we want to prove that $w = \bigcap Z$. Now $Z \subseteq |C|_R = \uparrow w$, so each $u \in Z$ has $w \subseteq u$, hence $w \subseteq \bigcap Z$. For the converse inclusion, let $D \in \bigcap Z$. Then as $(C \wedge A)^{\vdash} \in |C \wedge A|_R \subseteq Z$ we have $D \in (C \wedge A)^{\vdash}$ and so $C \wedge A \vdash D$. Likewise $C \wedge B \vdash D$. Hence $(C \wedge A) \vee (C \wedge B) \vdash D$, from Disjunction Elimination. But $C \wedge (A \vee B) \vdash (C \wedge A) \vee (C \wedge B)$ *by* $\wedge\vee$*-Distribution*, so this leads to $C \wedge (A \vee B) \vdash D$.

Now $A \vee B \in w = C^{\vdash}$, so $C \vdash A \vee B$. Hence $C \vdash C \wedge (A \vee B)$, and therefore $C \vdash D$, giving $D \in C^{\vdash} = w$.

This completes the proof that $w = \bigcap Z$, i.e. $w \lhd Z$, which completes the proof of (6.5.2) as explained. ⊣

Now define a model $\mathcal{M}_R$ on $\mathcal{S}_R$ by putting $|p|^{\mathcal{M}_R} = |p|_R$ for all propositional variables p. The Truth Lemma for this model is

THEOREM 6.5.5 (Truth Is Membership). *For any formula A in the language of* R, $|A|^{\mathcal{M}_R} = |A|_R$. *Thus for all* $w \in W_R$ *we have*

$$\mathcal{M}_R, w \models A \quad \textit{iff} \quad A \in w.$$

PROOF. When A is a variable this holds by definition. When A is the constant t, we have $|t|^{\mathcal{M}_R} = \uparrow\varepsilon = \uparrow(t^{\vdash}) = |t|_R$, as noted in (6.5.1).

The inductive cases of the connectives $\wedge$, $\neg$ and $\rightarrow$ follow by using parts (2)–(4) of Lemma 6.5.3 and the definition of $|-|^{\mathcal{M}_R}$ for these cases. ⊣

THEOREM 6.5.6 (Model System Characterisation of R). *For any formula A, the following are equivalent.*

(1) *A is a theorem of* R.
(2) *A is valid in all* R*-model systems.*
(3) *A is valid in the canonical model system* $\mathcal{S}_R$.
(4) *A is true in the model* $\mathcal{M}_R$.

PROOF. (1) implies (2) by the Soundness Theorem 6.4.4. (2) implies (3) as $\mathcal{S}_R$ is an R-model system (Theorem 6.5.4). (3) implies (4) by definition of validity in a model system.

(4) implies (1): if $\mathcal{M}_R \models A$, then $\mathcal{M}_R, \varepsilon \models A$, hence by the previous Theorem, $A \in \varepsilon = t^{\vdash}$. Thus $t \vdash A$, i.e. $\vdash_R t \to A$. But $\vdash_R (t \to A) \to A$ from the t-Axiom, so Modus Ponens then gives $\vdash_R A$. ⊣

It is noteworthy that although our definition of cover system only requires the refinement relation $\leq$ to be a preorder, in the canonical system $\mathcal{S}_R$ this relation is the *partial* order $\subseteq$. Hence R is also characterised by validity in the partially ordered R-model systems.

6.6. Modelling Fusion

In the "operational" semantics for a fusion-like connective $\circ$ in some substructural logics (Ono and Komori [1985], Došen [1989], Ono [1993]), the criterion for $\mathcal{M}, w \models A \circ B$ is

$$\text{there exist } u, v \text{ such that } u \cdot v \leq w,\ \mathcal{M}, u \models A \text{ and } \mathcal{M}, v \models B. \tag{6.6.1}$$

In this condition, $u \cdot v$ belongs to $|A|^{\mathcal{M}} \cdot |B|^{\mathcal{M}}$, so $w \in {\uparrow}(|A|^{\mathcal{M}} \cdot |B|^{\mathcal{M}})$. Hence this interpretation amounts to putting

$$|A \circ B|^{\mathcal{M}} = {\uparrow}(|A|^{\mathcal{M}} \cdot |B|^{\mathcal{M}}). \tag{6.6.2}$$

To characterise $|A \circ B|^{\mathcal{M}}$ in the cover semantics, define a binary operation $\otimes$ on subsets of a relevant cover system by

$$X \otimes Y = (X \Rightarrow Y^{\perp})^{\perp}.$$

Since $A \circ B$ is defined to be $\neg(A \to \neg B)$, this makes $|A \circ B|^{\mathcal{M}} = |A|^{\mathcal{M}} \otimes |B|^{\mathcal{M}}$ in any model on such a system.

LEMMA 6.6.1. *In any relevant model system, $X \Rightarrow Y^{\perp} = (X \cdot Y)^{\perp}$ and hence $X \otimes Y = (X \cdot Y)^{\perp\perp}$.*

PROOF. We use the fact that $v \cdot w \perp u$ iff $v \perp w \cdot u$, which follows by commutativity of $\cdot$, contraposition, and symmetry of $\perp$.

Now $v \in X \Rightarrow Y^{\perp}$ iff $v \cdot w \in Y^{\perp}$ for all $w \in X$, iff $v \cdot w \perp u$ for all $w \in X$ and $u \in Y$, iff $v \perp w \cdot u$ for all $w \cdot u \in X \cdot Y$, iff $v \in (X \cdot Y)^{\perp}$. That proves the first equation.

The second equation follows directly from the first by applying $(-)^{\perp}$. ⊣

From this Lemma we get that in a model on a relevant cover system, we always have

$$|A \circ B|^{\mathcal{M}} = (|A|^{\mathcal{M}} \cdot |B|^{\mathcal{M}})^{\perp\perp}.$$

But this may not give us (6.6.2). In general we have $X \subseteq {\uparrow}X \subseteq X^{\perp\perp}$, because $X^{\perp\perp}$ is an up-set (Lemma 6.4.1(2)) including X. Hence $({\uparrow}X)^{\perp\perp} = X^{\perp\perp}$ as

$X^{\perp\perp}$ is the least orthoclosed superset of X. But we will not have $\uparrow X = X^{\perp\perp}$ unless $\uparrow X$ is orthoclosed, and, for our case of interest, it is not evident that $X = \uparrow(|A|^{\mathcal{M}} \cdot |B|^{\mathcal{M}})$ must be orthoclosed.

However we do have the desired condition (6.6.2) in the canonical model $\mathcal{M}_{\mathrm{R}}$, as follows from

LEMMA 6.6.2. *For any formulas A and B,*

$$|A \circ B|_{\mathrm{R}} = \uparrow(|A|_{\mathrm{R}} \cdot |B|_{\mathrm{R}}).$$

PROOF. Let $C^{\vdash} \in |A \circ B|_{\mathrm{R}}$. Then $C \vdash A \circ B$. Define $u = A^{\vdash}$ and $v = B^{\vdash}$. Hence $A \in u$, so $u \in |A|_{\mathrm{R}}$, and likewise $v \in |B|_{\mathrm{R}}$. Thus $u \cdot v \in |A|_{\mathrm{R}} \cdot |B|_{\mathrm{R}}$. But $u \cdot v = (A \circ B)^{\vdash} \subseteq C^{\vdash}$ as $C \vdash A \circ B$, hence $C^{\vdash} \in \uparrow(|A|_{\mathrm{R}} \cdot |B|_{\mathrm{R}})$.

Conversely, let $C^{\vdash} \in \uparrow(|A|_{\mathrm{R}} \cdot |B|_{\mathrm{R}})$. Then there exist $D^{\vdash} \in |A|_{\mathrm{R}}$ and $E^{\vdash} \in |B|_{\mathrm{R}}$ with $D^{\vdash} \cdot E^{\vdash} \subseteq C^{\vdash}$. Hence $(D \circ E)^{\vdash} \subseteq C^{\vdash}$, so $C \vdash D \circ E$. But $D \vdash A$ and $E \vdash B$, so $D \circ E \vdash A \circ B$. Hence $C \vdash A \circ B$, implying $C^{\vdash} \in |A \circ B|_{\mathrm{R}}$. ⊣

Applying the Truth Lemma of $\mathcal{M}_{\mathrm{R}}$ (Theorem 6.5.5) to this result gives

$$|A \circ B|^{\mathcal{M}_{\mathrm{R}}} = \uparrow(|A|^{\mathcal{M}_{\mathrm{R}}} \cdot |B|^{\mathcal{M}_{\mathrm{R}}}).$$

Since $\mathcal{M}_{\mathrm{R}}$ characterises the logic R, we conclude that R is also characterised by validity in all models satisfying (6.6.2) on R-model structures.

6.7. Cover System Characterisation of RQ

The semantic analysis of R by cover relations $\rhd$ will now be lifted to a characterisation of the quantified logic RQ by structures based on cover systems. We define an RQ-*model system* to be a structure

$$\mathcal{S} = (W, \leq, \rhd, \cdot, \varepsilon, \perp, U, \mathit{Prop}, \mathit{PropFun}),$$

such that:

- $(W, \leq, \rhd, \cdot, \varepsilon, \perp, \mathit{Prop})$ is an R-model system.
- U is a set, the *universe of* $\mathcal{S}$.
- *PropFun* is a set of *admissible propositional functions* from U^{ω} to *Prop*, containing the function α_{ε} with constant value $\uparrow\varepsilon$, and closed under the operations $\cap$, $(-)^{\perp}$ and $\Rightarrow$ that are defined pointwise by lifting the corresponding operations on *Prop*. For $\alpha, \beta : U^{\omega} \to \wp W$, the functions $\alpha \cap \beta$, $\alpha^{\perp}$, $\alpha \Rightarrow \beta : U^{\omega} \to \wp W$ are defined by

$$(\alpha \cap \beta)f = \alpha f \cap \beta f$$
$$(\alpha^{\perp})f = (\alpha f)^{\perp}$$
$$(\alpha \Rightarrow \beta)f = \alpha f \Rightarrow \beta f.$$

- *PropFun* is closed under operations $\forall_x$ for all individual variables x, where, for each $\alpha : U^\omega \to \wp W$, the function $\forall_x\alpha : U^\omega \to \wp W$ is defined by

$$(\forall_x\alpha)f = \bigcap_{a\in U} \alpha(f[a/x]).$$

- *Cover-Join Property*: For each $\alpha \in PropFun$,

$$\sum_{a\in U} \alpha(f[a/x]) = j\Big(\bigcup_{a\in U} \alpha(f[a/x]\Big).$$

The Cover-Join Property uses the orthojoin operation $\sum_I X_i = (\bigcup_I X_i)^{\perp\perp}$ that was introduced in (6.4.5). This Property will ensure the validity of the $\wedge\exists$-Distribution axiom of RQ. It is an infinitary generalisation of the property $X + Y = j(X \cup Y)$ of members of *Prop* in an R-model system, which we showed to characterise the $\wedge\vee$-Distribution axiom of R. A full generalisation might ask that $\sum \mathcal{Z} = j(\bigcup \mathcal{Z})$ for any set $\mathcal{Z}$ of (admissible) propositions, but the Cover-Join Property requires this only when $\mathcal{Z} = \{\alpha(f[a/x]) : a \in U\}$ for some $\alpha \in PropFun$ and $f \in U^\omega$.

Since the existential quantifier $\exists x$ is defined as $\neg\forall x\neg$, it is natural to introduce an operation $\exists_x$ on propositional functions by defining $\exists_x\alpha = (\forall_x(\alpha^\perp))^\perp$. Then for any $f \in U^\omega$, with the help of (6.4.7) we obtain

$$(\exists_x\alpha)f = \Big(\bigcap_{a\in U} \alpha(f[a/x])^\perp\Big)^\perp = \sum_{a\in U} \alpha(f[a/x]). \tag{6.7.1}$$

Fix a signature $\mathcal{L}$ consisting of individual constants and predicate symbols. A *premodel* $\mathcal{M}$ for $\mathcal{L}$ on the system $\mathcal{S}$ assigns to each constant $\mathsf{c} \in \mathcal{L}$ an element $|\mathsf{c}|^\mathcal{M} \in U$, and to each n-ary predicate symbol $P \in \mathcal{L}$ a function $|P|^\mathcal{M} : U^\omega \to \wp W$. Then $\mathcal{M}$ assigns a function $|\varphi|^\mathcal{M} : U^\omega \to \wp W$ to each $\mathcal{L}$-formula φ, with the case of atomic φ being as usual (6.2.1), and the other cases given by:

- $|t|^\mathcal{M} = \alpha_\varepsilon$.
- $|\varphi \wedge \psi|^\mathcal{M} = |\varphi|^\mathcal{M} \cap |\psi|^\mathcal{M}$.
- $|\neg\varphi|^\mathcal{M} = (|\varphi|^\mathcal{M})^\perp$.
- $|\varphi \to \psi|^\mathcal{M} = |\varphi|^\mathcal{M} \Rightarrow |\psi|^\mathcal{M}$.
- $|\forall x\varphi|^\mathcal{M} = \forall_x|\varphi|^\mathcal{M}$.

The principal differences from the semantics for RQ sketched in Section 6.2 are that t is now interpreted as $\uparrow\varepsilon$, the * operator is replaced by the operator $(-)^\perp$ induced by the orthogonality relation, $\Rightarrow$ is now determined by the fusion operation as in (6.4.1), and $\forall_x$ is now defined independently of *Prop* by set intersection $\bigcap$, rather than $\bigsqcap$.

Writing $\mathcal{M}, w, f \models A$ to mean that $w \in |A|^\mathcal{M} f$, we get a truth relation satisfying the following conditions:

- $\mathcal{M}, w, f \models P\tau_1 \cdots \tau_n$ iff $w \in |P|^\mathcal{M}(|\tau_1|^\mathcal{M} f, \ldots, |\tau_n|^\mathcal{M} f)$.
- $\mathcal{M}, w \models t$ iff $\varepsilon \leq w$.

- $\mathcal{M}, w, f \models \varphi \wedge \psi$ iff $\mathcal{M}, w, f \models \varphi$ and $\mathcal{M}, w, f \models \psi$.
- $\mathcal{M}, w, f \models \neg\varphi$ iff for all u, $\mathcal{M}, u, f \models \varphi$ implies $w \perp u$.
- $\mathcal{M}, w, f \models \varphi \to \psi$ iff for all u, $\mathcal{M}, u, f \models \varphi$ implies $\mathcal{M}, w \cdot u, f \models \psi$.
- $\mathcal{M}, w, f \models \forall x \varphi$ iff for all $a \in U$, $\mathcal{M}, w, f[a/x] \models \varphi$.

Thus the universal quantifier $\forall x$ gets the simple classical semantics of a one-universal-domain Kripkean model, as Routley and Meyer proposed, but the model is based on a relevant cover system rather than a Routley-Meyer R-frame.

For disjunction, the truth sets have

$$|\varphi \vee \psi|^{\mathcal{M}} f = |\varphi|^{\mathcal{M}} f + |\psi|^{\mathcal{M}} f = j(|\varphi|^{\mathcal{M}} f \cup |\psi|^{\mathcal{M}} f),$$

and so as in (6.4.9),

$$\mathcal{M}, w, f \models \varphi \vee \psi \text{ iff for some } Z,\ w \lhd Z \subseteq |\varphi|^{\mathcal{M}} f \cup |\psi|^{\mathcal{M}} f.$$

For the existential quantifier,

$$|\exists x \varphi|^{\mathcal{M}} = |\neg\forall x \neg\varphi|^{\mathcal{M}} = \left(\forall_x (|\varphi|^{\mathcal{M}})^{\perp}\right)^{\perp} = \exists_x |\varphi|^{\mathcal{M}},$$

so for any $f \in U^{\omega}$, by (6.7.1) we have

$$|\exists x \varphi|^{\mathcal{M}} f = \sum_{a \in U} |\varphi|^{\mathcal{M}} f[a/x]. \tag{6.7.2}$$

$\mathcal{M}$ is called a *model* if $|\varphi|^{\mathcal{M}} \in \mathit{PropFun}$ for all *atomic* formulas φ. The closure conditions on *PropFun* ensure that if $\mathcal{M}$ is a model, then $|\varphi|^{\mathcal{M}} \in \mathit{PropFun}$ for every formula φ, and hence that $|\varphi|^{\mathcal{M}} f \in \mathit{Prop}$ for every φ and every $f \in U^{\omega}$.

LEMMA 6.7.1. *If $\mathcal{M}$ is a model, then*

$$\mathcal{M}, w, f \models \exists x \varphi \text{ iff for some } Z,\ w \lhd Z \subseteq \bigcup_{a \in U} |\varphi|^{\mathcal{M}} f[a/x].$$

PROOF. If $\mathcal{M}$ is a model, then $|\varphi|^{\mathcal{M}} \in \mathit{PropFun}$, so by (6.7.2) and the Cover-Join Property,

$$|\exists x \varphi|^{\mathcal{M}} f = j\Big(\bigcup_{a \in U} |\varphi|^{\mathcal{M}} f[a/x]\Big).$$

The definition of j then gives the result. ⊣

As with our other model theories, truth of a formula depends only on the assignment of values to its *free* variables. The following formulation of this is proven similarly to Lemma 1.6.1, and the proof is left to the reader.

LEMMA 6.7.2 (Free Assignment). *In any premodel $\mathcal{M}$ on an* RQ-*model system, for any formula φ, if assignments $f, g \in U^{\omega}$ agree on all free variables of φ, then $|\varphi|^{\mathcal{M}} f = |\varphi|^{\mathcal{M}} g$.* ⊣

This in turn is used to prove the usual result relating free substitution to assignment-updating:

LEMMA 6.7.3 (Substitution). *Let φ be any formula, and τ a term that is free for x in φ. Then in any premodel $\mathcal{M}$,*

$$\mathcal{M}, w, f \models \varphi(\tau/x) \quad \textit{iff} \quad \mathcal{M}, w, f[|\tau|^{\mathcal{M}} f/x] \models \varphi. \qquad \dashv$$

A formula φ is *valid in* premodel $\mathcal{M}$, written $\mathcal{M} \models \varphi$, if $\mathcal{M}, \varepsilon, f \models \varphi$ for all $f \in U^{\omega}$. This is equivalent to having $\uparrow\varepsilon \subseteq |\varphi|^{\mathcal{M}} f$ for all such f. φ is *valid in* the model system $\mathcal{S}$ if it is valid in all *models* on $\mathcal{S}$.

Our proof of soundness for R in cover system models depended on the Semantic Entailment Lemma 6.4.3. The corresponding result here, with essentially the same proof is:

LEMMA 6.7.4 (Semantic Entailment). *In any premodel $\mathcal{M}$, for any $f \in U^{\omega}$ we have: $\mathcal{M}, \varepsilon, f \models \varphi \to \psi$ iff $|\varphi|^{\mathcal{M}} f \subseteq |\psi|^{\mathcal{M}} f$ iff for all $w \in W$, $\mathcal{M}, w, f \models \varphi$ implies $\mathcal{M}, w, f \models \psi$.* $\dashv$

If we partially order propositional functions pointwise by inclusion, i.e. put

$$|\varphi|^{\mathcal{M}} \subseteq |\psi|^{\mathcal{M}} \quad \text{iff} \quad \text{for all } f \in U^{\omega},\ |\varphi|^{\mathcal{M}} f \subseteq |\psi|^{\mathcal{M}} f$$

as in (4.1.1), then this last Lemma becomes

$$\mathcal{M} \models \varphi \to \psi \quad \text{iff} \quad |\varphi|^{\mathcal{M}} \subseteq |\psi|^{\mathcal{M}}.$$

THEOREM 6.7.5 (Soundness for RQ). *Every* RQ-*theorem is valid in every* RQ-*model system.*

PROOF. Let $\mathcal{M}$ be a model on an RQ-model system $\mathcal{S}$. Since every truth set $|\varphi|^{\mathcal{M}} f$ belongs to *Prop*, and $\mathcal{S}$ is based on an R-model system, all instances of R-axioms in the present language are valid in $\mathcal{M}$ and the rules of Adjunction and Modus Ponens preserve this validity, by the arguments of the Soundness Theorem 6.4.4 for R. Thus it remains to deal with the quantificational postulates for RQ listed in Table 6.2.1.

Universal Instantiation: Let $\mathcal{M}, w, f \models \forall x\varphi$. Then for any term τ that is freely substitutable for x in φ, we have $\mathcal{M}, w, f[|\tau|^{\mathcal{M}} f/x] \models \varphi$ by the semantics of $\forall$ in $\mathcal{M}$, and hence $\mathcal{M}, w, f \models \varphi(\tau/x)$ by the Substitution Lemma. By the Semantic Entailment Lemma, this proves that the axiom $\forall x\varphi \to \varphi(\tau/x)$ is valid in $\mathcal{M}$.

$\forall$-Introduction Rule: Let $\mathcal{M} \models \varphi \to \psi$ with x not free in φ. Suppose $\mathcal{M}, w, f \models \varphi$. Then for any $a \in U$, the variable-assignments f and $f[a/x]$ agree on all free variables of φ, hence $\mathcal{M}, w, f[a/x] \models \varphi$ by the Free Assignment Lemma, so $\mathcal{M}, w, f[a/x] \models \psi$ by Semantic Entailment as $\varphi \to \psi$ is valid in $\mathcal{M}$. Therefore $\mathcal{M}, w, f \models \forall x\psi$. This shows that $\mathcal{M} \models \varphi \to \forall x\psi$. Hence $\forall$-Introduction preserves validity in $\mathcal{M}$.

$\wedge\exists$-Distribution: Suppose x not free in φ and $\mathcal{M}, w, f \models \varphi \wedge \exists x\psi$. Then $\mathcal{M}, w, f \models \exists x\psi$, so by Lemma 6.7.1 there is some Z such that

$$w \lhd Z \subseteq \bigcup_{a \in U} |\psi|^{\mathcal{M}} f[a/x].$$

Then $Z \subseteq \uparrow w$ as covers are refining. But $\uparrow w \subseteq |\varphi|^{\mathcal{M}} f$ as $\mathcal{M}, w, f \models \varphi$ and $|\varphi|^{\mathcal{M}} f$ is an up-set, so

$$Z \subseteq |\varphi|^{\mathcal{M}} f \cap \bigcup_{a \in U} |\psi|^{\mathcal{M}} f[a/x] = \bigcup_{a \in U} \left(|\varphi|^{\mathcal{M}} f \cap |\psi|^{\mathcal{M}} f[a/x]\right).$$

Now for any $a \in U$, $|\varphi|^{\mathcal{M}} f = |\varphi|^{\mathcal{M}} f[a/x]$ as x is not free in φ, so

$$|\varphi|^{\mathcal{M}} f \cap |\psi|^{\mathcal{M}} f[a/x] = |\varphi|^{\mathcal{M}} f[a/x] \cap |\psi|^{\mathcal{M}} f[a/x] = |\varphi \wedge \psi|^{\mathcal{M}} f[a/x].$$

Hence

$$w \lhd Z \subseteq \bigcup_{a \in U} |\varphi \wedge \psi|^{\mathcal{M}} f[a/x].$$

This implies $\mathcal{M}, w, f \models \exists x(\varphi \wedge \psi)$, by Lemma 6.7.1 again. Thus by Semantic Entailment we have shown that the axiom $\varphi \wedge \exists x \psi \to \exists x(\varphi \wedge \psi)$ is valid in the model $\mathcal{M}$. $\dashv$

To prove that RQ is complete for validity in RQ-model structures, we will construct a canonical model on the system

$$\mathcal{S}_{\mathrm{RQ}} = (W_{\mathrm{RQ}}, \leq, \rhd, \cdot, \varepsilon, \perp, U_{\mathcal{L}}, Prop_{\mathrm{RQ}}, PropFun_{\mathrm{RQ}}),$$

which is defined as follows:

- W_{RQ} is the set of principal theories $\varphi^{\vdash} = \{\psi : \varphi \vdash \psi\}$, where φ, ψ range over all $\mathcal{L}$-formulas, and $\vdash$ refers to deducibility in RQ.
- The universe $U_{\mathcal{L}}$ is the set of all individual constants of $\mathcal{L}$, which is assumed to be infinite.
- $(W_{\mathrm{RQ}}, \leq, \rhd, \cdot, \varepsilon, \perp)$ is the relevant cover system on W_{RQ} defined in the language of $\mathcal{L}$-formulas in the same way that the cover system of the canonical R-model structure $\mathcal{S}_{\mathrm{R}}$ was built from propositional formulas (see Theorem 6.5.4 and preceding material). Thus:
  - $w \leq u$ iff $w \subseteq u$.
  - $Z \rhd w$ iff $\bigcap Z = w$.
  - $\varphi^{\vdash} \cdot \psi^{\vdash} = (\varphi \circ \psi)^{\vdash}$.
  - $\varepsilon = t^{\vdash}$.
  - $\varphi^{\vdash} \perp \psi^{\vdash}$ iff $\varphi \vdash \neg\psi$.
- $Prop_{\mathrm{RQ}} = \{|\varphi|_{\mathrm{RQ}} : \varphi$ is an $\mathcal{L}$-sentence$\}$, where

$$|\varphi|_{\mathrm{RQ}} = \{w \in W_{\mathrm{RQ}} : \varphi \in w\} = \{\psi^{\vdash} : \psi \vdash \varphi\}.$$

- $PropFun_{\mathrm{RQ}} = \{\alpha_{\varphi} : \varphi$ is an $\mathcal{L}$-formula$\}$, where $\alpha_{\varphi} : U_{\mathcal{L}}{}^{\omega} \to \wp W_{\mathrm{RQ}}$ is defined by $\alpha_{\varphi} f = |\varphi^{f}|_{\mathrm{RQ}}$.

Since φ^{f} is a sentence, $|\varphi^{f}|_{\mathrm{RQ}}$ belongs to the set $Prop_{\mathrm{RQ}}$ of admissible propositions of $\mathcal{S}_{\mathrm{RQ}}$, so we have $\alpha_{\varphi} : U_{\mathcal{L}}{}^{\omega} \to Prop_{\mathrm{RQ}}$.

The structure $(W_{\mathrm{RQ}}, \leq, \rhd, \cdot, \varepsilon, \perp, Prop_{\mathrm{RQ}})$ is an R-model system. The proof of this is the same as the proof in Theorem 6.5.4 that $\mathcal{S}_{\mathrm{R}}$ is an R-model system. It uses the fact that the results of Lemma 6.5.3 all hold with $|A|_{\mathrm{R}}, |B|_{\mathrm{R}}$ replaced by $|\varphi|_{\mathrm{RQ}}, |\psi|_{\mathrm{RQ}}$ for any $\mathcal{L}$ formulas φ, ψ.

To show that $\mathcal{S}_{\mathrm{RQ}}$ is an RQ-model structure, we have some further preliminary work to do.

LEMMA 6.7.6. *For any $\mathcal{L}$-formula φ:*

(1) $|\forall x\varphi|_{\mathrm{RQ}} = \bigcap_{\tau\in U_{\mathcal{L}}} |\varphi(\tau/x)|_{\mathrm{RQ}}$.

(2) *For all $f \in U_{\mathcal{L}}{}^{\omega}$,* $|(\forall x\varphi)^f|_{\mathrm{RQ}} = \bigcap_{\tau\in U_{\mathcal{L}}} |\varphi^{f[\tau/x]}|_{\mathrm{RQ}}$.

(3) $|\exists x\varphi|_{\mathrm{RQ}} = \sum_{\tau\in U_{\mathcal{L}}} |\varphi(\tau/x)|_{\mathrm{RQ}}$.

(4) *For all $f \in U_{\mathcal{L}}{}^{\omega}$,* $|(\exists x\varphi)^f|_{\mathrm{RQ}} = \sum_{\tau\in U_{\mathcal{L}}} |\varphi^{f[\tau/x]}|_{\mathrm{RQ}}$.

PROOF. (1) asserts that for any $w \in W_{\mathrm{RQ}}$,

$$\forall x\varphi \in w \text{ iff for all } \tau \in U_{\mathcal{L}}, \varphi(\tau/x) \in w.$$

But if $\forall x\varphi \in w$, then for any $\tau \in U_{\mathcal{L}}$ we have $\forall x\varphi \vdash \varphi(\tau/x)$ by the Universal Instantiation axiom, so $\varphi(\tau/x) \in w$ as w is closed under deducibility. Conversely, suppose $\varphi(\tau/x) \in w$ for all $\tau \in U_{\mathcal{L}}$. Let $w = \psi^{\vdash}$, and choose a constant $\mathsf{c} \in \mathcal{L}$ that does not occur in φ or ψ. Then $\varphi(\mathsf{c}/x) \in w$, i.e. $\psi \vdash \varphi(\mathsf{c}/x)$. From this we get $\psi \vdash \forall x\varphi$ by the rule $\forall$GC of Lemma 6.2.1(5), hence $\forall x\varphi \in \psi^{\vdash} = w$ as required.

(2) follows from (1) using $(\forall x\varphi)^f = \forall x(\varphi^{f\backslash x})$ and $\varphi^{f\backslash x}(\tau/x) = \varphi^{f[\tau/x]}$.

(3) uses the definition of $\exists x$ as $\neg\forall x\neg$, together with (1) and the result that $|\neg\psi|_{\mathrm{RQ}} = |\psi|_{\mathrm{RQ}}{}^{\perp}$, to infer that

$$|\exists x\varphi|_{\mathrm{RQ}} = \Big(\bigcap_{\tau\in U_{\mathcal{L}}} |\varphi(\tau/x)|_{\mathrm{RQ}}{}^{\perp}\Big)^{\perp},$$

which is $\sum_{\tau\in U_{\mathcal{L}}} |\varphi(\tau/x)|_{\mathrm{RQ}}$ by (6.4.7).

(4) follows from (3) using $(\exists x\varphi)^f = \exists x(\varphi^{f\backslash x})$ etc. ⊣

LEMMA 6.7.7. (1) $\alpha_{\varepsilon} = \alpha_t$.

(2) $\alpha_{\varphi} \cap \alpha_{\psi} = \alpha_{\varphi\wedge\psi}$.

(3) $\alpha_{\varphi}{}^{\perp} = \alpha_{\neg\varphi}$.

(4) $\alpha_{\varphi} \Rightarrow \alpha_{\psi} = \alpha_{\varphi\to\psi}$.

(5) $\forall_x\alpha_{\varphi} = \alpha_{\forall x\varphi}$.

(6) $\exists_x\alpha_{\varphi} = \alpha_{\exists x\varphi}$.

PROOF. (1) holds because $\alpha_{\varepsilon}(f) = {\uparrow}\varepsilon = |t|_{\mathrm{RQ}} = |t^f|_{\mathrm{RQ}}$ for all $f \in U_{\mathcal{L}}{}^{\omega}$.

(2) asserts that for all f, $|\varphi^f|_{\mathrm{RQ}} \cap |\psi^f|_{\mathrm{RQ}} = |(\varphi\wedge\psi)^f|_{\mathrm{RQ}}$. This holds because $(\varphi\wedge\psi)^f = \varphi^f \wedge \psi^f$.

(3) and (4) hold similarly, as $|\varphi^f|_{\mathrm{RQ}}{}^{\perp} = |\neg\varphi^f|_{\mathrm{RQ}}$ and $|\varphi^f|_{\mathrm{RQ}} \Rightarrow |\psi^f|_{\mathrm{RQ}} = |\varphi^f \to \psi^f|_{\mathrm{RQ}} = |(\varphi\to\psi)^f|_{\mathrm{RQ}}$.

For (5) we use part (2) of the last Lemma in reasoning that

$$(\alpha_{\forall x\varphi})f = \bigcap_{\tau\in U_{\mathcal{L}}} |\varphi^{f[\tau/x]}|_{\mathrm{RQ}} = \bigcap_{\tau\in U_{\mathcal{L}}} \alpha_{\varphi}(f[\tau/x]) = (\forall_x\alpha_{\varphi})f.$$

(6) is similar to (5), but using part (4) of the last Lemma and (6.7.1). ⊣

This Lemma shows that $\mathit{PropFun}_{\mathrm{RQ}}$ contains the function α_{ε} with constant value ${\uparrow}\varepsilon$, and is closed under the operations $\cap$, $(-)^{\perp}$, $\Rightarrow$, and $\forall_x$ for all

$x \in \mathsf{InVar}$. So to show that $\mathcal{S}_{\mathrm{RQ}}$ is an RQ-model structure, it remains only to show that $PropFun_{\mathrm{RQ}}$ has the Cover-Join Property. This is where we use the $\wedge\exists$-Distribution axiom of RQ.

Lemma 6.7.8. *For any $\mathcal{L}$-formula φ,*

$$|\exists x\varphi|_{\mathrm{RQ}} = j\Big(\bigcup_{\tau\in U_{\mathcal{L}}} |\varphi(\tau/x)|_{\mathrm{RQ}}\Big).$$

Proof. We have to show that for any $w \in W_{\mathrm{RQ}}$,

$$\exists x\varphi \in w \quad \text{iff} \quad \text{there exists } Z \subseteq W_{\mathrm{RQ}} \text{ with } w \lhd Z \subseteq \bigcup_{\tau\in U_{\mathcal{L}}} |\varphi(\tau/x)|_{\mathrm{RQ}}.$$

From right to left first: for each $\tau \in U_{\mathcal{L}}$ we have $\varphi(\tau/x) \vdash \exists x\varphi$ by Existential Generalisation (Lemma 6.2.1(1)), so for any theory $u \in |\varphi(\tau/x)|_{\mathrm{RQ}}$ we have $\varphi(\tau/x) \in u$, hence $\exists x\varphi \in u$. This shows that

$$\exists x\varphi \in \bigcap\Big(\bigcup_{\tau\in U_{\mathcal{L}}} |\varphi(\tau/x)|_{\mathrm{RQ}}\Big).$$

But now if there exists Z with $w \lhd Z \subseteq \bigcup_{\tau\in U_{\mathcal{L}}} |\varphi(\tau/x)|_{\mathrm{RQ}}$, then

$$\bigcap\Big(\bigcup_{\tau\in U_{\mathcal{L}}} |\varphi(\tau/x)|_{\mathrm{RQ}}\Big) \subseteq \bigcap Z = w,$$

and so $\exists x\varphi \in w$.

For the converse, let $\exists x\varphi \in w$. Suppose that $w = \psi^{\vdash}$. We assume for now that x is not free in ψ, and defer the opposite case till the end. Let

$$Z = \bigcup_{\tau\in U_{\mathcal{L}}} |\psi \wedge \varphi(\tau/x)|_{\mathrm{RQ}}.$$

We show that Z is a w-cover meeting our requirements. First, if $u \in Z$, then $\psi \wedge \varphi(\tau/x) \in u$ for some $\tau \in U_{\mathcal{L}}$, hence $\psi \in u$ and so $w = \psi^{\vdash} \subseteq u$. This shows that $w \subseteq \bigcap Z$. For the converse inclusion, let $\chi \in \bigcap Z$. Choose a constant c that does not occur in χ or in $\psi \wedge \varphi$. Then as x is not free in ψ,

$$\big((\psi \wedge \varphi)(\mathsf{c}/x)\big)^{\vdash} = \big(\psi \wedge \varphi(\mathsf{c}/x)\big)^{\vdash} \in |\psi \wedge \varphi(\mathsf{c}/x)|_{\mathrm{RQ}} \subseteq Z.$$

Since $\chi \in \bigcap Z$, this gives $\chi \in \big((\psi \wedge \varphi)(\mathsf{c}/x)\big)^{\vdash}$, i.e. $(\psi \wedge \varphi)(\mathsf{c}/x) \vdash \chi$. By the rule $\exists$GC of Lemma 6.2.1(5), this implies $\exists x(\psi \wedge \varphi) \vdash \chi$.

Now as x is not free in ψ, by $\wedge\exists$-Distribution $\psi \wedge \exists x\varphi \vdash \exists x(\psi \wedge \varphi)$, so this leads to $\psi \wedge \exists x\varphi \vdash \chi$. But $\exists x\varphi \in w = \psi^{\vdash}$, so $\psi \vdash \exists x\varphi$. Hence $\psi \vdash \psi \wedge \exists x\varphi$, and therefore $\psi \vdash \chi$, giving $\chi \in \psi^{\vdash} = w$.

We have now shown that $w = \bigcap Z$, i.e. $w \lhd Z$. Since $|\psi \wedge \varphi(\tau/x)|_{\mathrm{RQ}} \subseteq |\varphi(\tau/x)|_{\mathrm{RQ}}$, we also have $Z \subseteq \bigcup_{\tau\in U_{\mathcal{L}}} |\varphi(\tau/x)|_{\mathrm{RQ}}$, completing the proof in this case.

It remains to explain what to do if it happens that x is free in ψ. In that case we take a bound alphabetical variant $\exists y\varphi(y/x)$ of $\exists x\varphi$ with y not occurring in φ or ψ. Then $\exists x\varphi \vdash \exists y\varphi(y/x)$ (Lemma 6.2.1(3)), so from $\exists x\varphi \in w$ we obtain $\exists y\varphi(y/x) \in \psi^{\vdash}$ with y not free in ψ. Hence we can apply the above argument, with $\varphi(y/x)$ in place of φ, to conclude that there exists some $Z \rhd w$ such that $Z \subseteq \bigcup_{\tau \in U_{\mathcal{L}}} |\varphi(y/x)(\tau/y)|_{\mathrm{RQ}}$. But as y is not in φ, $\varphi(y/x)(\tau/y) = \varphi(\tau/x)$, so again we have our desired conclusion that

$$w \lhd Z \subseteq \bigcup_{\tau \in U_{\mathcal{L}}} |\varphi(\tau/x)|_{\mathrm{RQ}}. \qquad \dashv$$

COROLLARY 6.7.9. *$\mathcal{S}_{\mathrm{RQ}}$ has the Cover-Join Property.*

PROOF. Take any $\alpha_\varphi \in \mathit{PropFun}_{\mathrm{RQ}}$ and $f \in U_{\mathcal{L}}{}^{\omega}$. We have to show that

$$\sum_{\tau \in U_{\mathcal{L}}} \alpha_\varphi(f[\tau/x]) = j\Big(\bigcup_{\tau \in U_{\mathcal{L}}} \alpha_\varphi(f[\tau/x]\Big).$$

Since $\alpha_\varphi(f[\tau/x]) = |\varphi^{f[\tau/x]}|_{\mathrm{RQ}} = |\varphi^{f \setminus x}(\tau/x)|_{\mathrm{RQ}}$, we can write the above equation as

$$\sum_{\tau \in U_{\mathcal{L}}} |\varphi^{f[\tau/x]}|_{\mathrm{RQ}} = j\Big(\bigcup_{\tau \in U_{\mathcal{L}}} |\varphi^{f \setminus x}(\tau/x)|_{\mathrm{RQ}}\Big).$$

The left side of this is equal to $|(\exists x\varphi)^f|_{\mathrm{RQ}}$ by Lemma 6.7.6(4), and the right side is equal to $|\exists x(\varphi^{f \setminus x})|_{\mathrm{RQ}}$ by the Lemma just proved. But $(\exists x\varphi)^f = \exists x(\varphi^{f \setminus x})$. $\dashv$

That completes the proof that $\mathcal{S}_{\mathrm{RQ}}$ is an RQ-model structure. Its canonical premodel $\mathcal{M}_{\mathrm{RQ}}$ is defined in the usual way, by putting $|\mathsf{c}|^{\mathcal{M}_{\mathrm{RQ}}} = \mathsf{c}$, and

$$|P|^{\mathcal{M}_{\mathrm{RQ}}}(\mathsf{c}_1, \dots, \mathsf{c}_n) = |P\mathsf{c}_1 \cdots \mathsf{c}_n|_{\mathrm{RQ}} \in \mathit{Prop}_{\mathrm{RQ}}.$$

THEOREM 6.7.10 (Truth Is Membership). *Let φ be any $\mathcal{L}$-formula. Then for all $f \in U_{\mathcal{L}}{}^{\omega}$,*

$$|\varphi|^{\mathcal{M}_{\mathrm{RQ}}} f = |\varphi^f|_{\mathrm{RQ}},$$

and hence for all $w \in W_{\mathrm{RQ}}$,

$$\mathcal{M}_{\mathrm{RQ}}, w, f \models \varphi \quad \textit{iff} \quad \varphi^f \in w.$$

PROOF. The case that φ is the atomic $P\tau_1 \cdots \tau_n$ is handled in the usual way, as in the proof of Theorem 1.9.4. When φ is t, we have

$$|t|^{\mathcal{M}_{\mathrm{RQ}}} f = {\uparrow}\varepsilon = {\uparrow}(t^{\vdash}) = |t|_{\mathrm{RQ}}.$$

The inductive case of negation, assuming the result for φ, is

$$|\neg\varphi|^{\mathcal{M}_{\mathrm{RQ}}} f = (|\varphi|^{\mathcal{M}_{\mathrm{RQ}}} f)^{\perp} = (|\varphi^f|_{\mathrm{RQ}})^{\perp} = |\neg(\varphi^f)|_{\mathrm{RQ}} = |(\neg\varphi)^f|_{\mathrm{RQ}}.$$

The cases for conjunction $\wedge$ and relevant implication $\rightarrow$ are similar.

For the case of the quantifier $\forall$, if the result holds for φ, then

$$\begin{aligned} |\forall x\varphi|^{\mathcal{M}_{\mathrm{RQ}}} f &= \bigcap_{\tau \in U_{\mathcal{L}}} |\varphi|^{\mathcal{M}_{\mathrm{RQ}}} f[\tau/x] && \text{semantics of } \forall \\ &= \bigcap_{\tau \in U_{\mathcal{L}}} |\varphi^{f[\tau/x]}|_{\mathrm{RQ}} && \text{induction hypothesis on } \varphi \\ &= |(\forall x\varphi)^f|_{\mathrm{L}} && \text{Lemma 6.7.6(2)}, \end{aligned}$$

so it holds for $\forall x\varphi$. $\dashv$

COROLLARY 6.7.11. *$\mathcal{M}_{\mathrm{RQ}}$ is a model.*

PROOF. We have to show that the propositional function $|\varphi|^{\mathcal{M}_{\mathrm{RQ}}}$ is admissible, i.e. is a member of $\mathit{PropFun}_{\mathrm{RQ}}$, whenever φ is atomic. But in fact, by the Theorem just proved, for an arbitrary φ and any $f \in U_{\mathcal{L}}{}^{\omega}$,

$$|\varphi|^{\mathcal{M}_{\mathrm{RQ}}} f = |\varphi^f|_{\mathrm{RQ}} = \alpha_\varphi f,$$

so $|\varphi|^{\mathcal{M}_{\mathrm{RQ}}}$ is just the function α_φ, which is admissible by definition of $\mathit{PropFun}_{\mathrm{RQ}}$. ⊣

Finally, all the pieces are in place to show that RQ is characterised by validity in its model systems.

THEOREM 6.7.12 (Model System Characterisation of RQ). *For any $\mathcal{L}$-formula φ, the following are equivalent.*

(1) *φ is a theorem of* RQ.
(2) *φ is valid in all* RQ-*model systems.*
(3) *φ is valid in the canonical model system $\mathcal{S}_{\mathrm{RQ}}$.*
(4) *φ is valid in the model $\mathcal{M}_{\mathrm{RQ}}$.*

PROOF. Given the Soundness Theorem 6.7.5 and the facts that $\mathcal{S}_{\mathrm{RQ}}$ is an RQ-model system and $\mathcal{M}_{\mathrm{RQ}}$ is a model on $\mathcal{S}_{\mathrm{RQ}}$, the implications from (1) to (2), (2) to (3) and (3) to (4) follow directly.

(4) implies (1): if φ is valid in $\mathcal{M}_{\mathrm{RQ}}$, then for any $f \in U_{\mathcal{L}}{}^{\omega}$, $\mathcal{M}_{\mathrm{RQ}}, \varepsilon, f \models \varphi$, hence by Theorem 6.7.10, $\varphi^f \in \varepsilon = t^{\vdash}$. Thus $t \vdash \varphi^f$, i.e. $\vdash_{\mathrm{RQ}} t \to \varphi^f$, and therefore $\vdash_{\mathrm{RQ}} \varphi^f$ by the t-Axiom.

In particular, if $x_1, \dots, x_n$ are the free variables of φ, and $\mathsf{c}_1, \dots, \mathsf{c}_n$ are distinct constants that do not occur in φ, then $\vdash_{\mathrm{RQ}} \varphi(\mathsf{c}_1/x_1, \dots \mathsf{c}_n/x_n)$. Hence $\vdash_{\mathrm{RQ}} \varphi$ by the rule GC*. ⊣

6.8. Heyting Implication for Full Systems

We are now going to show that the propositional logic R is characterised by model systems that are *full* in the sense that every proposition, i.e. every localised up-set, is admissible. In such systems, the localised up-sets are precisely the orthoclosed up-sets. The full systems will give us a characterisation of R by model systems that are completely determined by their underlying relevant cover system (see Theorem 6.8.7 and Corollary 6.9.5).

There is an analogy here with the situation in Boolean modal logic, where a proposition is an unstructured set of worlds, and a general frame is full if it contains all such sets. A Kripke frame (W, R) can be identified with the full general frame $(W, R, \wp W)$ having all propositions admissible (see Section 1.3). Some modal logics are characterised by validity in such full structures, and hence are viewed as being characterised by their Kripke frames. The corresponding situation holds for R with its cover systems and their notion of proposition. To show that, we will need to examine a relationship between R and the implication connective from Heyting's intuitionistic logic.

A relevant cover system $\mathcal{S} = (W, \leq, \triangleright, \cdot, \varepsilon, \perp)$ will be called *strongly relevant* if it satisfies the following additional postulates:

- *Refinement of ε is local*: $w \lhd Z \subseteq \uparrow\varepsilon$ implies $\varepsilon \leq w$.
- *Orthoclosure localisation*: $X^{\perp\perp} \subseteq jX$ for all up-sets X.

The first of these states that if w locally refines ε, in the sense that it has a cover consisting of points refining ε, then w itself refines ε. This means that the up-set $\uparrow\varepsilon$ of points refining ε is localised, i.e. $j\uparrow\varepsilon \subseteq \uparrow\varepsilon$, and therefore is a proposition.

The inclusion $jX \subseteq X^{\perp\perp}$ holds for arbitrary $X \subseteq W$, by Lemma 6.4.1(6). Hence the *orthoclosure localisation* postulate is equivalent to the assertion that each up-set X has $X^{\perp\perp} = jX$. This implies that if an up-set X is localised ($jX = X$), then it is orthoclosed ($X^{\perp\perp} = X$), and conversely. Actually, that converse holds for any relevant cover system, as was shown in Corollary 6.4.2 and used extensively subsequently. Now that Corollary is refined to

THEOREM 6.8.1. *In a strongly relevant cover system, the propositions are precisely the orthoclosed up-sets.* ⊣

For a system $\mathcal{S} = (W, \leq, \triangleright, \cdot, \varepsilon, \perp)$ we now define

$$\mathcal{S}^+ = (W, \leq, \triangleright, \cdot, \varepsilon, \perp, \mathit{Prop}(\mathcal{S})),$$

where $\mathit{Prop}(\mathcal{S})$ is the set of *all* localised up-sets of $\mathcal{S}$.

THEOREM 6.8.2. *If $\mathcal{S}$ is a strongly relevant cover system, then $\mathcal{S}^+$ is a full* R-*model system.*

PROOF. We need to show that $\mathit{Prop}(\mathcal{S})$ has the closure properties required by the definition of an R-model system.

First, since *refinement of ε is local* in $\mathcal{S}$, the up-set $\uparrow\varepsilon$ is localised, as noted above, so belongs to $\mathit{Prop}(\mathcal{S})$. It is readily seen that the intersection of two localised up-sets is a localised up-set. Also, $X^{\perp}$ is always a localised up-set, and so is $X \Rightarrow Y$ whenever Y is, by parts (2)–(4) of Lemma 6.4.1. So $\mathit{Prop}(\mathcal{S})$ is closed under $\cap$, $(-)^{\perp}$, and $\Rightarrow$.

Finally, we need to check that $X + Y = j(X \cup Y)$ for any $X, Y \in \mathit{Prop}(\mathcal{S})$. But $X + Y = (X \cup Y)^{\perp\perp}$, and $X \cup Y$ is an up-set for such X and Y, so the property follows immediately from orthoclosure localisation. ⊣

In view of this result, we can define a propositional formula to be *valid in* a strongly relevant cover system $\mathcal{S}$ if it is valid in the full R-model system $\mathcal{S}^+$. This means that the formula is true in any model on $\mathcal{S}$ no matter what localised up-sets are assigned to its variables. The logic R is sound for this notion of validity: every R-theorem is valid in all strongly relevant cover systems, since it is valid in all R-model structures.

In attempting to show that R is, conversely, complete for validity in strongly relevant cover systems, an immediate strategy is to consider the canonical model system $\mathcal{S}_{\mathrm{R}}$ of Section 6.5, and ask if its underlying relevant cover system is strongly relevant. Refinement of ε is local in $\mathcal{S}_{\mathrm{R}}$, since $\uparrow\varepsilon = |t|_{\mathrm{R}} \in Prop_{\mathrm{R}}$, and all members of $Prop_{\mathrm{R}}$ are localised. But orthoclosure localisation is problematic. The condition $X^{\perp\perp} \subseteq jX$ holds for X of the form $|A|_{\mathrm{R}}$, but when X is an arbitrary up-set of $(W_{\mathrm{R}}, \subseteq)$, it is not evident that it does hold.

To overcome this obstacle we introduce a new device. To motivate it, recall the important relationship

$$A \circ B \vdash C \quad \text{iff} \quad A \vdash B \to C,$$

which is expressed by saying that relevant implication $\to$ is *residual* to the fusion connective $\circ$ in the logic R. That property was used to show that the fusion operation in $\mathcal{S}_{\mathrm{R}}$ preserves covering (see the proof of Theorem 6.5.2) and to prove that $|A|_{\mathrm{R}} + |B|_{\mathrm{R}} = |A \vee B|_{\mathrm{R}}$ (Lemma 6.5.3(5)). Our new device is a binary connective $\supset$ which is to be analogously residual to conjunction , i.e. to satisfy the relationship

$$A \wedge B \vdash C \quad \text{iff} \quad A \vdash B \supset C. \tag{6.8.1}$$

This is a characteristic property of the implication connective of Heyting's intuitionistic propositional logic, so we will refer to $\supset$ as *Heyting implication*.

We extend the language of R by adding this new connective $\supset$, and work with the formulas generated from propositional variables and t by the connectives $\wedge$, $\neg$, $\to$, and $\supset$. Formulas of the original language, i.e. those without any occurrence of $\supset$, will be called *pure*. In the new language, a logic HR is defined by specifying its axioms and rules to be those of R, as in Table 6.1.1, together with the new $\wedge$-*Residuation* rules

$$\frac{A \wedge B \to C}{A \to (B \supset C)} \qquad \frac{A \to (B \supset C)}{A \wedge B \to C}\,. \tag{6.8.2}$$

Now we take $A \vdash B$ to mean that $\vdash_{\mathrm{HR}} A \to B$. Then the $\wedge$-Residuation rules ensure that (6.8.1) holds.

To interpret $\supset$ semantically, we would need a binary operation $X \Supset Y$ on up-sets that is residual to the set intersection operation $X \cap Y$, i.e. for all up-sets X, Y, Z we want

$$Z \cap X \subseteq Y \quad \text{iff} \quad Z \subseteq X \Supset Y. \tag{6.8.3}$$

Such an operation exists on the up-sets of any preordered set, and is defined by

$$X \Supset Y = \{w \in W : \uparrow w \cap X \subseteq Y\}. \tag{6.8.4}$$

Thus $w \in X \Supset Y$ iff $w \leq u \in X$ implies $u \in Y$.

The transitivity of $\leq$ alone makes $X \Supset Y$ an up-set, regardless of whether X and Y are, and (6.8.3) holds for any up-set Z. In fact the right-to-left part of (6.8.3) holds for any Z, by reflexivity of $\leq$, and putting $Z = X \Supset Y$ here gives the useful result that

$$(X \Supset Y) \cap X \subseteq Y. \tag{6.8.5}$$

We can interpret $\supset$ in any model $\mathcal{M}$ with a preorder, by putting

$$|A \supset B|^{\mathcal{M}} = |A|^{\mathcal{M}} \Supset |B|^{\mathcal{M}}. \tag{6.8.6}$$

In terms of the associated truth relation, this states that

$$\mathcal{M}, w \models A \supset B \text{ iff for all } u \geq w,\ \mathcal{M}, u \models A \text{ implies } \mathcal{M}, u \models B. \tag{6.8.7}$$

That is the semantic analysis of Heyting implication due to Kripke [1965].

On a more abstract level, an operation with the defining characteristic of $\Supset$ can be introduced on any set that has a complete partial order $\sqsubseteq$ for which finite meets $\sqcap$ distribute over arbitrary joins $\bigsqcup$, meaning that

$$a \sqcap (\bigsqcup \mathcal{Z}) = \bigsqcup\{a \sqcap b : b \in \mathcal{Z}\}. \tag{6.8.8}$$

Then putting $a \Supset b = \bigsqcup\{c : c \sqcap a \sqsubseteq b\}$ defines an operation satisfying

$$c \sqsubseteq a \Supset b \quad \text{iff} \quad c \sqcap a \sqsubseteq b. \tag{6.8.9}$$

Conversely, the existence of an operation $\Supset$ satisfying (6.8.9) ensures that distribution of $\sqcap$ over $\bigsqcup$ always holds. A complete partially ordered set satisfying (6.8.8) is sometimes called a *locale* or *complete Heyting algebra*.

The collection *Up* of up-sets of a preordered set is a locale, in which meets and joins are just set intersections and unions. So the induced operation on *Up* is given by

$$X \Supset Y = \bigcup\{Z \in Up : Z \cap X \subseteq Y\},$$

which is demonstrably equivalent to (6.8.4). The question of whether localised up-sets form a locale will be considered in the next section.

The crucial fact about HR that helps us solve our completeness problem for R is that it is a *conservative* extension of R. This means that it has no new $\supset$-free theorems. If a pure formula is a theorem of HR, then it must already have been a theorem of R. In other words, if a formula is an HR-theorem but not an R-theorem, then it contains $\supset$. This observation, which is due to Restall [1998], can be proved by using the Routley-Meyer models for R described in Section 6.1, which can be expanded to HR models by interpreting $\supset$ as in (6.8.6). The details are as follows.

THEOREM 6.8.3. HR *is a conservative extension of* R.

PROOF. Let A_0 be a pure formula that is not an R-theorem. We have to show that A_0 is not an HR-theorem.

Since R is complete for its Routley-Meyer semantics, there is a model $\mathcal{M}$ on a Routley-Meyer R-frame having a base world w such that $\mathcal{M}, w \not\models A_0$.

Define a new model $\mathcal{M}'$ for the language with $\supset$, on the same frame, by putting $|A|^{\mathcal{M}'} = |A|^{\mathcal{M}}$ if $A \in$ PropVar or $A = t$, then inductively:

$$|A \wedge B|^{\mathcal{M}'} = |A|^{\mathcal{M}'} \cap |B|^{\mathcal{M}'}.$$

$$|\neg A|^{\mathcal{M}'} = (|A|^{\mathcal{M}'})^*.$$

$$|A \to B|^{\mathcal{M}} = |A|^{\mathcal{M}'} \Rightarrow |B|^{\mathcal{M}'}.$$

$$|A \supset B|^{\mathcal{M}} = |A|^{\mathcal{M}'} \Supset |B|^{\mathcal{M}'}.$$

Comparing these equations with (6.1.1), a straightforward induction shows that if A is any pure formula, then $|A|^{\mathcal{M}'} = |A|^{\mathcal{M}}$. Hence in particular, $\mathcal{M}', w \not\models A_0$. Thus if we can show that all the HR-theorems are true in $\mathcal{M}'$, it will follow that A is not an HR-theorem, as required.

Since $\mathcal{M}'$ is based on an R-frame, the proof of the soundness theorem for R of Routley and Meyer [1973] applies here to show that all R-axioms in our expanded language are true in $\mathcal{M}$, and the rules of R preserve this truth. The essential prerequisite for that soundness proof is that every truth set $|A|^{\mathcal{M}'}$ be an up-set. But the class of up-sets is closed under $\Supset$ as well as $\cap$, * and $\Rightarrow$, so this does hold. Indeed, as Restall [1998, p. 185] puts it, "we do not gain any new propositions on a frame by enriching our expressive powers to include intuitionistic implication".

It remains only to show that the new $\wedge$-Residuation rules defining HR preserve truth in $\mathcal{M}'$. The fact that all truth sets are up-sets also ensures that the semantic entailment property (6.1.2) holds for $\mathcal{M}'$, so the requirement comes down to showing that

$$|A \wedge B|^{\mathcal{M}'} \subseteq |C|^{\mathcal{M}'} \quad \text{iff} \quad |A|^{\mathcal{M}'} \subseteq |B \supset C|^{\mathcal{M}'}.$$

But this follows from the residuation property (6.8.3) and the definition of $|-|^{\mathcal{M}'}$. ⊣

A model $\mathcal{M}_{\mathrm{HR}}$ on a canonical system $\mathcal{S}_{\mathrm{HR}}$ can be constructed by repeating the construction of $\mathcal{M}_{\mathrm{R}}$ and $\mathcal{S}_{\mathrm{R}}$, but using the formulas from the language including $\supset$, and taking $\vdash$ as the deducibility relation of HR. $\mathcal{S}_{\mathrm{HR}}$ is based on the set W_{HR} of principal theories $A^{\vdash} = \{B : A \vdash B\}$, where A, B range over all formulas of the expanded language. Its admissible propositions are the sets $|A|_{\mathrm{HR}} = \{w \in W_{\mathrm{HR}} : A \in w\}$. Since HR includes all instances of the R-axioms and is closed under the R-rules, $\mathcal{S}_{\mathrm{HR}}$ is an R-model system, by the same proof as for $\mathcal{S}_{\mathrm{R}}$. The purpose of introducing Heyting implication here is to prove the following result. Note that both $\wedge$-Residuation rules are used in the proof.

THEOREM 6.8.4. *$\mathcal{S}_{\mathrm{HR}}$ is based on a strongly relevant cover system.*

PROOF. $\mathcal{S}_{\mathrm{HR}}$ is based on a relevant cover system, by the same argument as for $\mathcal{S}_{\mathrm{R}}$ (Theorem 6.5.2). Then refinement of ε is local in $\mathcal{S}_{\mathrm{R}}$, since $\uparrow\varepsilon$ is the localised set $|t|_{\mathrm{HR}}$.

The burden of the proof is to show that $X^{\perp\perp} \subseteq jX$ for any given up-set X of $\mathcal{S}_{\mathrm{HR}}$. To facilitate this, we choose from each $w \in W_{\mathrm{HR}}$ a fixed (but arbitrary) generator A_w, so that $w = A_w^{\vdash}$.

Now suppose that $w \in X^{\perp\perp}$. Define $Z = \{(A_v \wedge A_w)^{\vdash} : v \in X\}$. Then $Z \subseteq X$, for if $v \in X$, then since $A_v \wedge A_w \vdash A_v$ we get $v = A_v^{\vdash} \subseteq (A_v \wedge A_w)^{\vdash}$, so $(A_v \wedge A_w)^{\vdash} \in X$ as X is an up-set.

Next we will show that Z is a cover of w. This will give $w \lhd Z \subseteq X$ and hence $w \in jX$, proving that $X^{\perp\perp} \subseteq jX$, so completing the Theorem. First, for each $v \in X$ we have $w = A_w^{\vdash} \subseteq (A_v \wedge A_w)^{\vdash}$. Hence $w \subseteq \bigcap Z$. To prove the converse inclusion, take any $B \in \bigcap Z$. Then for each $v \in X$, $B \in (A_v \wedge A_w)^{\vdash}$, so $A_v \wedge A_w \vdash B$, hence $A_v \vdash A_w \supset B$ by the first $\wedge$-Residuation rule (6.8.2), and so $\neg(A_w \supset B) \vdash \neg A_v$. Thus if we define $u = (\neg(A_w \supset B))^{\vdash}$ we have $u \perp (A_v)^{\vdash} = v$. As this holds for all $v \in X$ we now have $u \in X^{\perp}$. But $w \in X^{\perp\perp}$, so then $w \perp u$. This means that $A_w \vdash \neg\neg(A_w \supset B)$. Hence $A_w \vdash A_w \supset B$ by Double Negation Elimination. Therefore $A_w \wedge A_w \vdash B$ by the other $\wedge$-Residuation rule, leading to $A_w \vdash B$, and therefore $B \in w$.

That completes the proof that $w = \bigcap Z$, i.e. $w \lhd Z$, which completes the Theorem as explained. ⊣

Here is how Heyting implication impacts on the algebra of admissible propositions in $\mathcal{S}_{\mathrm{HR}}$:

LEMMA 6.8.5. $|A|_{\mathrm{HR}} \Supset |B|_{\mathrm{HR}} = |A \supset B|_{\mathrm{HR}}$.

PROOF. Take any element $C^{\vdash}$ of W_{HR}.

Let $C^{\vdash} \in |A|_{\mathrm{HR}} \Supset |B|_{\mathrm{HR}}$. Now $C^{\vdash} \subseteq (C \wedge A)^{\vdash} \in |A|_{\mathrm{HR}}$, so then $(C \wedge A)^{\vdash} \in |B|_{\mathrm{HR}}$ by definition of $\Supset$. This implies that $C \wedge A \vdash B$. Hence $C \vdash A \supset B$ by $\wedge$-Residuation, giving $C^{\vdash} \in |A \supset B|_{\mathrm{HR}}$.

Conversely, let $C^{\vdash} \in |A \supset B|_{\mathrm{HR}}$. Then $C \vdash A \supset B$, hence $C \wedge A \vdash B$ by the other by $\wedge$-Residuation rule. Now for any $u \in W_{\mathrm{HR}}$, if $C^{\vdash} \subseteq u \in |A|_{\mathrm{HR}}$, then $C \in u$ and $A \in u$, so $C \wedge A \in u$ and hence $B \in u$ as u is closed under conjunction and deducibility, therefore $u \in |B|_{\mathrm{HR}}$. This proves that $C^{\vdash} \in |A|_{\mathrm{HR}} \Supset |B|_{\mathrm{HR}}$. ⊣

The model $\mathcal{M}_{\mathrm{HR}}$ is defined by putting $|p|^{\mathcal{M}_{\mathrm{HR}}} = |p|_{\mathrm{HR}}$ for $p \in \mathsf{PropVar}$, and then using $\uparrow\varepsilon$, $\cap$, $(-)^{\perp}$, $\Rightarrow$ and $\Supset$ to extend the definition of $|A|^{\mathcal{M}_{\mathrm{HR}}}$ to all formulas.

THEOREM 6.8.6 (Truth Is Membership). *$|A|^{\mathcal{M}_{\mathrm{HR}}} = |A|_{\mathrm{HR}}$ for any formula A in the language including $\supset$. Thus for all $w \in W_{\mathrm{HR}}$ we have*

$$\mathcal{M}_{\mathrm{HR}}, w \models A \quad \textit{iff} \quad A \in w.$$

PROOF. The result holds for $A \in \mathsf{PropVar}$ by definition. The equations of Lemma 6.5.3 concerning the function $|-|_{\mathrm{R}}$ all hold for $|-|_{\mathrm{HR}}$, and this takes care of the case that A is t, and the inductive case of the connectives $\wedge$, $\neg$ and $\rightarrow$. The last Lemma takes care of the inductive case for $\supset$. ⊣

We are now in a position to prove that R is complete for truth in all models on strongly relevant cover systems.

THEOREM 6.8.7 (Cover System Characterisation of R). *For any $\supset$-free formula A, the following are equivalent.*

(1) *A is a theorem of* R.
(2) *A is valid in all strongly relevant cover systems.*
(3) *A is valid in the relevant cover system of $\mathcal{S}_{\mathrm{HR}}$.*
(4) *A is true in the model $\mathcal{M}_{\mathrm{HR}}$.*

PROOF. The only part requiring more work is that (4) implies (1). But if $\mathcal{M}_{\mathrm{HR}}, \varepsilon \models A$, then $A \in \varepsilon = t^{\vdash}$ by the Theorem just proved. This implies that $\vdash_{\mathrm{HR}} t \to A$, and hence $\vdash_{\mathrm{HR}} A$ by the t-Axiom. But if A is pure, it then follows that $\vdash_{\mathrm{R}} A$, because HR is a conservative extension of R. $\dashv$

We will strengthen this result in the next section, by showing that R is characterised by a special class of strongly relevant cover systems (see Corollary 6.9.5).

6.9. Localic Cover Systems for HR and HRQ

The logic HR was introduced as an auxiliary mechanism to derive the characterisation of R just given. But we might also ask for a semantic analysis of HR itself. This requires a further condition on cover systems.

The point here is that, whereas the propositions in a Routley-Meyer frame are the up-sets, in a cover system they are the *localised* up-sets. In any model on a relevant cover system, the truth of the Disjunction Elimination scheme depends on all truth sets being localised (see the proof of Theorem 6.4.4). For the language including $\supset$, this will require sets of the form $|A|^{\mathcal{M}} \Supset |B|^{\mathcal{M}}$ to be localised. That is true for the model $\mathcal{M}_{\mathrm{HR}}$, but may not be true for other models on relevant cover systems.

We will say that set Y *refines* set X if $Y \subseteq \uparrow X$, i.e. for all $y \in Y$ there is some $x \in X$ such that $x \leq y$. In other words, every member of Y refines some member of X. A cover system will be called *localic* if it satisfies the following postulate:

- *Refinement of covers*: if $w \leq u$, then every w-cover can be refined to a u-cover, i.e. if $Z \rhd w$, then there exists a $Z' \rhd u$ with $Z' \subseteq \uparrow Z$.

LEMMA 6.9.1. *In a localic cover system, if X is an up-set and Y is localized, then $X \Supset Y$ is a proposition.*

PROOF. $X \Supset Y$ is an up-set regardless of the status of X and Y. To show it is localized, let $w \lhd Z \subseteq X \Supset Y$. We want $w \in X \Supset Y$, i.e. $\uparrow w \cap X \subseteq Y$. So let $u \in \uparrow w \cap X$, in order to show $u \in Y$. We have $\uparrow Z \subseteq X \Supset Y$ as $X \Supset Y$ is an up-set, and $\uparrow u \subseteq X$ as X is an up-set.

Since $w \leq u$, by *refinement of covers* there is a u-cover $Z' \subseteq \uparrow Z$, hence $Z' \subseteq X \Supset Y$. Also, as $u \lhd Z'$ we get $Z' \subseteq \uparrow u \subseteq X$. Thus $Z' \subseteq (X \Supset Y) \cap X$. But $(X \Supset Y) \cap X \subseteq Y$ (6.8.5), so have $u \lhd Z' \subseteq Y$. Since Y is localized, this implies $u \in Y$ as required. ⊣

This result ensures that the set $Prop(\mathcal{S})$ of propositions of a localic cover system $\mathcal{S}$ is closed under the operation $\Supset$. Since this operation satisfies

$$Z \cap X \subseteq Y \quad \text{iff} \quad Z \subseteq X \Supset Y$$

for all propositions X, Y, Z, it follows that $Prop(\mathcal{S})$ is a *locale*, as defined in our discussion of (6.8.8) and (6.8.9). Hence our use of the adjective "localic" for $\mathcal{S}$. In this locale $Prop(\mathcal{S})$, the meet $\bigsqcap \mathcal{Z}$ of a set $\mathcal{Z}$ of propositions is the set intersection $\bigcap \mathcal{Z}$. If $\mathcal{S}$ is also a strongly relevant cover system, so that the propositions are precisely the orthoclosed up-sets (Theorem 6.8.1), then the join $\bigsqcup \mathcal{Z}$ is the orthojoin $(\bigcup \mathcal{Z})^{\perp\perp}$, which is equal to $j(\bigcup \mathcal{Z})$.

THEOREM 6.9.2 (Soundness for HR). *Every* HR-*theorem is valid in every localic strongly relevant cover system.*

PROOF. Let $\mathcal{S}$ be a localic strongly relevant cover system. By strong relevance, the full system $\mathcal{S}^+$ is an R-model system (Theorem 6.8.2), so $Prop(\mathcal{S})$ contains $\uparrow\varepsilon$ and is closed under $\cap$, $(-)^\perp$ and $\Rightarrow$. Since $\mathcal{S}$ is localic, $Prop(\mathcal{S})$ is also closed under $\Supset$, as just observed.

It follows that in any model $\mathcal{M}$ on $\mathcal{S}$, the truth sets $|A|^{\mathcal{M}}$ are all localised up-sets. The proof of Soundness Theorem 6.4.4 for R then applies here to show that all R-axioms in the expanded language are true in $\mathcal{M}$, and the rules of R preserve this truth. Thus it remains only to show that the $\wedge$-Residuation rules defining HR preserve truth in $\mathcal{M}$. But the fact that $\Supset$ is residual to $\cap$ ensures that

$$|A \wedge B|^{\mathcal{M}} \subseteq |C|^{\mathcal{M}} \quad \text{iff} \quad |A|^{\mathcal{M}} \subseteq |B \supset C|^{\mathcal{M}}. \qquad ⊣$$

To prove completeness for HR, we need the following fact.

LEMMA 6.9.3. *$\mathcal{S}_{\mathrm{HR}}$ is based on a localic strongly relevant cover system.*

PROOF. We already know from Theorem 6.8.4 that $\mathcal{S}_{\mathrm{HR}}$ is based on a strongly relevant cover system. So we have to show that refinement of covers holds for this system. Choose from each $w \in W_{\mathrm{HR}}$ a fixed generator A_w, so that $w = A_w^{\vdash}$.

Now suppose $w \leq u$ and Z is a cover of w. We need to refine Z to a cover of u. Put $Z' = \{(A_v \wedge A_u)^{\vdash} : v \in Z\}$. Then for each member $(A_v \wedge A_u)^{\vdash}$ of Z' we have $v = A_v^{\vdash} \subseteq (A_v \wedge A_u)^{\vdash}$ with $v \in Z$. Hence $Z' \subseteq \uparrow Z$, i.e. Z' refines Z.

Next, since $u = A_u^{\vdash} \subseteq (A_v \wedge A_u)^{\vdash}$ for every $v \in Z$, we have $u \subseteq \bigcap Z'$. To prove the converse inclusion, take any $B \in \bigcap Z'$. Then for any $v \in Z$ we have $B \in (A_v \wedge A_u)^{\vdash}$, so $A_v \wedge A_u \vdash B$, and hence $A_v \vdash A_u \supset B$, making $A_u \supset B \in v$. Thus $A_u \supset B \in \bigcap Z$. But $\bigcap Z = w$, as $Z \rhd w$, so

$A_w \vdash A_u \supset B$, implying $A_w \wedge A_u \vdash B$. Now $A_u \vdash A_w$, as $w \subseteq u$, so this yields $A_u \vdash B$, and hence $B \in u$.

Altogether we have $u = \bigcap Z'$, i.e. $Z' \rhd u$. Thus Z' is a cover of u refining Z, as required. ⊣

THEOREM 6.9.4 (Cover System Characterisation of HR). *For any formula A of the language including $\supset$, the following are equivalent.*

(1) *A is a theorem of* HR.
(2) *A is valid in all localic strongly relevant cover systems.*
(3) *A is valid in the relevant cover system of $\mathcal{S}_{\mathrm{HR}}$.*
(4) *A is true in the model $\mathcal{M}_{\mathrm{HR}}$.*

PROOF. (1) implies (2) by the Soundness Theorem 6.9.2. (2) implies (3) by the last Lemma. (3) implies (4) by definition of validity in a relevant cover system.

(4) implies (1): that $\mathcal{M}_{\mathrm{HR}} \models A$ implies $\vdash_{\mathrm{HR}} A$ was shown in the proof of Theorem 6.8.7. ⊣

COROLLARY 6.9.5. *The logic* R *is characterised by validity in all localic strongly relevant cover systems.*

PROOF. If A is pure, then $\vdash_{\mathrm{R}} A$ iff $\vdash_{\mathrm{HR}} A$ by conservativity of HR over R. Hence the Corollary holds by equivalence of (1) and (2) in the Theorem. ⊣

REMARK 6.9.6. Localic cover systems give a natural modelling for intuitionistic logic itself. This can be extended to a semantics for intuitionistic *modal* logics that provides the Kripkean relational modelling of $\Diamond$, namely

$$w \models \Diamond\varphi \quad \text{iff} \quad \text{there is some } v \text{ with } wRv \text{ and } v \models \varphi,$$

without validating the scheme

$$\Diamond(\varphi \vee \psi) \leftrightarrow \Diamond\varphi \vee \Diamond\psi.$$

A full discussion of this development is given in Goldblatt [2010]. ⊣

Turning now to quantificational logic, for a given signature $\mathcal{L}$ we add the Heyting implication connective to the syntax used for RQ, and consider all the formulas generated as a result. Again, a formula will be called *pure* if it does not contain $\supset$. A new logic HRQ is defined to have all the axioms and rules of RQ together with the $\wedge$-Residuation rules (6.8.2) for $\mathcal{L}$-formulas in this expanded language. Closure properties of the set of HRQ-theorems are given by Lemma 6.2.1, whose results all hold when L is HRQ.

Consider structures of the form $\mathcal{S} = (W, \leq, \rhd, \cdot, \varepsilon, \perp, U)$, comprising a relevant cover system with a specified set U of individuals. We refer to such an $\mathcal{S}$ as an *individualised* cover system. Define

$$\mathcal{S}^{++} = (W, \leq, \rhd, \cdot, \varepsilon, \perp, U, \mathit{Prop}(\mathcal{S}), \mathit{PropFun}(\mathcal{S})),$$

where $\mathit{Prop}(\mathcal{S})$ is the set of *all* localised up-sets of $\mathcal{S}$, and $\mathit{PropFun}(\mathcal{S})$ is the set of *all* functions from U^ω to $\mathit{Prop}(\mathcal{S})$.

THEOREM 6.9.7. *If $\mathcal{S}$ is a strongly relevant individualised cover system, then $\mathcal{S}^{++}$ is a full* RQ-*model system.*

PROOF. By Theorem 6.8.2, $(W, \leq, \rhd, \cdot, \varepsilon, \perp, Prop(\mathcal{S}))$ is an R-model system. Since $\uparrow\varepsilon \in Prop(\mathcal{S})$, the function α_ε with constant value $\uparrow\varepsilon$ belongs to $PropFun(\mathcal{S})$. Since $Prop(\mathcal{S})$ is closed under $\cap$, $(-)^\perp$ and $\Rightarrow$, we get that $PropFun(\mathcal{S})$ is closed under the corresponding pointwise defined operations on propositional functions. For instance, if $\alpha \in PropFun(\mathcal{S})$, then for all $f \in U^\omega$, $(\alpha^\perp)^f = (\alpha f)^\perp \in Prop(\mathcal{S})$, since $\alpha f \in Prop(\mathcal{S})$ and $Prop(\mathcal{S})$ is closed under $(-)^\perp$; hence $\alpha^\perp \in PropFun(\mathcal{S})$.

$PropFun(\mathcal{S})$ is also closed under the operations $\forall_x$, defined by $(\forall_x\alpha)f = \bigcap_{a\in U} \alpha(f[a/x])$. If each $\alpha(f[a/x])$ belongs to $Prop(\mathcal{S})$, then so too does the set $(\forall_x\alpha)f$, since $Prop(\mathcal{S})$ is closed under $\bigcap$. Hence $\forall_x\alpha \in PropFun(\mathcal{S})$.

It remains to prove the Cover-Join Property

$$\sum_{a\in U} \alpha(f[a/x]) = j(\bigcup_{a\in U} \alpha(f[a/x])$$

for $\alpha \in PropFun(\mathcal{S})$. Let $X = \bigcup_{a\in U} \alpha(f[a/x])$, which is an up-set, since each $\alpha(f[a/x])$ is an up-set. Then the Cover-Join Property asserts that $X^{\perp\perp} = jX$, an instance of orthoclosure localisation, which holds because $\mathcal{S}$ is strongly relevant. ⊣

If $\mathcal{S}$ is any *localic* strongly relevant individualised cover system, then $Prop(\mathcal{S})$ will be closed under $\Supset$, and hence $PropFun(\mathcal{S})$ will be closed under the operation $\alpha \Supset \beta$ defined pointwise by $(\alpha \Supset \beta)f = \alpha f \Supset \beta f$. We will define an $\mathcal{L}$-formula to be *valid in* $\mathcal{S}$ if it is valid in the full RQ-model system $\mathcal{S}^{++}$. This means that the formula is valid in all models on $\mathcal{S}^{++}$. In view of the fullness, we can take any such model $\mathcal{M}$ to be simply given by an interpretation assigning to each n-ary predicate symbol $P \in \mathcal{L}$ a function of the form $|P|^\mathcal{M} : U^\omega \to Prop(\mathcal{S})$ (and to each constant $\mathsf{c} \in \mathcal{L}$ an element $|\mathsf{c}|^\mathcal{M} \in U$). This will ensure that the truth-sets $|\varphi|^\mathcal{M} f$ always belong to $Prop(\mathcal{S})$ when φ is atomic, and so $|\varphi|^\mathcal{M}$ belongs to $PropFun(\mathcal{S})$. The closure properties of $PropFun(\mathcal{S})$ will then guarantee that this remains true for all formulas of the language including $\supset$, so that we have $|\varphi|^\mathcal{M} : U^\omega \to Prop(\mathcal{S})$ for every formula φ.

THEOREM 6.9.8 (Cover System Characterisation of HRQ). *An $\mathcal{L}$-formula is a theorem of* HRQ *iff it is valid in all localic strongly relevant individualised cover systems.*

PROOF. *Soundness*: if $\mathcal{S}$ is a localic strongly relevant individualised cover system, then $\mathcal{S}^{++}$ is an RQ-model system, so validates the axioms of RQ, while the rules of RQ preserve this validity. Also $Prop(\mathcal{S})$ is closed under the operation $\Supset$ residual to $\cap$, so validity in $\mathcal{S}^{++}$ preserves the $\wedge$-Residuation rules. Hence all HRQ-theorems are valid in $\mathcal{S}$.

Completeness: HRQ has a canonical system $\mathcal{S}_{\mathrm{HRQ}}$, defined just like that for RQ, which is based on a cover system that is localic and strongly relevant, by the same proof as for $\mathcal{S}_{\mathrm{HR}}$ (Theorem 6.9.3). Thus if a formula φ is valid in all localic strongly relevant individualised cover systems, then it is valid in the canonical model $\mathcal{M}_{\mathrm{HRQ}}$. But then $\vdash_{\mathrm{HRQ}} \varphi$, by the argument of the "(4) implies (1)" case of Theorem 6.7.12. $\dashv$

In the light of our results, it is natural to conjecture that RQ itself is characterised by validity in all (localic) strongly relevant individualised cover systems. This would follow if HRQ is a conservative extension of RQ. In [Goldblatt 2009] we have shown that HRQ is conservative over an extension of RQ by axioms for an identity predicate. This reduces the question of conservativity of HRQ over RQ to the question of conservativity of the axioms for identity over RQ. For now that remains an open question.

In conclusion, it is noteworthy that the method of cover system semantics can be applied to other kinds of substructural logic, such as linear logic and the Lambek calculus, that lack the $\wedge\vee$-Distribution axiom. In such applications we can retain the *refinement of covers* postulate, allowing w-covers to be refined to u-covers whenever $w \leq u$, but must abandon the assumption that *covers are refining*, i.e. $Z \rhd w$ implies $Z \subseteq {\uparrow}w$. Writing $X \sqcup Y$ for $j(X \cup Y)$, the distribution law

$$X \cap (Y \sqcup Y') = (X \cap Y) \sqcup (X \cap Y')$$

can be readily derived, for any up-set X, if covers are refining.

Abandoning this assumption looks anomalous from the point of view of Kripke-Joyal semantics, where the requirement that covers are refining is built into the definition of "cover" itself, rather than being a postulate to be considered. But its absence makes model constructions easier. For instance, in building canonical models out of principal theories, the cover relation $Z \rhd w$ can be defined to mean the weaker $\bigcap Z \subseteq w$, rather than $\bigcap Z = w$, if we do not require that $Z \subseteq {\uparrow}w$. Also, in the non-distributive setting new algebraic techniques become available, in particular the *MacNeille completion* of a partially ordered set [MacNeille 1937]. This construction does not preserve distributivity of meets over joins, but it embeds into a complete partial ordering in a way that preserves any existing meets and joins, providing an effective tool for modelling the quantifiers on cover systems based on such complete orderings.

Some work in this direction can be found in [Goldblatt 2006a].

REFERENCES

ALAN R. ANDERSON AND NUEL D. BELNAP

[1975] ***Entailment: The Logic of Relevance and Necessity***, vol. I, Princeton University Press.

ALAN R. ANDERSON, NUEL D. BELNAP, AND J. MICHAEL DUNN

[1992] ***Entailment: The Logic of Relevance and Necessity***, vol. II, Princeton University Press.

RUTH C. BARCAN

[1946a] *A functional calculus of first order based on strict implication*, ***The Journal of Symbolic Logic***, vol. 11, no. 1, pp. 1–16.

[1946b] *The deduction theorem in a functional calculus of first order based on strict implication*, ***The Journal of Symbolic Logic***, vol. 11, no. 4, pp. 115–118.

[1947] *The identity of individuals in a strict functional calculus of second order*, ***The Journal of Symbolic Logic***, vol. 12, no. 1, pp. 12–15.

J. L. BELL

[1988] ***Toposes and Local Set Theories***, Oxford University Press.

JOHN L. BELL

[2005] *Cover schemes, frame-valued sets and their potential uses in spacetime physics*, ***Spacetime Physics Research Trends, Horizons in World Physics*** (Albert Reimer, editor), vol. 248, Nova Science Publishers. Manuscript at `http://publish.uwo.ca/~jbell`.

E. W. BETH

[1956] *Semantic construction of intuitionistic logic*, ***Mededelingen der Koninklijke Nederlandse Akademie van Wetenschappen, Afd. Letterkunde. Nieuwe Reeks***, vol. 19, no. 11, pp. 357–388.

PATRICK BLACKBURN, MAARTEN DE RIJKE, AND YDE VENEMA

[2001] ***Modal Logic***, Cambridge University Press.

W. J. BLOK

[1980] *The lattice of modal logics: an algebraic investigation*, ***The Journal of Symbolic Logic***, vol. 45, pp. 221–236.

GEORGE BOOLOS

[1979] ***The Unprovability of Consistency***, Cambridge University Press.

[1993] ***The Logic of Provability***, Cambridge University Press.

GERALDINE BRADY

[2000] ***From Peirce to Skolem: A Neglected Chapter in the History of Logic***, North-Holland/Elsevier.

ALDO BRESSAN

[1972] ***A General Interpreted Modal Calculus***, Yale University Press.

R. A. BULL

[1968] *An algebraic study of tense logics with linear time*, ***The Journal of Symbolic Logic***, vol. 33, pp. 27–38.

RUDOLPH CARNAP

[1947] ***Meaning and Necessity***, The University of Chicago Press, Enlarged Edition 1956.

B. F. CHELLAS

[1980] ***Modal Logic: An Introduction***, Cambridge University Press.

ALONZO CHURCH

[1944] ***Introduction to Mathematical Logic, Part I***, Princeton University Press.

[1956] ***Introduction to Mathematical Logic, Volume I***, Princeton University Press.

NINO B. COCCHIARELLA

[2001] *Philosophical perspectives on quantification in tense and modal logic*, ***Handbook of Philosophical Logic*** (D. M. Gabbay and F. Guenthner, editors), vol. 7, Kluwer Academic Publishers, 2nd ed., pp. 235–275.

GIOVANNA CORSI

[2006] *A unified completeness theorem for quantified modal logics*, ***The Journal of Symbolic Logic***, vol. 67, no. 4, pp. 1483–1510.

M. J. CRESSWELL

[1995] *Incompleteness and the Barcan Formula*, ***Journal of Philosophical Logic***, vol. 24, pp. 379–403.

[1997] *Some incompletable modal predicate logics*, ***Logique et Analyse***, vol. 160, pp. 321–334.

[2001] *How to complete some modal predicate logics*, ***Advances in Modal Logic, Volume 2*** (Michael Zakharyaschev, Krister Segerberg, Maarten de Rijke, and Heinrich Wansang, editors), CSLI Publications, pp. 155–178.

RANDALL R. DIPERT

[1994] *The life and logical contributions of O. H. Mitchell, Peirce's gifted student*, ***Transactions of the Charles S. Peirce Society***, vol. 30, no. 3, pp. 515–542.

KOSTA DOŠEN

[1989] *Sequent systems and groupoid models II*, ***Studia Logica***, vol. 48, no. 1, pp. 41–65.

J. MICHAEL DUNN

[1993] *Star and perp: Two treatments of negation*, ***Philosophical Perspectives***, vol. 7, pp. 331–357.

HERBERT B. ENDERTON

[1972] ***A Mathematical Introduction to Logic***, Academic Press.

KIT FINE

[1975] *Some connections between elementary and modal logic*, ***Proceedings of the Third Scandinavian Logic Symposium*** (Stig Kanger, editor), North-Holland, pp. 15–31.

[1983] *The permutation principle in quantificational logic*, ***Journal of Philosophical Logic***, vol. 12, pp. 33–37.

[1988] *Semantics for quantified relevance logic*, ***Journal of Philosophical Logic***, vol. 17, pp. 22–59, Reprinted in Anderson, Belnap, and Dunn [1992, §53].

[1989] *Incompleteness for quantified relevance logics*, ***Directions in Relevant Logic*** (Jean Norman and Richard Sylvan, editors), Kluwer Academic Publishers. Summarised in Anderson, Belnap, and Dunn [1992, §52], pp. 205–225.

MELVIN FITTING AND RICHARD L. MENDELSOHN

[1998] ***First-Order Modal Logic***, Kluwer Academic Publishers.

DAGFINN FØLLESDAL

[2004] ***Referential Opacity and Modal Logic***, Routledge.

GRAEME FORBES

[1985] ***The Metaphysics of Modality***, Oxford University Press.

GOTTLOB FREGE

[1879] ***Begriffsschrift, eine der arithmetischen nachgebildete Formelsprache des reinen Denkens***. English translation in van Heijenoort [1967], 1–82.

DOV M. GABBAY, IAN HODKINSON, AND MARK REYNOLDS

[1994] ***Temporal Logic. Mathematical Foundations and Computational Aspects, Volume 1***, Oxford University Press.

DOV M. GABBAY, VALENTIN SHEHTMAN, AND DMITRIJ SKVORTSOV

[2009] ***Quantification in Nonclassical Logic***, Studies in Logic and the Foundations of Mathematics, vol. 153, Elsevier.

JAMES W. GARSON

[2001] *Quantification in modal logic*, ***Handbook of Philosophical Logic*** (D. M. Gabbay and F. Guenthner, editors), vol. 3, Kluwer Academic Publishers, 2nd ed., pp. 267–323.

SILVIO GHILARDI

[1989] *Presheaf semantics and independence results for some non-classical first-order logics*, ***Archive for Mathematical Logic***, vol. 29, no. 2, pp. 125–136.

[1991] *Incompleteness results in Kripke semantics*, ***The Journal of Symbolic Logic***, vol. 56, no. 2, pp. 517–538.

JEAN-YVES GIRARD

[1987] *Linear logic*, ***Theoretical Computer Science***, vol. 50, pp. 1–102.

ROBERT GOLDBLATT

[1974] *Semantic analysis of orthologic*, ***Journal of Philosophical Logic***, vol. 3, pp. 19–35. Reprinted in Goldblatt [1993].

[1982] ***Axiomatising the Logic of Computer Programming***, Lecture Notes in Computer Science, vol. 130, Springer-Verlag.

[1992] ***Logics of Time and Computation***, second ed., CSLI Lecture Notes No. 7, CSLI Publications, Stanford University.

[1993] ***Mathematics of Modality***, CSLI Lecture Notes No. 43, CSLI Publications, Stanford University.

[2006a] *A Kripke-Joyal semantics for noncommutative logic in quantales*, ***Advances in Modal Logic, Volume 6*** (Guido Governatori, Ian Hodkinson, and Yde Venema, editors), College Publications, London, pp. 209–225. `www.aiml.net/volumes/volume6/`.

[2006b] *Mathematical modal logic: A view of its evolution*, ***Logic and the Modalities in the Twentieth Century*** (Dov M. Gabbay and John Woods, editors), Handbook of the History of Logic, vol. 7, Elsevier, pp. 1–98. Manuscript available at `www.msor.vuw.ac.nz/~rob`.

[2009] *Conservativity of Heyting implication over relevant quantification*, ***The Review of Symbolic Logic***, vol. 2, no. 2, pp. 310–341.

[2010] *Cover semantics for quantified lax logic*, ***Journal of Logic and Computation***, `doi:10.1093/logcom/exq029`.

ROBERT GOLDBLATT AND IAN HODKINSON

[2009] *Commutativity of quantifiers in varying-domain Kripke models*, ***Towards Mathematical Philosophy*** (David Makinson, Jacek Malinowski, and Heinrich Wansing, editors), Trends in Logic, vol. 28, Springer, pp. 9–30.

ROBERT GOLDBLATT, IAN HODKINSON, AND YDE VENEMA

[2003] *On canonical modal logics that are not elementarily determined*, ***Logique et Analyse***, vol. 181, pp. 77–101. Published October 2004.

[2004] *Erdős graphs resolve Fine's canonicity problem*, ***The Bulletin of Symbolic Logic***, vol. 10, no. 2, pp. 186–208.

ROBERT GOLDBLATT AND EDWIN D. MARES

[2006] *A general semantics for quantified modal logic*, ***Advances in Modal Logic, Volume 6*** (Guido Governatori, Ian Hodkinson, and Yde Venema, editors), College Publications, `http://www.aiml.net/volumes/volume6/`, pp. 227–246.

PAUL R. HALMOS

[1954] *Polyadic Boolean algebras*, ***Proceedings of the National Academy of Sciences of the United States of America***, vol. 40, no. 5, pp. 296–301.

[1962] ***Algebraic Logic***, Chelsea, New York.

CHARLES HARTSHORNE AND PAUL WEISS

[1933] ***Collected Papers of Charles Sanders Peirce***, Vol. III: Exact Logic, Harvard University Press, Cambridge, Massachusetts.

L. HENKIN, J. D. MONK, A. TARSKI, H. ANDRÉKA, AND I. NÉMETI

[1981] ***Cylindric Set Algebras***, Lecture Notes in Mathematics, vol. 883, Springer-Verlag.

LEON HENKIN

[1950] *Completeness in the theory of types*, ***The Journal of Symbolic Logic***, vol. 15, pp. 81–91.

[1957] *A generalisation of the concept of ω-completeness*, ***The Journal of Symbolic Logic***, vol. 22, pp. 1–14.

LEON HENKIN, J. DONALD MONK, AND ALFRED TARSKI

[1971] ***Cylindric Algebras I***, North-Holland.

[1985] ***Cylindric Algebras II***, North-Holland.

JAAKKO HINTIKKA

[1969] ***Models for Modalities: Selected Essays***, D. Reidel.

G. E. HUGHES AND M. J. CRESSWELL

[1968] ***An Introduction to Modal Logic***, Methuen.

[1984] ***A Companion to Modal Logic***, Methuen.

[1996] ***A New Introduction to Modal Logic***, Routledge.

I. L. HUMBERSTONE

[1988] *Operational semantics for positive R*, ***Notre Dame Journal of Formal Logic***, vol. 29, no. 1, pp. 61–80.

STEPHEN COLE KLEENE

[1952] ***Introduction to Metamathematics***, Van Nostrand.

SAUL A. KRIPKE

[1963a] *Semantical analysis of modal logic I. Normal modal propositional calculi*, ***Zeitschrift für mathematische Logik und Grundlagen der Mathematik***, vol. 9, pp. 67–96.

[1963b] *Semantical considerations on modal logic*, ***Acta Philosophica Fennica***, vol. 16, pp. 83–94.

[1965] *Semantical analysis of intuitionistic logic I*, ***Formal Systems and Recursive Functions*** (J. N. Crossley and M. A. E. Dummett, editors), North-Holland, pp. 92–130.

[1967] *Review of Lemmon* [1966], ***Mathematical Reviews***, vol. 34, pp. 1021–1022, MR0205835 (MR 34 #5661).

[1971] *Identity and necessity*, ***Identity and Individuation*** (Milton K. Munitz, editor), New York University Press, pp. 135–164.

[1980] ***Naming and Necessity***, Harvard University Press.

[1992] *Individual concepts: Their logic, philosophy, and some of their uses*, ***Proceedings and Addresses of the American Philosophical Association***, vol. 66, no. 2, pp. 70–73. Summary of invited paper given at American Philosophical Association Eastern Division Meeting, Washington, D.C., December 1992.

J. LAMBEK AND P. J. SCOTT

[1986] ***Introduction to Higher Order Categorical Logic***, Cambridge University Press.

E. J. LEMMON

[1960] *Quantified S4 and the Barcan formula*, ***The Journal of Symbolic Logic***, vol. 25, no. 4, pp. 391–392.

[1966] *Algebraic semantics for modal logics II*, ***The Journal of Symbolic Logic***, vol. 31, pp. 191–218.

[1977] ***An Introduction to Modal Logic***, American Philosophical Quarterly Monograph Series, vol. 11, Basil Blackwell, Oxford. Written in 1966 in collaboration with Dana Scott. Edited by Krister Segerberg.

LEONARD LINSKY

[1971] ***Reference and Modality***, Oxford University Press.

SAUNDERS MAC LANE AND IEKE MOERDIJK
[1992] ***Sheaves in Geometry and Logic: A First Introduction to Topos Theory***, Springer-Verlag.

H. M. MACNEILLE
[1937] *Partially ordered sets*, ***Transactions of the American Mathematical Society***, vol. 42, pp. 416–460.

D. C. MAKINSON
[1971] *Some embedding theorems for modal logic*, ***Notre Dame Journal of Formal Logic***, vol. 12, pp. 252–254.

RUTH BARCAN MARCUS
[1993] ***Modalities: Philosophical Essays***, Oxford University Press.

EDWIN D. MARES
[2004] ***Relevant Logic: A Philosophical Interpretation***, Cambridge University Press.

EDWIN D. MARES AND ROBERT GOLDBLATT
[2006] *An alternative semantics for quantified relevant logic*, ***The Journal of Symbolic Logic***, vol. 71, no. 1, pp. 163–187.

J. C. C. MCKINSEY AND ALFRED TARSKI
[1944] *The algebra of topology*, ***Annals of Mathematics***, vol. 45, pp. 141–191.
[1946] *On closed elements in closure algebras*, ***Annals of Mathematics***, vol. 47, pp. 122–162.

O. H. MITCHELL
[1883] *On a new algebra of logic*, ***Studies in Logic, by Members of the John Hopkins University*** (C. S. Peirce, editor), Little, Brown and Company, Boston. Facsimile reprint, with an Introduction by Max H. Fisch and a preface by Achim Eschbach, John Benjamins Publishing Company, Amsterdam/Philadelphia, 1983, pp. 72–106.

FRANCO MONTAGNA
[1984] *The predicate modal logic of provability*, ***Notre Dame Journal of Formal Logic***, vol. 25, pp. 179–189.

ANDRJEZ MOSTOWSKI
[1948] *Proofs of non-deducibility in intuitionistic functional calculus*, ***The Journal of Symbolic Logic***, vol. 13, no. 4, pp. 204–207.

HIROAKIRA ONO
[1973] *Incompleteness of semantics for intermediate predicate logics I. Kripke's semantics*, ***Proceedings of the Japan Academy***, vol. 49, no. 9, pp. 711–713.

[1983] *Model extension theorem and Craig's interpolation theorem for intermediate predicate logics*, ***Reports on Mathematical Logic***, vol. 15, pp. 41–58.

[1993] *Semantics for substructural logics*, ***Substructural Logics*** (Peter Schroeder-Heister and Kosta Došen, editors), Oxford University Press, pp. 259–291.

HIROAKIRA ONO AND YUICHI KOMORI

[1985] *Logics without the contraction rule*, ***The Journal of Symbolic Logic***, vol. 50, no. 1, pp. 169–201.

C. S. PEIRCE

[1885] *On the algebra of logic: A contribution to the philosophy of notation*, ***American Journal of Mathematics***, vol. 7, no. 2, pp. 180–196. Continued as Vol. 7, No. 3, pp. 197–202. Reprinted as pp. 210–249 of Hartshorne and Weiss [1933].

GRAHAM PRIEST

[2008] ***An Introduction to Non-Classical Logic: From If to Is***, Cambridge University Press.

ARTHUR N. PRIOR

[1956] *Modality and quantification in S5*, ***The Journal of Symbolic Logic***, vol. 21, no. 1, pp. 60–62.

H. RASIOWA

[1951] *Algebraic treatment of the functional calculi of Heyting and Lewis*, ***Fundamenta Mathematicae***, vol. 38, pp. 99–126.

HELENA RASIOWA AND ROMAN SIKORSKI

[1950] *A proof of the completeness theorem of Gödel*, ***Fundamenta Mathematicae***, vol. 37, pp. 193–200.

[1963] ***The Mathematics of Metamathematics***, PWN–Polish Scientific Publishers, Warsaw.

GREG RESTALL

[1998] *Displaying and deciding substructural logics* 1: *Logics with contraction*, ***Journal of Philosophical Logic***, vol. 27, pp. 179–216.

RICHARD ROUTLEY AND ROBERT K. MEYER

[1973] *The semantics of entailment*, ***Truth, Syntax and Modality*** (Hughes Leblanc, editor), North-Holland, pp. 199–243.

RICHARD ROUTLEY, ROBERT K. MEYER, VAL PLUMWOOD, AND ROSS T. BRADY

[1982] ***Relevant Logics and Their Rivals 1***, Ridgeview Publishing Co., Atascadero, California.

BERTRAND RUSSELL

[1975] ***My Philosophical Development***, Unwin Books.

GERHARD SCHURZ

[1997] ***The Is-Ought Problem***, Kluwer Academic Publishers.

DANA SCOTT

[1967] *Existence and description in formal logic*, ***Bertrand Russell: Philosopher of the Century*** (Ralph Schoenman, editor), George, Allen and Unwin, pp. 181–200.

[1970] *Advice on modal logic*, ***Philosophical Problems in Logic*** (K. Lambert, editor), Reidel, pp. 143–173.

VALENTIN SHEHTMAN AND DMITRIJ SKVORTSOV

[1990] *Semantics of non-classical first-order predicate logics*, ***Mathematical Logic*** (P. P. Petkov, editor), Plenum Press, pp. 105–116.

ROBERT C. STALNAKER

[2003] ***Ways a World Might Be: Metaphysical and Anti-Metaphysical Essays***, Oxford University Press.

R. H. THOMASON

[1970] *Some completeness results for modal predicate calculi*, ***Philosophical Problems in Logic*** (K. Lambert, editor), D. Reidel, pp. 56–76.

S. K. THOMASON

[1972] *Semantic analysis of tense logic*, ***The Journal of Symbolic Logic***, vol. 37, pp. 150–158.

A. S. TROELSTRA AND P. VAN ULSEN

[1999] *The discovery of E. W. Beth's semantics for intuitionistic logic*, ***JFAK. Essays dedicated to Johan van Benthem on the occasion of his 50th birthday*** (Jelle Gerbrandy, Maarten Marx, Maarten de Rijke, and Yde Venema, editors), Vossiuspers, Amsterdam University Press, ISBN 90 5629 104 1, `www.illc.uva.nl/j50/contribs/troelstra/`.

ALASDAIR URQUHART

[1972] *Semantics for relevant logics*, ***The Journal of Symbolic Logic***, vol. 37, no. 1, pp. 159–169.

[1984] *The undecidability of entailment and relevant implication*, ***The Journal of Symbolic Logic***, vol. 49, no. 4, pp. 1059–1073.

J. F. A. K. VAN BENTHEM

[1975] *A note on modal formulas and relational properties*, ***The Journal of Symbolic Logic***, vol. 40, no. 1, pp. 55–58.

[1977] *Modal logic as second-order logic*, ***Report 77–04***, Department of Mathematics, University of Amsterdam, March.

[1979] *Canonical modal logics and ultrafilter extensions*, ***The Journal of Symbolic Logic***, vol. 44, pp. 1–8.

JEAN VAN HEIJENOORT

[1967] ***From Frege to Gödel: A Source Book in Mathematical Logic, 1879–1931***, Harvard University Press.

INDEX

□∀-Intro, 123
∃GC, 210
∀GC, 7
∀∧-Distribution, 106
∀∨-Distribution, 209
∧-Residuation rules, 241
∧∃-Distribution, 209

actual individuals, 1, 15
Actual Instantiation, 1, 4
actualist interpretation, xi, 1, 4, 167
Adjunction, 204
admissibility, 59
admissible
  conjunction, x, 17
  disjunction, x, 18
  E-, 103
  formula, 29, 168
  individual concept, 165
  object, 165
  proposition, x, 14, 23
  propositional function, xii, 129
  semantics, x
  set, x, 23
AI, 4
atom, 40
atomic formula, 2, 162
AtSI, 176
axiom
  AI, 4
  AtSI, 176
  B, 58
  BF, 16
  CBF, 14
  CQ, 5
  DD, 194
  EI, 104
  EI^{+}, 169, 176
  EX, 176
  FNE, 188
  K, 4
  McKinsey, 14, 62–63
  NE, 116
  NI, 159, 176
  NNE, 116
  NNI, 176
  SI, 160
  UD, 4
  UI, 5
  VQ, 4
  WR, 176

Barcan Formula, xi, 16, 61, 155–157
Barcan shape, 154
base world, 206
BF, 16
  independence from NNE, 121–122
  relation to T∀-Intro+NNE, 123–125
BF-Lemma, 85
body of a term, 193
Boolean implication, 18
Boolean tautology, 3
bound alphabetic variant, 6
bounded morphism, 55
Brouwerian Axiom, 58
Brouwerian lattice, 20

C-complete set of formulas, 84
canonical
  functional model structure, 132
  functionally – logic, 152
  general frame, 47, 53, 57
    for identity, 179
  Kripke frame, 54
  Kripkean model
    for $\mathcal{E}$-logics, 112

model
nature of, 45
on $\mathcal{F}_{\mathrm{R}}$, 208
model structure
for a Kripkean identity logic, 189
for a Kripkean identity logic with definite descriptions, 198
for $\mathcal{E}$-logics, 107
for identity, 181
for QR, 213
varying-domains, 47
model system
for HR, 243
for R, 226
for RQ, 235
premodel, 47
for a Kripkean identity logic with definite descriptions, 199
for identity, 182
for QR, 213
on a functional structure, 133
propositional modal logic, 54
R-frame, 208
S-model, 54
varying-domain model structure
for $\mathcal{E}$-logics, 107
CBF, 14
characterisation of
HR, 247
HRQ, 248
QEK+T∀-Intro, 115
QEK+T∀-Intro+NE, 117
QES, 108
QES with S canonical, 108
QES+BF, 155
with S canonical, 156
QES+NE, 117
QES+NE with S canonical, 117
QES+T∀-Intro, 115
with canonical S preserved by subframes, 116
QES+T∀-Intro+NE, 117
QES+T∀-Intro+NE with canonical S preserved by subframes, 117
QES+T∀-Intro+NNE, 123
QES+UI+BF, 157
QK, 52
by contracting-domains structures, 72
QK$^{\approx}$, 186
QK+CBF
by constant-domains structures, 76
QK+CQ+CBF+BF, 92
QK+CQ+UI+BF, 97
QKB, 58
QR, 213
QS, 54
by contracting-domains structures, 74
with S canonical, 58
QS$^{\approx}$, 186
with canonical S preserved by subframes, 186
QS$^{\approx}$ + T∀-Intro + FNE, 190
QS$^{\approx}$ + T∀-Intro + FNE + DD, 201
QS+BF, 156
QS+CBF, 69
by constant-domains structures, 77
with S canonical, 69
with S canonical by constant-domains structures, 79
QS+CQ+BF, 157
QS+CQ+CBF+BF, 94
with canonical S preserved by subframes, 101
QS+CQ+UI+BF, 98
with canonical S preserved by subframes, 101
QS+UI, 81
with S canonical, 81
QS+UI+BF, 157
QS4M+CQ+CBF+BF, 99
QS4M+CQ+UI+BF, 99
R
by cover systems, 245
by model systems, 229, 231
by Routley-Meyer frames, 209
RQ
by model structures, 213
by model systems, 239
S4, 12
S5, 12
characterised by a class, 32
characteristic
Kripkean model
for logics with CBF and BF, 89
for logics with UI and BF, 95
model, 45
closed term, 2
with definite descriptions, 193
commutativity of quantification, xii
Commuting Quantifiers, 5
complete for (validity in) a class, 32
complete Heyting algebra, 242

complete lattice, 217
complete partial order, 20, 217
completeness for
  first-order logic, 21
  QK, 52
  $QK^{\approx}$, 186
  QS, 54
  $QS^{\approx}$, 186
  QS+CBF, 69
components of a frame, 80, 164
Con, 2, 162
concept variables, 162
conclusion, 4
cone, 216
congruent, 6, 141
connected worlds, 80
conservative extension, 43, 242
consistent set of formulas, 46
constant domains, xi, 67
constant partial function, 164
constants
  individual, 2
  individual concept, 162
  object, 162
continuous time, 16
contracting domains, xi, 67
Converse Barcan Formula, xi, 14, 61
co-tenability, 205
cover relation, 216
cover system, 216
  individualised, 247
  localic, 245
  relevant, 220
  strongly relevant, 240
Cover-Join Property, 232
CQ, 5, 40–43, 59–61
Cterm, 198
cylindric algebra, 21

DD, 194
deducible, 45, 208
definite description, 193
  elimination of, 195
definite set of formulas, 199
discrete time, 16
disjoint, for tuples, 136
domain
  of a world, 23
  of existing individuals, ix, 1, 15
  one universal –, 80
domain of a function, 163

E-admissible, 103
$\mathcal{E}$-logic, 105
$\mathcal{E}\forall$-complete, 111
  with identity, 187
$\mathcal{E}\forall$-Intro, 104, 169
EI, 104
EI^{+}, 169, 176
elimination of definite descriptions, 195
empty function, 31, 163
entailment, x, 17
  in a model on an R-frame, 207
EX, 176
existence predicate, xii, 103, 162
Existence property of $|P|^{\mathcal{M}}$, 166
existence proposition, xi, 23, 163
existence set, 23, 103
Existential Generalisation, 210
Existing Instantiation, 104, 169
expanding domains, xi, 15, 67
extension, 160
extensional, 160
Extensionality property of $|P|^{\mathcal{M}}$, 166

F (Falsum), 2
$\mathcal{F}_{\mathrm{L}}$, 56
$\mathcal{F}_{\mathrm{S}}$, 54
FNE, 188
formula
  pure, 241
  admissible, 29, 168
  atomic, 2, 162
  identity, 162
  of a signature, 2
  of a signature with identity, 162
  propositional modal, 2
  self-identity, 162
frame, 11
  general, 13
  R–, 206
  S–, 12
Free Assignment Lemma, 28, 168, 233
free at a formula, 137
  strongly, 143
  sufficiently, 141
free for
  a predicate symbol, 137
  a variable, 4, 136
freely substitutable for, 4
full
  general frame, 14
  model structure, 31

model system, 239
full of objects, 165, 175
functional model structure, xii, 129
functionally canonical, 152
fusion
connective, 205
semantics of, 207, 230
operation, 218

$\mathcal{G}_L$, 53, 179
$\mathcal{G}_S$, 57
GC, 5
GC*, 7
general frame, 13
canonical, 47, 57
for identity, 179
full, 14
general model, 13
Generalisation on Constants, 5
generated subframe, 55
GL, 15, 63–66
Gödel-Löb logic, 15, 63–66
greatest lower bound, x, 17, 24

hold locally, 214

identical at w, 163
identity, 162
identity formula, 162
identity set, 163
implication operation
Boolean, 18
defined by fusion, 218
incomplete logic
for frame semantics, 12
inconsistent logic, 12
index, 261–268
Indistinguishability Lemma, 170
indistinguishability set, 164
indistinguishable at w, 164
semantically, 170
individual concept, xii, 161
admissible, 165
individual concept constants, 162
individual constants, 2
individual variables, 1
individualised cover system, 247
inference rule, 4
$\Box\forall$-Intro, 123
$\exists$-Elimination, 6
$\exists$-Equivalence, 6
$\exists$-Monotonicity, 5
$\exists$GC, 210
$\forall$-Equivalence, 5
$\forall$-Introduction, 5
modified, 104
$\forall$-Monotonicity, 5
$\forall$GC, 7
$\wedge$-Residuation, 241
Adjunction, 204
$\mathcal{E}\forall$-Intro, 104, 169
GC, 5
GC*, 7
MP, 4
N, 4
Relettering of Bound Variables, 6
Replacement of Provable Equivalents, 6
Sub, 7
Sub*, 7
substitution for predicate letters, 146
T$\forall$-Intro, 109, 187
TI, 4
TI*, 7
TNE, 177
UG, 4
infinitely many constants, 43, 48, 52
inflationary operation, 216
inner subframe, 55
point-generated, 78
instance of P, 137
intension, 160
intensional, 160
intensional conjunction, 204
InVar, 2, 161

join, x, 18, 24

K, 4, 9
KB, 58
Kripke frame, 11
canonical, 54
Kripke-Joyal semantics, 217, 249
Kripkean
canonical model, 112
characteristic model, 89, 95
Kripkean identity logic, 188
Kripkean premodel, 30
with identity, 187

L-model, 32
L-structure, 32
least upper bound, x, 18, 24
Lemma

BF-, 85
Free Assignment, 28, 168, 233
Indistinguishability, 170
Lindenbaum's, 46
Rasiowa-Sikorski, 21
Semantic Entailment, 223, 234
Simultaneous Substitution, 139
Substitution, 29, 168, 234
Lindenbaum algebra, 20
Lindenbaum's Lemma, 46, 84, 112, 179, 208
local character, 215
local members, 216
locale, 242
localic cover system, 245
localised set, 216
locally
constant, 215
determined, 215
equal, 215
holding property, 214
true, 215
logic
$\mathcal{E}$-, 105
inconsistent, 12
Kripkean identity –, 188
propositional modal, 8
quantified modal, 5
relevant, 203
second-order, 13
with identity, 177

$\mathcal{M}_{\mathrm{HR}}$, 243
$\mathcal{M}_{\mathrm{HRQ}}$, 249
$\mathcal{M}_{\mathrm{L}}$, 47, 107, 182
$\mathfrak{M}_{\mathrm{L}}$, 133
$\mathcal{M}_{\mathrm{L}}^{\flat}$, 71
$\mathcal{M}_{\mathrm{L}}^{\circ}$, 75
$\mathcal{M}_{\mathrm{L}}^{\natural}$, 89
$\mathcal{M}_{\mathrm{L}}^{K}$, 95, 112, 190, 199
$\mathfrak{M}_{\mathrm{QR}}$, 213
$\mathcal{M}_{\mathrm{R}}$, 229
$\mathcal{M}_{\mathrm{RQ}}$, 238
$\mathcal{M}_{\mathrm{S}}$, 54
MacNeille completion, 21, 249
maximal set of formulas, 46
McKinsey axiom, 14, 62–63
meet, x, 17, 24
model
canonical, nature of, 45
characteristic, 45
general, 13
of a logic (L-model), 32
on a functional model structure, 130
on a Kripke frame, 11
on a model structure, 29
on a model structure for identity, 168
on a QR-model structure, 211
on an R-frame, 206
on an R-model system, 222
on an RQ-model system, 233
model structure, 22
canonical
for a Kripkean identity logic, 189
for $\mathcal{E}$-logics, 107
for identity, 181
for QR, 213
varying-domains, 47
for language with identity, 165
for QR, 211
for RQ, 213
functional, xii, 129
model system
for R, 222
canonical, 226
for RQ, 231
canonical, 235
Modus Ponens, 4
monotonic under refinement, 220
MP, 4

N, 4
NE, 116
nearness, 214
Necessitation, 4
Necessity of Existence, 116
Necessity of Identity, 160
Necessity of Non-Existence, 116
negation complete set of formulas, 46
NI, 159, 176
NNE, 116, 118
NNI, 176
null object, 197

Ob, 162
ObCon, 162
object
admissible, 165
null, 197
object constants, 162
object rich, 178
object variables, 161
ObVar, 162

one universal domain, 80
open cover, 214
open set, 214
operational interpretation of relevant implication, 208
orthoclosed set, 219
orthoclosure, 219
orthoclosure localisation, 240
orthocomplement, 219
orthogonality relation, 218
orthojoin, 219

P-instance, 137
p-morphism, 55
parameter, 137
parameterless substitution, 137, 145
partial function, 163
  constant, 164
  rigid, 164
partial order, 206
  complete, 20, 217
PC, 5
φ^f, 49
$\varphi(\tau/x)$, 3
$\varphi(\tau'//\tau)$, 160
point-generated inner subframe, 78
polyadic algebra, 22
possibilist interpretation, x, 5
possible individuals, 15
powerset $\wp$, 11
premiss, 4
premodel, 25, 129
  canonical, 47
    for identity, 182
    for QR, 213
    on a functional structure, 133
  for language with identity, 165
  Kripkean, 30
    with identity, 187
  on a QR-model structure, 211
  on an RQ-model system, 232
preorder, 206
preserved by subframes, 100
prime theory, 208
principal theory, 225
proof sequence, 8, 43
Prop, x, 13, 23
PropFun, xii, 129
proposition
  admissible, x, 14, 23
  as localised up-set, 217
  as set of worlds, ix, 13, 160
  existence, xi, 23
  strongest, 18
  weakest, 17
propositional function, xii, 20, 22, 26, 127, 128, 166
  admissible, 129
propositional modal formula, 2
propositional modal logic, 8
propositional variables, 2
PropVar, 2
pure formula, 241

QES, 108, 154
QK, 9
QR-model structure, 211
QS, 9
QS$^{\approx}$, 177
QS4M, 14
quantification
  actualist interpretation of, xi, 1, 4, 167
  as joins and meets, 20
  commutativity of, xii
  possibilist interpretation of, x, 5
quantified modal logic, 5, 105

R, 203
  axioms and rules for, 204
  theorems of, 205
R-frame, 206
R-model system, 222
R-theory, 208
  prime, 208
  principal, 225
  regular, 208
Rasiowa-Sikorski Lemma, 21
refinement
  as increase of information, 217
  of a cover, 215
  of a set by a set, 245
  relation, 216
reflexive-transitive closure, 78
regular theory, 208
Relettering of Bound Variables, 6
relevant cover system, 220
relevant logic, 203
repairable incompleteness, 15
Replacement of Provable Equivalents, 6
residual to
    the fusion connective, 241
    conjunction, 241

  set intersection, 241
  the fusion operation, 218
rich (=object rich), 178
rigid concept, xii
rigid designator, 160
rigid partial function, 164
RQ, 209
RQ-model structure, 213
RQ-model system, 231

$\mathcal{S}_{\mathrm{HR}}$, 243
$\mathcal{S}_{\mathrm{HRQ}}$, 249
$\mathcal{S}_{\mathrm{L}}$, 47, 107, 181
$\mathfrak{S}_{\mathrm{L}}$, 133
$\mathcal{S}^{\flat}_{\mathrm{L}}$, 70
$\mathcal{S}^{\circ}_{\mathrm{L}}$, 75
$\mathcal{S}^{\natural}_{\mathrm{L}}$, 89
$\mathcal{S}^{K}_{\mathrm{L}}$, 95, 112, 189, 199
$\mathfrak{S}_{\mathrm{QR}}$, 213
$\mathcal{S}_{\mathrm{R}}$, 226
$\mathcal{S}_{\mathrm{RQ}}$, 235
S-frame, 12
S4, 12, 20
S4.2, 16
S4M, 14
S5, 12
satisfaction relation, 27
second-order logic, 13
self-identity formula, 162
Semantic Entailment Lemma, 223, 234
semantically indistinguishable at w, 170
sentence, 3
sequential updating, 138
shape, 154
SI, 160
signature, 2, 162
simultaneous substitution, 3, 137
Simultaneous Substitution Lemma, 139
simultaneous updating
  of a propositional function, 140
  of a variable assignment, 138
sort, 162
sound for (validity in) a class, 32
soundness for
  HR, 246
  QK, 38
  $\mathrm{QK}^{\approx}$, 177
  R, 223
  RQ, 234
  substitution for predicate letters, 151
  T∀-Intro, 111
square-decreasing, 220
strict implication, 16, 109
strongest proposition, 18
strongly free at, 143
strongly relevant cover system, 240
Sub, 7
Sub*, 7
subframe, 55
  inner, 55
  generated, 55
substitution
  for free variables, 3
  for predicate letters, 127, 137, 146
  for propositional variables, 3
  simultaneous, 3, 137
Substitution Lemma, 29, 168, 234
substitution operator, 3, 137
  free at a formula, 137
    strongly, 143
    sufficiently, 141
  parameter of, 137
  parameterless, 137, 145
substitution-instance, 2
Substitutivity of Identicals, 160, 177
sufficiently free at, 141

⊤ (Verum), 2
t, 203
tautology, 3
Tem, 109
template, 109, 175
Template ∀-Introduction, 109, 187
Template Non-Existence, 177
term, 2, 162
  closed, 2
    with definite descriptions, 193
  object –, 162
Term Instantiation, 4
theorem of a logic, 5
TI, 4
TI*, 7
TNE, 177
Trivial system, 60
true in a model on
  a Kripke frame, 11
  an R-frame, 207
  an R-model system, 223
Truth Lemma for
  $\mathcal{M}_{\mathrm{HR}}$, 244
  $\mathcal{M}_{\mathrm{L}}$, 51
    with $\mathcal{E}$, 108

with identity, 184
$\mathfrak{M}_{\mathrm{L}}$, 133
$\mathcal{M}_{\mathrm{L}}^{\flat}$, 72
$\mathcal{M}_{\mathrm{L}}^{\circ}$, 75
$\mathcal{M}_{\mathrm{L}}^{\natural}$, 91
$\mathcal{M}_{\mathrm{L}}^{K}$, 96
with $\mathcal{E}$, 114
with identity, 199
$\mathfrak{M}_{\mathrm{QR}}$, 213
$\mathcal{M}_{\mathrm{R}}$, 229
$\mathcal{M}_{\mathrm{RQ}}$, 238
$\mathcal{M}_{\mathrm{S}}$, 54
truth relation, 11, 27, 207
truth set, 11
two-sorted language, 161

UD, 4
UG, 4
UI, 5
underlying
(Kripke) frame, 23
canonical general frame, 53
general frame, 23
Universal
Distribution, 4
Generalisation, 4
Instantiation, 1, 5
universe, 23
up-set, 206
updatable structure, 129
updating, 26
sequential, 138
simultaneous, 138

Vacuous Quantification, 4
valid in a
class, 11, 32
frame, 11
functional model structure, 130
general frame, 14
model structure, 32
model structure for identity, 168
premodel, 32
premodel for identity, 167
premodel on an RQ-model system, 234
QR-premodel, 212
R-frame, 207
R-model system, 223
RQ-model system, 234
strongly relevant cover system, 240
valuation, 166
variable-assignment, 26
updating of, 26
variables
individual, 1
propositional, 2
concept, 162
object, 161
varying domains, 1
Verum, 2
Verum system, 60
VQ, 4

Weak Reflexivity, 176
weakest proposition, 17
WR, 176